Rural Gerontology

This book provides the first foundation of knowledge about the intellectual traditions, contemporary scope and future prospects for the interdisciplinary field of rural gerontology.

With a focus on rural regions, small towns and villages, which have the highest rates of population ageing worldwide, *Rural Gerontology* is aimed at understanding what it means for rural people, communities and institutions to be at the forefront of twenty-first-century demographic change. The book offers important insights from rural ageing studies into today's most pressing gerontological problems. With chapters from more than 65 established and emerging rural ageing researchers, it is the first synthesis of knowledge about rural gerontology, harnessing a burgeoning interdisciplinary scholarship on the rural dimensions of ageing, old age and older populations. With a view to advancing a critical understanding of rural ageing populations, this book will have an overreaching impact across the social sciences by drawing on advancements in understandings of rural ageing from social, environmental, geographical and critical gerontology to facilitate a comprehensive exploration of the diversity, complexity and implications of the ageing process in rural settings.

Bringing together valuable international perspectives, this book makes a timely contribution to gerontology, rural studies and the social sciences, and will appeal to scholars and researchers across USA and Canada, UK and Ireland, Australia and New Zealand, Europe, China and countries in Africa, South America and South-East Asia.

Mark Skinner is Dean of Social Sciences at Trent University, Canada, where he is also Professor of Geography and holds the Canada Research Chair in Rural Aging, Health and Social Care.

Rachel Winterton is Senior Research Fellow at the John Richards Centre for Rural Ageing Research, La Trobe Rural Health School, La Trobe University, Australia.

Kieran Walsh is Professor of Ageing & Public Policy in the Discipline of Economics and Director of the Irish Centre for Social Gerontology at the National University of Ireland Galway.

Perspectives on Rural Policy and Planning
Series Editors: Andrew Gilg and Mark Lapping

This well-established series offers a forum for the discussion and debate of the often-conflicting needs of rural communities and how best they might be served. Offering a range of high-quality research monographs and edited volumes, the titles in the series explore topics directly related to planning strategy and the implementation of policy in the countryside. Global in scope, contributions include theoretical treatments as well as empirical studies from around the world and tackle issues such as rural development, agriculture, governance, age and gender.

European Integration and Rural Development
Actors, Institutions and Power
Michael Kull

Evaluating the European Approach to Rural Development
Grass-roots Experiences of the LEADER Programme
Edited by Leo Granberg, Kjell Andersson and Imre Kovách

Globalization and Europe's Rural Regions
Edited by John McDonagh, Birte Nienaber and Michael Woods

Service Provision and Rural Sustainability
Infrastructure and Innovation
Edited by Greg Halseth, Sean Markey and Laura Ryser

The Changing World of Farming in Brexit UK
Edited by Matt Lobley, Michael Winter and Rebecca Wheeler

Rural Gerontology
Towards Critical Perspectives on Rural Ageing
Edited by Mark Skinner, Rachel Winterton and Kieran Walsh

For more information about this series, please visit: www.routledge.com/Perspectives-on-Rural-Policy-and-Planning/book-series/ASHSER-1035

Rural Gerontology

Towards Critical Perspectives on Rural Ageing

Edited by Mark Skinner, Rachel Winterton and Kieran Walsh

Routledge
Taylor & Francis Group
LONDON AND NEW YORK

First published 2021
by Routledge
2 Park Square, Milton Park, Abingdon, Oxon OX14 4RN

and by Routledge
52 Vanderbilt Avenue, New York, NY 10017

Routledge is an imprint of the Taylor & Francis Group, an informa business

British Library Cataloguing-in-Publication Data
A catalogue record for this book is available from the British Library

Library of Congress Cataloging-in-Publication Data
A catalog record for this book has been requested

ISBN: 978-0-367-89479-5 (hbk)
ISBN: 978-1-003-01943-5 (ebk)

Typeset in Times
by Apex CoVantage, LLC

Contents

Figures

Tables

Contributors

Padmore Adusei Amoah is Research Assistant Professor in the School of Graduate Studies, Asia Pacific Institute of Ageing Studies and Institute of Policy Studies of Lingnan University, Hong Kong. His research centres on social epidemiology and social policy issues on health-related wellbeing among children, youth, older persons and migrants.

Gavin J. Andrews is Professor in the Department of Health, Aging and Society, McMaster University, Canada. A leading health geographer, his wide-ranging research explores the dynamics between space/place and ageing, holistic medicine, health care work, phobias, sports and fitness, health histories and popular music. Much of his work is positional and considers the progress, state-of-the-art and future of health geography. In recent years, he has developed an interest in non-representational theory. This approach challenges the very nature and purpose of social science – which has tended to dig deeper and deeper for meaning – by animating the immediacy and "taking place" of health and well-being. He has edited the pioneering Routledge books *Ageing and Place: Perspectives, Policy, Practice* (with David R. Phillips, 2005), and *Geographical Gerontology: Perspectives, Concepts, Approaches* (with Mark Skinner and Malcolm P. Cutchin, 2018).

Liat Ayalon, PhD, is Professor in the School of Social Work, at Bar Ilan University, Israel. She coordinates an international EU funded PhD program on the topic of ageism (EuroAgeism.eu). She is also the Israeli Principal Investigator of the EU funded MascAge program to study ageing masculinities in literature and cinema. She has led an international research network on the topic of ageism, funded through COST (Cooperation in Science and Technology; COST IS1402, notoageism.com). She consults both national and international organisations concerning the development and evaluation of programs and services for older adults.

Rachel J. Bar is Director of Health and Research at Canada's National Ballet School. She completed her PhD in psychology as a Vanier Scholar at Ryerson University, Canada, and currently is a Postdoctoral Fellow at the Trent Centre for Aging & Society, Trent University, Canada. Her research explores the benefits of dance for older adult populations and the utility of arts-based knowledge translation of health research.

Laura Bates is Research Assistant in the School of Environment, University of Auckland. Her Master's thesis and associated work has centred on the connections between place attachment, experiences of home and well-being.

E. Helen Berry is an alumna of Ohio State University who completed postdoctoral work at the University of Michigan and is Professor of Sociology at Utah State University. Her research mainly examines migration at different points of the life course, including among youth, persons with disabilities and racial and ethnic groups. Most recently, she has focused on the demography of ageing and its associated concerns, particularly in rural places. Her primary teaching interests are in urban and rural sociology, population studies, social statistics and graduate student training. She is a Women and Gender Research Institute Distinguished Professor and Director of the Yun Kim Population Research Laboratory. She is also a past President of the Rural Sociological Society and is active in the Population Association of America.

Vanessa Burholt is Professor in Gerontology at the Faculty of Medical and Health Sciences, jointly in the School of Nursing and School of Population Health, at The University of Auckland, New Zealand. She completed her PhD at Bangor University, where she became Deputy Director of the Centre for Social Policy Research and Development in 1998, and in 2004 took over as Director. She was awarded a personal Chair at Bangor in 2007, and shortly after moved to Swansea University, as Director of the Centre for Innovative Ageing. In 2016 she was also appointed Director of the pan-Wales Centre for Ageing and Dementia Research.

Brídín Carroll is Postdoctoral Researcher in the Irish Centre of Social Gerontology (ICSG) at the National University of Ireland Galway, Ireland. She completed her PhD in 2013 in the discipline of Geography at NUI Galway on the topic of food sustainability. She then spent three years a postdoctoral researcher in University College Dublin, Ireland, and one year as a postdoctoral researcher in the Healthy and Positive Ageing Initiative, a research initiative based in Ireland's Department of Health. Her research interests in marginalised populations have pivoted away from food poverty to the health and well-being of older people. At the ICSG, she is working currently on the *Older Traveller and Older Homeless* study.

Matthew Carroll is Senior Research Fellow in the School of Rural Health at Monash University, Australia and an adjunct Senior Research Fellow at Federation University, Australia. He is one of the lead investigators of the Hazelwood Health Study, investigating the impacts of the 2014 Hazelwood mine fire on residents in adjacent communities, lead of the Psychological Impacts stream, co-lead on the Older Person's stream and an investigator of the Community Wellbeing and Adult Survey streams. His research focuses on evidence-based ageing policy, older people and heat, impacts of climate change and preventing social isolation. He is National Convenor of the Emerging Researchers in Ageing initiative, which provides support and encouragement to students undertaking higher degrees by research in the field of ageing in Australia and internationally.

Yang Cheng is Associate Professor in the Faculty of Geographical Science at Beijing Normal University, China. She is interested in combining the use of qualitative and quantitative methods to investigate the geography of population ageing and elderly care in the economic, political, social and cultural context of a rapidly changing China, especially in mega cities. Her current research focuses on the intergenerational differences in social support, migration to urban new areas and access to health care services for older population living in communities in Beijing, and maternal health risks and access to maternity care.

Amber Colibaba is Research Coordinator of the Rural Aging Canada Research Chair program at Trent University, Canada, where she is also Research Associate, Trent Centre for Aging & Society. She recently graduated with the Collaborative Specialization in Aging Studies from the Master of Arts in Sustainability Studies program at Trent University with a research interest in rural ageing, voluntarism and community development.

Malcolm P. Cutchin is Professor at the Institute of Gerontology and the Department of Health Care Sciences, Wayne State University, USA. A primary focus of his work is human-place relationships and well-being, and his studies have often focused on residential environments. Project foci have included rural physician retention; community based care settings for seniors; stress and health of ethnic populations living near petrochemical plants; preventive home visits for at-risk older adults; neighborhood stressors, daily activities and the stress response of older African-Americans; and Detroit patrol officers' neighbourhood-related stress. He has an abiding interest in the philosophical pragmatism, especially the writings of John Dewey and those who study Dewey. He is a Fellow of the Gerontological Society of America and was the Editor-in-Chief of the *Journal of Applied Gerontology* (2006–2011).

Stefanie Doebler is Lecturer in Social Statistics and Sociology at Lancaster University, UK. Her main research interest is in social inequalities over the life course and ageing, using quantitative methods of analysis. After completing her PhD at the University of Manchester in 2013, she worked as a research Fellow at the Queen's University Belfast, UK, where she came into contact with large-scale population data linkage studies, which have since inspired her research.

Nata Duvvury is Senior Lecturer and Director, Center for Global Women's Studies and Leader of Gender and Public Policy Cluster in the Whitaker Institute at the National University of Ireland (NUI) Galway, Ireland. Before

NUI Galway, she was Director, Gender, Violence and Rights Team at the International Center for Research on Women (ICRW). She is an international development expert with more than 30 years of experience in gender, development and empowerment. Her work includes research and advocacy on gender-based violence, women's property rights and gender and pensions in a variety of settings including conflict and post-conflict contexts. She has several peer reviewed journal articles, book chapters and research reports on these themes.

Jacquie Eales is Research Manager with the Research on Aging, Policies and Practice (RAPP) team in the Department of Human Ecology at the University of Alberta, Canada. Having grown up in a rural farming community, she is particularly interested in the interface between older adults and the family and community environments in which they live, and the process of creating a best person-environment fit. As research manager and knowledge mobilisation expert, she has focused on bridging research, policy and practice to make a meaningful difference in the lives of older adults and their families. She has worked collaboratively with older adults, family caregivers, government and community partners to ensure research projects are relevant to their needs and findings are translated in effective and creative ways.

Nina Glasgow is Senior Research Associate (Emerita) in the Department of Global Development, Cornell University, USA. She received a PhD in Sociology from the University of Illinois, Urbana-Champaign in 1982 and was employed as a sociologist in the Economic Research Service, U.S. Department of Agriculture in Washington, D.C. from 1980–1987. During her career, she co-authored and co-edited four books published numerous journal articles, book chapters and many policy and research briefs on several topics related to ageing in rural environments. She remains active in research and publishing and is on the International Advisory Committee for the Centre for Rural Economy, Newcastle University, UK.

Amanda Grenier is Professor in the Factor-Inwentash Faculty of Social Work at the University of Toronto, Canada, the Norman and Honey Schipper Chair in Gerontological Social Work, and a senior scientist with Baycrest's Rotman Research Institute. She has carried out research on life course transitions, social constructs of frailty, ageing with a disability, homelessness among older people, social isolation among seniors and precarious ageing. She is the author of *Transitions and the LifeCourse: Challenging the Constructions of Growing Old*. Her new books include an edited collection on *Precarious Aging* and a book on *Homelessness in Late Life*.

Alisa Grigorovich is CIHR Health System Impact Postdoctoral Fellow at KITE-Toronto Rehabilitation Institute, University Health Network. Her research program is interdisciplinary and centres on exploring social, ethical and policy issues related to health, ageing, technology and care. A central focus of her research concerns the use of arts to challenge stigma associated with dementia and to improve health and social inclusion across community-based and institutional care environments.

Greg Halseth is Professor in the Geography Program at the University of Northern British Columbia, Canada, where he is also the Canada Research Chair in Rural and Small Town Studies and Co-Director of UNBC's Community Development Institute. His research examines rural and small town community development, and community strategies for coping with social and economic change, all with a focus upon northern B.C.'s resource-based towns. He has served on the governing council of the Social Sciences and Humanities Research Council of Canada, the Advisory Committee on Rural Issues for the Federal Secretary of State for Rural Development, the Community Advisory Committee for the BC Ministry of Forests Mountain Pine Beetle Task Force and other advisory committees.

Neil Hanlon is Professor of Geography at the University of Northern British Columbia, Canada. His research interests include rural health care delivery, age- and able-minded community development, the place-embeddedness of care networks and regional health governance. With Mark Skinner, he co-edited *Ageing Resource Communities: New Frontiers of Rural Population Chance, Community Development, and Voluntarism* (Routledge, 2016). His work appears in leading journals, such as *Health & Place*, *Social Science & Medicine* and *Health & Social Care in the Community.*

Stine Hansen recently completed her PhD in the School of Earth, Environment & Society at McMaster University, Canada. Her interests include ageing and disability. She is currently working as an online educational developer with McMaster's MacPherson Institute for Leadership, Innovation and Excellence in Teaching.

Jesse Heley is Senior Lecturer in the Department of Geography and Earth Sciences at Aberystwyth University, UK, and Research Associate of the Wales Institute of Social & Economic Research, Data & Methods (WISERD). He is the Director of Employability & Enterprise for the Institute of Geography, History, Politics and Psychology. Having completed his PhD at Aberystwyth University on the subject of 'Rurality, Class, Aspiration and the Emergence of a New Squirearchy', he was appointed as a Research Associate in DGES in 2009.

Catherine Hagan Hennessy is Professor of Ageing at the University of Stirling, UK, where she co-leads the Ageing and Dementia Research Programme. Her work focuses on older people's health, well-being and social inclusion. At the University of Sheffield she held the post of Deputy Director of the Economic and Social Research Council's Growing Older Programme on Extending Quality Life and the UK National Collaboration on Ageing Research. From 2005–2015 she was Professor of Public Health and Ageing at Plymouth University, where she led the interdisciplinary "Grey and Pleasant Land" study on rural ageing. She is co-editor and co-author of *Countryside Connections: Older People, Community and Place in Rural Britain* (Policy Press, 2014).

Rachel Herron is Associate Professor in the Department of Geography and Environment at Brandon University, Canada, and the Canada Research Chair in Rural and Remote Mental Health. Her current research examines the vulnerability and complexity of care relationships, social inclusion and meaningful engagement for people with dementia and the diversity of lived experiences of rural mental health. Her research has been published in peer-reviewed journals such as *Health & Place*, *Social Science & Medicine*, *Gender, Place & Culture* and *Environment and Planning*, as well as in edited volumes by international academic publishers.

Jaco Hoffman is Associate Professor at North-West University, South Africa. He has been cross-disciplinarily involved in African gerontology over the past 25 years through research, practice and advocacy as well as in management. He currently leads the Optentia Research Focus sub-programme: Ageing and Generational Dynamics in Africa (AGenDA) with a visiting appointment as Professorial Fellow in the Oxford Institute of Population Ageing/Oxford Martin School at the University of Oxford, UK. He is also an Honorary Professor in the Institute of Ageing in Africa, Medical Faculty at the University of Cape Town, South Africa. He serves as co-director of the International Longevity Centre, South Africa based at both North-West University and the University of Cape Town.

Kai Huang is Associate Professor of Operations Management in the DeGroote School of Business, McMaster University, Canada. He received his PhD from the School of Industrial and Systems Engineering at Georgia Institute of Technology, USA. He specialises in the optimisation under uncertainty and data-driven optimisation techniques with applications in business analytics and supply chain management. His research has appeared in journals as *Operations Research*, *European Journal of Operational Research*, *Information Systems* and *Operational Research.*

Anthea Innes is the first Professor of Dementia at Salford University, UK, the Coles Medlock Director for the Salford Institute for Dementia. She was the founder of the Bournemouth University Dementia Institute and led the first online Masters Programme in Dementia Studies when working at the University of Stirling. At the core of her work is a concern to engage with the lived experience of those impacted by dementia, namely, those diagnosed with the condition, their family members and professional care supporters. Her research interests span the care continuum from pre-diagnosis to end of life. Her particular research topics include rurality, technology and creative approaches to supporting persons with dementia.

Robin Kearns is Professor of Geography in the School of Environment, University of Auckland, New Zealand. He has researched across a wide range of themes in social and cultural geography with a particular interest in the links between place experience and well-being. He is an editor of the international journal *Health & Place* and his most recent book is *Children's Health and Wellbeing in Urban Environments* (Routledge, 2017).

Norah Keating is Professor at Swansea University, UK and Chair, Centre for Innovative Ageing. She is a social gerontologist whose theoretical and empirical research has created evidence, challenged discourses and influenced policy in social, physical and global contexts of ageing. Her program of theoretical, empirical and policy work on physical contexts of ageing has challenged beliefs about rural communities as good places to grow old and informed the global agenda on age-friendly communities. She challenges gerontologists to focus on macro trends such as climate change, economic recession and political instability and their disproportionate influence on older persons. She leads the Global Social Issues on Ageing which fosters collaboration on issues of ageing at the interfaces of regional issues and global trends.

Sinéad Keogh is Postdoctoral Researcher at the Irish Centre for Social Gerontology (ICSG) at the National University of Ireland Galway (NUIG), Ireland. Sinéad completed her PhD in 2014 on "The Allocation of Time to Non-Market Off-Farm Activities by Farm Household Members in the West of Ireland" at the Discipline of Economics at NUI Galway. In 2015, Sinéad took a position as a postdoctoral researcher in the Centre for Disability Law and Policy in NUI Galway, where she worked primarily on research that explored developments in independent living and personal assistance for persons with disabilities. Since joining the ICSG in 2018, Sinéad has worked on the PLACED-Lives project, examining the influence of critical transitions on the accumulation of multidimensional social exclusion in later life.

Pia Kontos is Senior Scientist at Toronto Rehabilitation Institute, University Health Network and an Associate Professor in the Dalla Lana School of Public Health at the University of Toronto, Canada. In her research she focuses on structural and relational vulnerability to stigma associated with dementia in long-term care settings; the development of theories, policies and practices that support ethical care relationships and the development and evaluation of art-based initiatives to reduce stigma and improve social inclusion and quality of care.

An Kosurko is Research Associate with the Rural Aging Canada Research Chair program and the Trent Centre for Aging & Society at Trent University, Canada, where she recently graduated from the Master of Arts in Sustainability Studies Program. Her areas of interest include communications, community-based research, voluntarism, placemaking and artistic practice. She is undertaking PhD studies in social sciences at the University of Helsinki, Finland.

Mayeso Lazaro is a Lecturer in the Department of Human Ecology at the University of Malawi, Malawi, and is the Faculty Representative for Postgraduate Studies and the Faculty Representative for Research and Publications. He holds an interdisciplinary PhD in Human Geography and Social Sciences from the University of Hull, UK. Currently, he is interested in researching grandparenting in rural contexts. Over the past six years, he has also been involved in several consultancies with the Government of Malawi as well as various non-governmental organisations such as Food and Agriculture Organization (FAO), Mathematica Policy Research (US), Project Concern Universal (PCI), Save the Children International, United Nations Children Education Fund (UNICEF) and World Vision International (WVI).

Andrew S. Maclaren is Research Fellow in the Health Services Research Unit at the University of Aberdeen, UK, having previously completed his doctorate and held a teaching fellowship at the university in the Department of Geography and Environment. His interests lie in cultural, political and social geographies with a particular focus on everyday life, drawing on contemporary theories of place. His research has engaged with the lived experience, conceptualisation and understanding of rural spaces and places, with his doctoral research considering this in relation to the lives of older people in rural communities. Maclaren's research has been published in leading geography and interdisciplinary journals including *Progress in Human Geography* and *Sociologia Ruralis*.

Sean Markey is Professor with the School of Resource and Environmental Management, Simon Fraser University, Canada. His research concerns issues of local and regional economic development, rural and small-town development, community sustainability and natural infrastructure. He continues to work with municipalities, non-profit organisations, Indigenous communities and the business community to promote and develop sustainable forms of community and regional development. He is also an Adjunct Professor with the Department of Geography at the University of Northern British Columbia, Canada.

Elizabeth McCrillis is Assistant Professor in the Department of Psychology and Director of the Trent Centre for Aging & Society at Trent University, Canada. She teaches psychology courses in ageing, health, qualitative research methods and the history of psychology, and supervises undergraduate and graduate students studying the psychology of ageing and applied health psychology topics more broadly. Her research is focused on the sustainability of rural age-friendly communities programs, having worked in collaboration with communities in various Canadian provinces who have implemented age-friendly programs. Her research takes a collaborative, interdisciplinary and community-based approach to studying a variety of ageing and health topics focused within rural communities.

Verena Menec is Professor in the Department of Community Health Sciences at the University of Manitoba, Canada. She has been working in the area of age-friendly communities since 2006, when she conducted the focus group research for the WHO Age-Friendly Cities project in one of the participating cities in the province of Manitoba. She was also part of the research team for the Canadian Age-Friendly Rural and Remote Communities project. As part of a 5-year program of research, she subsequently conducted many projects on age-friendliness in partnership with the Province of Manitoba's Age-Friendly Manitoba Initiative.

Paul Milbourne is Professor of Human Geography at in the School of Geography and Planning at Cardiff University, UK, where he is also the Head of School and Director of the Centre for Research on Environment, Society and Space. His main research interest lie in the field of social geography and, more specifically, the geographies of welfare and poverty. He is also interested in environmental geographies, particularly the interplay between social and environmental forms of (in)justice within contemporary societies.

Verónica Montes de Oca is Senior Researcher at the Social Research Institute of the National University Autonomous of Mexico (UNAM), Mexico and National research level III of the National System of Researcher by National Council of Science and Technology (SNI-CONACYT) and was president of the Latin American Association of Population (ALAP). Since 2012, she has been coordinator of the Teaching and Research University Network of Ageing and Old Age at UNAM (SUIEV in Spanish). Her research program includes demographic ageing in Mexico, Latin American and Caribbean regions, social support networks, households and families, impact on older persons of transnational migration between Mexico and US, active ageing from a comparative perspective in different countries and strategies of care for older persons with dementia in urban and rural areas in Mexico.

Áine Ní Léime is Senior Researcher and Deputy Director at the Irish Centre for Social Gerontology, National University of Ireland Galway, Ireland. Her recent research is in the area of ageing, gender, employment and retirement from a life-course perspective with a recent focus on extending working life. She has conducted research on ageing and creativity, ageing and volunteering and on public policy in relation to pensions and employment in later life. She has several publications in these areas including articles in peer-reviewed journals, edited books and book chapters and, most recently has co-edited a special issue of *Ageing & Society* on Gender and Extended Working Life and a book on Gender, Health and Extended Working Life Policy.

K. Bruce Newbold is Professor of Geography in the School of Earth, Environment & Society at McMaster University, Canada. Trained as a population geographer, he has taught at the University of Illinois, USA, and has held guest scholar positions at the University of Glasgow, UK, and the University of California San Diego, USA. He has authored or co-authored over 150 articles, along with editing or co-editing books. His research interests include population ageing, immigration, migration, health and transportation issues.

Sheila Novek is currently a PhD student in the Department of Community Health Sciences at the University of Manitoba, Canada. Her research has focused on age-friendly communities, as well as health and support services for people with early onset dementia.

Hanna Ojala is University Lecturer in Gender Studies at the Faculty of Social Sciences, Tampere University, Finland. Her research has focused on issues of ageing men related to health, retirement, anti-ageing, sports and communities.

Eamon O'Shea is Professor in the School of Business and Economics, founding Director of the Irish Centre for Social Gerontology (ICSG) and Director of the Centre for Economic & Social Research on Dementia (CESRD) at the National University of Ireland Galway, Ireland. His research interests are focused on the economics of ageing, rural gerontology and dementia. His work has been influential in setting the agenda for the reform of services and policies for older people in Ireland. He is the holder of a Health Research Board Leader award in Dementia in Ireland.

Maree Petersen is Senior Lecturer at the University of Queensland, Australia. She has a background in social work and has a particular interest in the lives and experiences of older people, encompassing areas of homelessness, housing and well-being. Her research focuses on developing knowledge to assist in addressing disadvantage among older people and the design of strategies to address their housing and support needs.

David R. Phillips is Professor Emeritus in the Department of Sociology and Social Policy at Lingnan University, Hong Kong, China and Adjunct Professor in the Department of Health, Aging and Society at McMaster University, Canada and in the Department of Geography and Planning at Macquarie University, Australia. He has extensive research and teaching experience in social gerontology, health and epidemiology in the Asia-Pacific region and elsewhere. He has been an adviser to the WHO Western Pacific Region on its reviews of the frameworks on ageing and health in the region. His most recent book is the second edition of *Global Health* (Routledge, 2017, co-author Kevin McCracken).

Judith Phillips is Deputy Principal (Research) and Professor of Gerontology at the University of Stirling, Scotland, UK. She is the UK Research Director for "Healthy Ageing" under the ESRC/AHRC Industrial Strategy Challenge Fund. A geographer by background, she continues to focus her research on spatial aspects of ageing and has researched and published widely on urban and rural contexts of older people, such as assessing the needs of older people in urban settings and establishing a critical human ecology perspective on rural ageing. She has also worked closely with policy makers and social work practitioners attempting to ensure her research has impact on the lives of older people.

Ilkka Pietilä is Assistant Professor of Social Gerontology at the Faculty of Social Sciences, University of Helsinki, Finland. His work has covered topics such as ageing men and masculinities, ageing and embodiment and retired men's communities.

Laura Poulin is a PhD candidate in Canadian Studies at Trent University, Canada, where she holds a Frederick Banting and Charles Best Doctoral Scholarship funded by the Canadian Institutes of Health Research (CIHR). She is interested in the dynamic interplay between older adults, health and rural communities. The main focus of her research is the transitional care of rural older adults. Her most recent publications in *Social Science & Medicine* and *Healthcare Management Forum* emphasise the importance of contextually sensitive approaches to older adult health and care.

Patricia Rea is CÁTEDRAS CONACYT researcher at the Social Research Institute, National University Autonomous of Mexico (UNAM), Mexico. She is a member of the National System of Researchers of the National Council of Science and Technology (SNI-CONACYT), the Interdisciplinary University Seminar on Aging and Old Age (SUIEV) at UNAM and was a visiting researcher at the Institute of Latin American Studies, University of Texas. Her areas of research include indigenous and intercultural education and development policies implemented among different ethnic groups of Mexico, with special emphasis on indigenous women from Oaxaca. In recent years, her research themes include demographic ageing, public policies and citizenship and indigenous and intercultural old age.

Vera Roos is Professor in the Ageing and Generational Dynamics in Africa (AGenDA) programme of the Optentia Research Focus Area at North-West University, South Africa, and an affiliate research fellow of the Oxford Institute of Population Ageing at the University of Oxford, UK. She focuses on relational experiences and on the contribution of older individuals in challenged contexts. She developed the Mmogo-method®, a visual data-collection tool, to enable research participation despite age, language or cultural barriers, and to obtain layered information of personal, relational and group experiences. This informed relational theory development (Self-Interactional Group Theory) and intergenerational interventions and led to the development of an information and communication eDirectory system, Yabelana, with context-specific information for use on smart and older generation mobile phone devices.

Mark W. Rosenberg is Professor in the Department of Geography and Planning and cross-appointed as Professor in the Department of Public Health Science at Queen's University, Canada. He is the Tier 1 Canada Research Chair in Development Studies. Over the course of his career, Professor Rosenberg has carried out a wide range of studies on access to health care and their utilisation with a particular emphasis on understanding how barriers to access affect the utilisation of health services among vulnerable populations, especially the older population, women in need of diagnostic service or reproductive health services and Aboriginal Peoples.

Graham D. Rowles is Professor of Gerontology (Emeritus) and was Founding Director of the Graduate Center for Gerontology at the University of Kentucky, USA. His research focuses on the lived experience of ageing with a central theme of this work is exploration of the changing relationship between older adults and their environments and implications of this relationship for health, well-being and environmental design. He has conducted in-depth ethnographic research with elderly populations in inner city, rural (Appalachian) and nursing facility environments. Recent research has focused on long-term care and the meaning of place in old age. His publications include the book *Prisoners of Space? Exploring the Geographical Experience of Older People*, seven co-edited books and numerous articles and book chapters. A Fellow of the Gerontological Society of America and the Association for Gerontology in Higher Education, he has served as President of the Southern Gerontological Society and the Association for Gerontology in Higher Education.

Laura Ryser is Research Manager of the Rural and Small Town Studies Program at the University of Northern British Columbia, Canada. She has worked extensively across northern BC and throughout Canada to explore rural restructuring and community transition, best practices guiding industry, work camp and community relationships, seniors' needs and implications for ageing resource towns, rural poverty, winter city design, labour mobility, the transformation of the voluntary sector, innovation and smart services and rural partnership development. She has served on the board of the Canadian Rural Revitalization Foundation and has served the Public Advisory Group for Canfor's Sustainable Forest Management Plan Certification Process and UNBC's Field Safety Committee. She is co-author of *Towards a Political Economy of Resource-dependent Regions* with Greg Halseth (Routledge, 2017).

Thomas Scharf is Professor of Social Gerontology in the Institute of Health & Society at Newcastle University, UK, President of the British Society of Gerontology and chair of the Age-Friendly City group in Newcastle. Before moving to Newcastle in 2016, he was Director of the Irish Centre for Social Gerontology at the National University of Ireland Galway (2010–2015) and Director of the Centre for Social Gerontology, Keele University (2006–2010). He is a Fellow of the Academy of Social Sciences and has held visiting professorships at the universities in Manchester, Keele, Vienna and Galway. His research focus is on topics relating to social exclusion and inequalities in later life and to the ways in which experiences of ageing differ across different types of community, including rural and disadvantaged urban communities.

Darren M. Scott is Professor of Geography at McMaster University, Canada. As a transportation geographer, his research interests include active transportation modes, activity-based travel behaviour, critical transportation infrastructure, disruptive mobility technologies, geographic information science, sustainable transportation and time geography.

Marjaana Seppänen is Professor of Social Work at the University of Helsinki, Finland. Coming from social gerontology with a background in social work, her many publications deal with older adults' social relationships, well-being and social exclusion, among other things.

Mark Skinner is Dean of Social Sciences at Trent University, Canada where he is also Professor of Geography and holds the Canada Research Chair in

Rural Aging, Health and Social Care. He was founding Director of the Trent Centre for Aging & Society (2013–2018) and is appointed to the Royal Society of Canada's College of New Scholars, Artists and Scientists. His previous books, *Ageing Resource Communities: New Frontiers of Rural Population Change, Community Development and Voluntarism* (2016, with Neil Hanlon) and *Geographical Gerontology: Perspectives, Concepts, Approaches* (2018, with Gavin Andrews and Malcolm Cutchin) are published by Routledge.

Elisa Tiilikainen is Assistant Professor in the Department of Social Sciences, University of Eastern Finland, Finland, working on loneliness and social relationships in later life, life course transitions and the effectiveness of elder care services.

Prakash Tyagi is Executive Director of GRAVIS, Founder-Director of GRAVIS Hospital, in India and is a visiting faculty at Jodhpur School of Public Health and a Clinical Associate Professor at the School of Global Health and Medicine of University of Washington, USA. In his academic work, he has studied internal medicine, geriatrics and global health. As the Executive Director of GRAVIS in India, he has been leading the organisation since last 14 years. GRAVIS works extensively in the Thar Desert of India, with key focuses on water security, food security and community health. GRAVIS reaches out to about 1.3 million people living in poverty with its work and has helped formation of over 3,000 Community Based Organizations.

Anna Urbaniak is Postdoctoral Researcher at the Irish Centre for Social Gerontology at National University of Ireland Galway, Ireland. She specialises in population ageing, life courses and people's interactions with and within place (rural and urban). She has a strong background in qualitative research methods. She has worked primarily in the areas of environmental gerontology, place and qualitative sociology. Her research interests focus on the relationship between older people and their surrounding environment within institutional and community settings.

Brenda Vrkljan is Professor in the Occupational Therapy program in the School of Rehabilitation Science and is an executive member of the McMaster Institute for Research on Aging (MIRA), McMaster University, Canada. Her research focuses on transportation mobility, medical risk and ageing. She is a co-lead researcher with the Canadian Driving Research Initiative for Vehicular Safety in the Elderly (Candrive), which includes a large, prospective multi-site cohort study that tracked the health and driving patterns of Canadians aged 70 and older.

Judi Walker holds a part-time professorial position in the Faculty of Medicine, Nursing and Health Sciences at Monash University, Australia, and adjunct professorial positions at University of Tasmania and Federation University Australia. She is Principal Co-Investigator of the Hazelwood Health Study, Lead of the Older Persons' research stream and has responsibilities for engagement with the study community and key stakeholders. She was CIA on an NHMRC Partnership Project Grant – *Aged Support and Aged Care: program and policy structures to support ageing well in rural and regional Australia* – and CIA for an ARC Linkage grant investigating the mechanisms behind age-related network shrinkage and social disengagement in order to design service models that can assist older rural people to stay socially engaged despite the challenges posed by increasing age.

Kieran Walsh is Professor of Ageing & Public Policy and Director of the Irish Centre for Social Gerontology, National University of Ireland Galway, Ireland. His research interests and expertise focus on social exclusion in later life; the relative nature of disadvantage in cross-national contexts; place and life course transitions; and informal and formal infrastructures of care. Kieran is also Chair of the European COST Action CA15122 on "Reducing Old-Age Social Exclusion" (ROSEnet – www.rosenetcost.com), which has over 180 members from 41 different countries. With objectives that address critical gaps in research, policy, and international interdisciplinary research capacity, ROSEnet aims to overcome fragmentation in conceptual innovation on old-age exclusion across the life course, in order to address the research-policy disconnect and tackle social exclusion among older people.

Jeni Warburton is Professor Emeritus and the John Richards Chair in Rural Aged Care Research at La Trobe University in Wodonga, Victoria, Australia. She has 20 years' experience of research into social policy, particularly relating to issues associated with an ageing population. Her main areas of expertise are in healthy and productive ageing as well as volunteering and community, with her current research focusing on ageing in rural communities. She has published widely both nationally and internationally on these topics, and her research has played a key role in the development of practice and policy around volunteering and social inclusion, particularly relating to older people.

Tanya Watson is a PhD graduate from the National University of Ireland Galway, School of Political Science and Sociology. Her PhD research, entitled "Altering Legacies as 'A Farmer in My Own Right': Married Women's Experiences of Farm Property Ownership in Ireland", uses narratives of personal experiences of farm women property owners to explore the extent to which women's agency in the rural economy is leveraged by property ownership. She received the Teagasc Walsh Fellowship in support of this research. Her research interests include women and agriculture, women's property ownership, changing gender relations on family farms, biographical narrative methodologies and identity.

Janine Wiles is Associate Professor in the School of Population Health, University of Auckland, New Zealand. Her research encompasses social/health geographies, critical social gerontologies, and community health; and links three themes: care, place and ageing. She has published numerous articles, book chapters and reports on topics including older people's resilience, well-being and identity and strengths-based approaches to ageing, comes and communities as sites or landscapes of care, the contextualised experiences of family or "lay" caregivers at home and in communities – in light of social and political changes and health care access for marginalised groups.

Rachel Winterton is Senior Research Fellow at the John Richards Centre for Rural Ageing Research, La Trobe Rural Health School, La Trobe University, Australia. Her research focuses on how rural communities, governments and organisations are managing and responding to challenges posed by population ageing through systems of governance, health and social infrastructure. She is internationally recognised for her work on rural ageing and voluntarism and is currently completing a series of projects with international collaborators exploring critical perspectives on volunteering in ageing rural communities. Other research interests include the implications of rural retirement migration for rural service provision, rural organisational capacity to facilitate age-friendly communities and the role of rural systems and structures in facilitating wellness for rural ageing populations.

Michael Woods is Professor and Personal Chair in the Department of Geography and Earth Sciences at Aberystwyth University, UK, as well as Co-Director of the Centre for Welsh Politics and Society/WISERD@Aberystwyth. His research interests focus on rural and political geography. Mike holds a prestigious European Research Council Advanced Grant, GLOBAL-RURAL, and is Co-ordinator of the €5m IMAJINE project, funded by the European Union's Horizon 2020 programme to examine territorial inequalities and spatial justice (2017–2021). He is editor of the *Journal of Rural Studies* and co-editor of the *Policy Press Book Series on Civil Society and Social Change*.

Preface

With the highest rates of population ageing worldwide, where better than in rural regions, small towns and villages to explore the processes and outcomes, experiences and representations, challenges and opportunities of twenty-first-century demographic change? Underlying this assertion is the growing importance of *rural gerontology* as an interdisciplinary, policy oriented and applied field of scholarship on the rural dimensions of ageing that, surprisingly, has received relatively little consideration to date. The aim of this book is to provide, for the first time, a landmark compendium that charts the intellectual evolution, contemporary scope and future prospects of rural gerontology. The book brings together contributions from leading scholars with a shared goal of advancing a more critical understanding of rural ageing and, in doing so, highlighting how rural ageing insights can inform many of the most important gerontological questions.

Rural Gerontology: Towards Critical Perspectives on Rural Ageing builds on a longstanding interest in ageing in rural contexts, which can be traced from the initial studies of the lives of older rural people in the early- and mid-twentieth century through to the first international conference on rural ageing at the turn of the twenty-first century. The chapters of this book inform the reader about the progression of this scholarship and acknowledge prominent rural ageing scholars along the way, many of whom are contributors to this volume. Of particular significance in prefacing the critical aspirations of this book is the ground breaking work of Norah Keating who, along with her colleagues, compelled us to think critically about rural ageing in her landmark Policy Press (2008) volume *Rural Ageing: A Good Place to Grow Old?*

This book began as a series of conversations among the three co-editors about the potential for an authoritative volume that would bring coherence to the burgeoning but seemingly disparate rural ageing literature. In the best of rural research traditions, it was on our long and winding road trips to meet with ageing communities throughout Connemara, the Kawarthas and Yarriambiack, for instance, that we came to recognize the need to showcase the intellectual breadth and impact of rural gerontology. And, in the best of interdisciplinary collaborative traditions, such conversations were only made possible because of the visiting scholar funding programs at our universities that brought us together in the mid-2010s, including La Trobe University's Transforming Human Societies Research

Focus Area Program, NUI Galway's Irish Centre for Social Gerontology Rural Ageing Fellowship Program and Trent University's Aging & Society Visiting Scholars Program. Indeed, this book is indebted to the foresight of Jeni Warburton, John Richards Centre for Rural Ageing Research (La Trobe University), Eamon O'Shea and Thomas Scharf, Irish Centre for Social Gerontology (National University of Ireland Galway), and Stephen Katz, Trent Centre for Aging & Society (Trent University) for creating new opportunities for international rural ageing research. The book is also inspired by the example of excellence in rural ageing scholarship, collaboration and mentorship set by University Professor Emeritus Alun Joseph (University of Guelph).

As co-editors, we are grateful to the contributing authors for their dedication to this book project, especially given the uncertain working conditions created by the COVID-19 global pandemic. We owe specific acknowledgements to Amber Colibaba (Trent University) for her editorial assistance without which this book would not have come to fruition and to Faye Leerink at Routledge for supporting the advancement of rural gerontology. This book was undertaken, in part, thanks to funding from the Canada Research Chairs program.

Mark Skinner,
Rachel Winterton,
Kieran Walsh
2021

Part I
Introduction

1 Introducing rural gerontology

Mark Skinner, Rachel Winterton and Kieran Walsh

Introduction

Rural ageing has become a defining demographic narrative of the twenty-first century. The enduring out-migration of rural youth, ageing in place of rural older residents and in-migration of retired older populations have awarded rural regions, small towns and villages with the highest rates of population ageing worldwide. With the global trend in rural population ageing, understanding what it means for rural people, households, communities, institutions, governments, economies, societies and cultures to be at the forefront of demographic change has become crucial for the development of informed research, policy and programmes. Though rural questions are not new in the scholarship on ageing, the aim of this book is to bring greater attention to the interdisciplinary field of study – *rural gerontology* – that offers important insights into today's most important gerontological problems. In doing so, and with a view to advancing a much needed critical understanding of rural ageing, the book provides the first comprehensive foundation of knowledge about rural gerontology, harnessing a burgeoning body of interdisciplinary scholarship on the rural dimensions of ageing.

Focusing on rural gerontology is timely because an increasingly robust international body of knowledge is emerging about the rural dimensions of, and approaches to, the study of ageing, old age and older populations. Indeed, a "rural turn" is already underway in gerontology. This is influenced by advancements within social, cultural, environmental, geographical and, more recently, critical gerontology calling for greater attention to rurality as both a context and, more importantly, a subject of ageing research. Notwithstanding that some gerontologists, as well as scholars of ageing from rural studies and across the broader social and health sciences, have long been interested in the lives of older rural people and rural environments of ageing, it is only in the past decade or so, that rural gerontology has emerged as a coherent field. What has been missing is an authoritative volume that describes, explains and expands inquiry into the constitution of the field and charts the evolution, contemporary scope and future prospects that distinguish rural gerontology as a body of scholarship. In taking up this opportunity, this book captures the momentum of recent calls by leading rural ageing scholars for greater engagement with critical perspectives and the development of a critical rural gerontology that, when combined with its foundational and thematic emphases, lays the groundwork for future developments in the field.

Rural gerontology

> *Decades of research on ageing in rural environments has resulted in limited progress towards developing a distinctive rural gerontology.*
>
> Rowles (1988: 115) in *Journal of Rural Studies*

By aiming to address gaps in knowledge about the scope and depth, perspectives and approaches that define rural gerontology as a field of study, this book seeks to respond to Rowles' (1988) observation that greater engagement with the diversity and complexity of experiences of rural ageing is needed – a view of the field that is still prescient today. More than three decades on, as evident in the program of the first *International Conference on Rural Aging* (Hermanova et al., 2001) and the international contributions to subsequent special issues and edited collections on rural ageing themes (e.g., Wenger, 2002; Brown and Glasgow, 2008, Keating, 2008; Milbourne, 2012; Glasgow and Berry, 2013; Hash, Jurkowski, and Krout, 2014; Hennessy, Means and Burholt, 2014; Skinner and Hanlon, 2016; Naskali, Harbison and Begum, 2019; Spelten and Burmeister, 2019), gerontologists and allied rural ageing scholars have begun to close this gap by expanding the theoretical breadth of rural ageing research to account for exclusion, identity and representation, among other identifiers of change and difference, as well as to strengthen the applied focus of gerontological scholarship on the issues facing rural populations (Currie and Phillip, 2019). With the growing momentum of rural ageing scholarship led by established and early career scholars, especially in the past decade, rural gerontology is finally emerging as a distinctive field of inquiry. To advance the field further, and what lies at the aspirational core of this book, it is essential to articulate, comprehensively and critically, the interdisciplinary development, contemporary scope and future prospects of rural gerontology.

Emergence of a distinctive rural gerontology

In their contemporary review of the field of rural gerontology, Scharf, Walsh and O'Shea (2016) observed that academic interest in "rural" as a context of ageing is long-standing. This is evident in the agrarian emphasis within historical studies of ageing – for instance, Cole's (1992) influential *The Journey of Life: A Cultural History of Aging in America* – as well as in the pioneering of social science research into the lives of older rural people that began in the mid-twentieth century (e.g., Youmans, 1967; Arensberg and Kimball, 1968; Coward and Lee, 1985; Wenger, 1988; Keating, 1991) and has led to the burgeoning of gerontological interest in the trends, processes and outcomes of rural demographic change that we see around the world today (Scharf, Walsh and O'Shea, 2016).

Propelled by the development and application of ecological perspectives within environmental gerontology (Kendig, 2003; Greenfield, 2012; Rowles and Bernard, 2013; Scheidt and Schwarz, 2013) and, more recently, spatial perspectives within geographical gerontology (Andrews et al., 2007; Skinner et al., 2018), rural has become a focal point for research on ageing contexts. Indeed, recent publications in the field argue for greater attention to rural ageing studies for their transferable insights into pressing program and policy issues facing contemporary ageing societies (e.g., retirement migration, social isolation and age-friendly communities) (Currie and Phillip, 2019). However, notwithstanding this intellectual trajectory, and the valuable contribution of the *International Rural Ageing Project* (1999) as well as the few but increasingly prominent national rural ageing research initiatives (e.g., Australia's *John Richards Centre for Rural Ageing Research* at La Trobe University and Ireland's Atlantic Philanthropies funded *Rural Ageing Observatory* at NUI Galway), rural gerontology is still considered by many, particularly those in the area of critical social sciences, to be in its early stages as a coherent field of inquiry (Cutchin, 2009; Burholt and Dobbs, 2012; Scharf, Walsh and O'Shea, 2016; Currie and Phillip, 2019). Within most scholarship and policy, for instance, there remains a failure to actively interrogate contemporary issues in ageing societies from a rural perspective and to understand how these issues themselves can emerge from within rural environments. There also remains an insufficient focus on adopting critical approaches in the analysis of the construction and propagation of rural ageing social phenomena.

As observed by Scharf, Walsh and O'Shea (2016), despite some gerontological studies continuing to view rural as merely a setting for empirical research rather than an ever-changing context that can potentially shape experiences and outcomes for older adults, an increasingly robust body of multi-disciplinary rural ageing scholarship – if not the foundations of a distinctive rural gerontology – is now beginning to emerge. Acknowledging the need to move the field beyond acting simply as a geographical location where people age, recent work by Skinner and Winterton (2018a, 2018b) explored specific aspects that are relevant to understanding a distinctive rural sub-strand of gerontology. The first relates to the mutually constitutive relationship between ageing and rural community sustainability, where older people are both reliant on, and simultaneously support, the capacity of their rural communities to enable ageing in place (Joseph and Skinner, 2012; Warburton and Winterton, 2017; Walsh et al., 2014). The second relates to the more radical transformation of rural settings, in response to trends relating to globalisation and counter-urbanisation. These dual trends have differential impacts on the capacity of rural communities to support population ageing and are part of an expanded depth of understanding rural approaches to gerontology (Winterton and Warburton, 2014; Skinner and Winterton, 2018b).

A third aspect of a distinctly rural gerontology is its interdisciplinarity, with rural studies of ageing drawn from geographical, sociological, demographic and health perspectives. Geographical perspectives on rural ageing have provided an understanding of how the various spaces, places and scales associated with the rural both shape the experiences of, and are shaped by, older adults (Skinner,

Andrews and Cutchin, 2018). Application of a sociological (particularly social gerontological) perspective has enabled a more critical approach in terms of how population ageing across and within these different settings is impacted by social factors, such as class, power, gender and culture, and national and international trends, such as globalisation and privatisation processes (e.g., Keating, 2008; Burholt and Dobbs, 2012). Demographic perspectives have allowed for the development of a more nuanced view of the dynamics of not just population ageing in rural settings, but other population processes, such as counter migration and outward migration, that shape older people's environments and their relational communities (e.g., Brown and Glasgow, 2008). The integration of health perspectives have provided a multi-level view of how health and ageing intersect in rural communities, including rural variations in health status, patterns of health service utilisation and a critical analysis of health infrastructure development (e.g., Gesler, Rabiner and Defriese, 2019).

Consequently, perspectives on rural gerontology are broad, and consider social, environmental and economic factors. However, despite the interdisciplinary benefits of bringing together these different disciplinary perspectives and their clear contribution to our understanding of rural ageing, there is still a tendency to think about these perspectives as distinct channels of research and scholarship (Scharf, Walsh and O'Shea, 2016). Thus, our capacity to fully embrace the interdisciplinarity that a rural ageing focus offers remains underdeveloped. Integrating these perspectives into a coherent foundational knowledge is key to addressing this deficit and fundamental for the continued development of the field of rural gerontology.

Recent developments towards a critical rural gerontology

Of particular importance for building our understanding of rural gerontology is the influence of three recent developments aimed at addressing the lack of engagement with contextually sensitive, policy oriented and critical perspectives within gerontology more broadly. First, is the "spatial turn" underway in social gerontology, led by greater engagement with geographical concepts of space and place in the study of older people's interactions with their environments, communities and societies (Andrews, Cutchin and Skinner, 2018; Andrews, Evans and Wiles, 2013). This work builds on and intersects with extensive and long-standing literatures on environmental gerontology and ageing in place (for an overview, see Wahl and Weisman, 2003; Rowles and Bernard, 2013; Peace, 2019) albeit with sometimes less of an emphasis on rural settings. Work within geographical gerontology is shedding light on how rural ageing contexts (spaces, places, environments, landscapes) are produced through the interactions of older people and the settings within which they experience and respond to ageing (Cutchin, 2009). The implication for rural gerontology is an increased recognition of the importance of context for understanding older adults' experiences of the processes and outcomes of ageing (Wiles et al., 2012). For instance, the various international contributions to Naskali, Harbison and Begum's (2019) recent volume on rural ageing in the circumpolar north highlight the challenges associated with ageing in not just geographical, but in particular socio-political and historical-cultural contexts (in their case the peripheral north) and how these lead to particular outcomes and opportunities for older rural residents and their communities.

Second, is the increasingly global perspective of ageing promoted by research and, especially, international policy developments that acknowledge the implications of globalisation and attendant challenges of rural decline for older rural populations and ageing rural communities (Scharf, Walsh and O'Shea, 2016). The "global turn" in rural ageing studies is, perhaps, best illustrated in the explicit engagement with rural settings within the WHO age-friendly communities framework (WHO, 2018), for instance, with the development and publication of national rural and remote age-friendly policies in Canada and elsewhere (Federal/Provincial/Territorial Ministers Responsible for Seniors, 2007). The implications for rural gerontology include a greater engagement with international contexts (migration, globalisation) in developing our understanding of rural ageing processes, outcomes and experiences as well as with rural experiences of global trends such as age-friendliness (Menec et al., 2015). An important development is the recognition of persistent gaps in gerontological scholarship among international contexts of rural ageing (e.g., global north versus global south), and the ways in which contemporary issues arising from and impacting on these contexts intersect with and are manifest within rural older people's lives. A notable example is the emphasis on understanding and addressing the implications of global pandemics, such as COVID-19, for older rural populations (Henning-Smith, 2020).

Third, is the overarching influence of culturally informed critical approaches to ageing studies that advocate for increasing attention to the diversity and complexity of ageing experiences, and how these are socially constructed (Katz, 2005). Work within critical gerontology is illuminating the ways in which gender, sexuality, race, class and ability relate to processes and outcomes, experiences and representations of ageing (Twigg and Martin, 2015). Implications for rural gerontology include the need to understand and challenge the construction (and contestation) of environments of rural ageing, the intersection of structural factors with everyday lived experiences and the myths and stereotypes that are associated with prevailing conceptions of ageing and rurality. Key studies highlight the increasing diversity of life histories, experiences and expectations of older adults in rural areas (Keating and Phillips, 2008) and how this diversity creates the likelihood that older rural adults will challenge the implicit and explicit rights and responsibilities associated with being a rural citizen (Chalmers and Joseph, 2006; Skinner and Winterton, 2018a). Nevertheless, notwithstanding these significant contributions, research on the rural ageing experience as a critical scholarly exploration is significantly underdeveloped. As part of gerontology's "critical turn", this limits the ability of the scientific community to illuminate complex interrelationships between structural change, discourses and environmental processes in rural people's lives and has very real implications for the development of meaningful policy and practice.

Taken together, the key developments outlined earlier point to the need for better understanding of the scope and depth of rural gerontology as a means of further advancing a contextually sensitive, globally applied and, increasingly, critically informed "rural turn" in gerontology more broadly as advocated for by Scharf, Walsh and O'Shea (2016) and Skinner and Winterton (2018a), among others (e.g., see Sidney, 2008). In seeking to not only coalesce but to also stimulate the growing coherency of the field, it is both important and timely to advance a critical approach in understanding rural gerontology. This endeavour requires careful attention to addressing fundamental gaps in our understanding of the diversity of older people's experiences and their interactions with and influence on the complexity of ageing societies. It asks critical questions, for instance, about the impacts of and responses to macro processes, from globalisation to global pandemics, on older rural residents and ageing rural communities; the influence (and idyll) of rurality within structures that empower and disenfranchise older people as well as their households, communities and economies; the differential roles and capacity of rural older people in creating and contesting their environments; and, perhaps most fundamentally, the absence of marginalised voices of older people within rural gerontology and rural ageing policy, public and popular discourses. In an effort to accelerate our collective momentum *towards critical perspectives on rural ageing*, this book gathers a selection of just some of the key contributors in this emerging field.

Foundations, scope and perspectives

The central argument of this book is that rural gerontology, particularly critical rural gerontology, provides constructive, distinct and essential contributions to the interdisciplinary study of ageing; contributions that advance our understanding of the relationship between rurality and ageing as well as the insights rural ageing bring to broader gerontological questions. To advance our knowledge about rural gerontology, the book draws together leading rural ageing scholars to explore the intellectual foundations, contemporary issues and future prospects of the field. International in reach, the 28 chapters that comprise the remainder of the book feature contributions from more than 65 established and emerging scholars at universities in 14 countries across five continents including Australia, Canada, China, Finland, Ireland, Israel, Malawi, Mexico, New Zealand, South Africa, the UK and the US. The chapters are illuminating in their focus on advancements in understandings of rural ageing from within gerontology (social, cultural and geographical) and from across rural studies and the social and health sciences (anthropology, geography, psychology, sociology, gender studies, planning and community development, social work and social policy, nursing and public health) and as such facilitate a comprehensive exploration of the diversity, complexity and implications of growing older in rural settings.

To achieve the foundational, scoping and critical objectives of the book, the chapters are organised into five major parts encompassing an introduction (Part I), interdisciplinary foundations (Part II), contemporary scope (Part III), emerging critical perspectives (Part VI) and a conclusion (Part V). The first part, featuring this introductory chapter (Skinner et al., Chapter 1), provides a brief historiography of the emergence of a "rural turn" in gerontology and the traditions and intellectual developments involved in the constitution of the field. The opening chapter sets the backdrop for the exploration of the foundations, scope and critical perspectives explored throughout the book.

The second part familiarises the reader with the broad areas of scholarship that constitute the foundations of the study of rural ageing. The emphasis is on introducing and explaining the contributions of the core fields that have informed the development of rural gerontology and critical rural gerontology over time, beginning with rural demography (Berry, Chapter 2), rural studies (Heley and Woods, Chapter 3), rural health (Hanlon and Poulin, Chapter 4), human ecology (Keating et al., Chapter 5) and social gerontology (Burholt and Scharf, Chapter 6). The five chapters in Part II set out the interdisciplinary foundation for understanding the growing coherence of rural gerontology as a field of inquiry with a particular view to providing insights into the increasingly global contexts of rural ageing and the influence of critical gerontology that are featured in the following parts of the book.

Expanding on the foundational chapters, the third part of the book explores the scope of contemporary rural gerontology from the global to local scales, emphasising key themes that represent the conceptual and empirical breadth of the field. The focus is on providing the reader a comprehensive overview of rural gerontology as a field of study, beginning with a global view of rural ageing issues, specifically from low- and middle-income countries (LMICs) of the global south (Amoah and Phillips, Chapter 7). This is followed by chapters that focus on national and regional (sub-national) issues, many of which have international significance, including ageing, gender and retirement (Duvvury et al., Chapter 8), rural-urban migration of older people (Cheng et al., Chapter 9), health and social care services (Glasgow and Doebler, Chapter 10), housing and homelessness (Petersen, Chapter 11), and transportation (Hansen et al., Chapter 12). Next are community, household and individual issues with chapters focusing on rural community development (Ryser et al., Chapter 13), age-friendly communities (Menec and Novek Chapter 14), ageing in place (Wiles et al., Chapter 15), inclusion and place-boundedness (Seppänen et al., Chapter 16), social connectivity and loneliness (Hennessy and Innes, Chapter 17), and care and caregiving (Urbaniak, Chapter 18). The 12 chapters in Part III provide a comprehensive view of the most salient contemporary issues that matter to rural gerontologists, encompassing demographic, socio-economic and cultural dimensions of rural ageing.

The fourth part examines the ways in which rural gerontology has engaged with critical perspectives, providing an account of the range of critical perspectives and approaches that have emerged in the field. This part begins with expositions on the application of postcolonial perspectives (Hoffman and Roos, Chapter 19), posthumanist traditions (MacLaren and Andrews, Chapter 20) and pragmatist approaches (Rowles and Cutchin, Chapter 21), followed by inquiries into emerging critical perspectives on social exclusion (Walsh et al., Chapter 22), active citizenship (Winterton and Warburton, Chapter 23), voluntarism and volunteering

(Colibaba et al., Chapter 24), poverty (Milbourne, Chapter 25), mental health and dementia (Herron and O'Shea, Chapter 26) as well as case studies of recent critical approaches to gerontechnology (Kosurko et al., Chapter 27) and older people, climate change and disasters (Carroll and Walker, Chapter 28). The ten chapters in Part IV elucidate the potential for rural critical gerontology to inform some of the most important questions within and beyond gerontology, including those related to decolonisation, stigma, environmental change and pandemics.

The fifth (and final) part features a concluding chapter (Winterton et al., Chapter 29) that draws together key debates and critiques from the contributions to the book and presents a new theoretically and conceptually informed framework that is representative of the twenty-first-century "rural turn" within gerontological studies. The closing chapter lays the foundations for a new critical rural gerontology that considers the diversity across, and within rural settlements globally, while also highlighting emergent issues that remain neglected within the literature in an effort to conclude the book with a critically informed agenda for future research that speaks to the need for reinforcing the importance of rural gerontology and the rural dimensions of ageing.

Concluding comments

Rural Gerontology: Towards Critical Perspectives on Rural Ageing makes several important and timely contributions to scholarship in gerontology as well as in rural studies and across the social and health sciences. As outlined in this introductory chapter, the chapters featured in the book inform our understanding of the "rural turn" in gerontology and what this means in terms of the evolution, scope and prospects for studying the rural dimensions of ageing. Taken together, the interdisciplinary contributions take up the challenge initially set by Rowles (1988) and address the more recent calls for inquiry into not only the constitution of a distinctive rural gerontology but also, and perhaps most importantly, the engagement of this field with critical perspectives. In doing so, they advance our theoretical and analytical framing of the complex ways in which the processes and outcomes, experiences and representations of ageing and rurality are entwined, thereby establishing at foundation of knowledge and approaches that will inform future developments in the field.

While certainly incomplete in covering of the full range of topics and time horizons of interest to scholars of rural gerontology, the foundational emphasis of the book is designed to be a useful point of departure for subsequent endeavours to uncover the full scope of the field. The emphasis within various chapters on under-researched contexts of rural ageing in the global south, for instance, provides an opportunity to move beyond the prevalence of rural ageing research in higher-income countries of the global north, towards greater inclusion of the diversity of voices of rural older adults, not to mention rural ageing researchers, from LMICs. Further underlying the aim to advance the field critically is a shared concern for the nascent application of rural gerontology in relation to under-considered experiences of rural ageing, particularly for older rural people who are racialised, Indigenous and lesbian, gay, bisexual, transgender, intersex or queer (LGBTIQ). With this in mind, and acknowledging that many of these voices are missing from the book, as they are from much of the current body of literature (for recent exceptions, see Grant and Walker, 2020; Ferreira, Leeson and Melhado, 2019; Pelcastre-Villafuerte et al., 2017), there is a need to further articulate the breadth of critical rural gerontology more comprehensively. Future dedicated volumes and/or special issues on the applications of critical rural gerontology in under-researched voices and contexts, akin to Naskali, Harbison and Begum's (2019) recent edited collection on rural ageing in the circumpolar north, would be welcome subsequent additions to the literature.

Looking forward, the ultimate aspiration of this book is to create a platform for new knowledge, opportunities and audiences for rural ageing scholarship – in effect to make a call to those who wonder why the experiences of people living in rural areas, small towns and villages matter so much to twenty-first-century demographic change – and, in doing so, to accelerate the development of a critical rural gerontology that will be evermore important in answering questions about the diversity of ageing, old age and older populations.

References

Andrews, G. J., Cutchin, M., McCracken, K., Phillips, D. R. and Wiles, J. (2007). Geographical gerontology: the constitution of a discipline. *Social Science & Medicine*, *65*(1), pp. 151–168.

Andrews, G. J., Cutchin, M. P, and Skinner, M. W. (2018). Space and place in geographical gerontology: Theoretical traditions, formations of hope. In: M. W. Skinner, G. J. Andrews and M. P. Cutchin, eds., *Geographical Gerontology: Perspectives, Concepts, Approaches*. London: Routledge.

Andrews, G. J., Evans, J. and Wiles, J. L. (2013). Re-spacing and re-placing gerontology: relationality and affect. *Ageing & Society*, 33(8), pp. 1339–1373.

Arensberg, C. M. and Kimball, S. T. (1968). *Family and Community in Ireland*, 2nd ed. Cambridge: Harvard University Press.

Brown, D. L. and Glasgow, N., (eds). (2008). *Rural Retirement Migration*. New York: Springer.

Burholt, V. and Dobbs, C. (2012). Research on rural ageing: where have we got to and where are we going in Europe? *Journal of Rural Studies*, 28(4), pp. 432–446.

Chalmers, A. I. and Joseph, A. E. (2006). Rural change and production of otherness: the elderly in New Zealand. In: P. Cloke, T. Marsden, P. and Mooney, eds., *Handbook of Rural Studies*. London: SAGE Publications, p. 388.

Cole, T. R. (1992) *The Journey of Life: A Cultural History of Aging in America*. Cambridge: Cambridge University Press.

Coward, R. T. and Lee, G. R., (eds). (1985). *The Elderly in Rural Society*. New York: Springer.

Currie, M. and Phillip, L. (2019). Rural ageing. In: D. Gu and M. E. Dupre, eds., *Encyclopedia of Gerontology and Population Aging*. New York: Springer, p. 9.

Cutchin, M. (2009). Geographical gerontology: new contributions and spaces for development. *The Gerontologist*, 49(3), pp. 440–444.

Federal/Provincial/Territorial Ministers Responsible for Seniors. (2007). *Age-Friendly Rural and Remote Communities: A Guide*. Ottawa, ON: Public Health Agency of Canada, Division of Aging and Seniors.

Ferreira, J. P., Leeson, G. and Melhado, V. R. (2019). Cartographias do envelhecimento em context rural: notas sobre raca/etnia, genero, classe e escolaridada [Cartographies of aging in the rural context: race/ethnicity, gender, schooling and social class]. *Trabalho, Educação e Saúde*, 17(1), e0017612. DOI: 10.1590/1981-7746-sol00176.

Gesler, W. M., Rabiner, D. G. and Defriese, G. H., (eds). (2019). *Rural Health and Aging Research: Theory, Methods, and Practical Applications*, 2nd ed. New York: Routledge.

Glasgow, N. and Berry, E. H., (eds.). (2013). *Rural Aging in 21st Century America*. Dordrecht: Springer.

Grant, R. and Walker, B. (2020). Older Lesbians' experiences of ageing in place in rural Tasmania, Australia: an exploratory qualitative investigation. *Health and Social Care Community*. Epub ahead of print 22 May. DOI: 10.1111/hsc.13032.

Greenfield, E. A. (2012). Using ecological frameworks to advance a field of research, practice, and policy on aging-in-place initiatives. *The Gerontologist*, 52(1), pp. 1–12.

Hash, K. M., Jurkowski, E. T. and Krout, J. A., (eds.). (2014). *Aging in Rural Places: Programs, Policies, and Professional Practice*. New York: Springer.

Hennessy, C., Means, R. and Burholt, V., (eds.). (2014) *Countryside Connections: Older People, Community and Place in Rural Britain*. Bristol: Policy Press.

Henning-Smith, C. (2020). The unique impact of COVID-19 on older adults in rural areas. *Journal of Aging and Social Policy*, 32, pp. 396–402.

Hermanova, H., Brown, D. K., Goins, R. T. and Briggs, R. (2001). The first international conference on rural aging: a global challenge. *The Journal of Rural Health*, 17, pp. 303–304.

International Rural Aging Project. (1999). *Shepherdstown report on rural aging: the result of the expert group meeting*, May 22–25. Shepherdstown, University of West Virginia.

Joseph, A. E. and Skinner, M. W. (2012). Voluntarism as a mediator of the experience of growing old in evolving rural spaces and changing rural places. *Journal of Rural Studies* 28, pp. 380–388.

Katz, S., (ed.) (2005). *Cultural Aging: Life Course, Lifestyle and Senior Worlds*. Peterborough: Broadview Press.

Keating, N. C. (1991). *Aging in Rural Canada*. Toronto, ON: Butterworths Canada.

Keating, N., (ed.). (2008). *Rural Ageing: A Good Place to Grown Old?* Bristol: Policy Press.

Keating, N. and Phillips, J. (2008). A critical human ecology perspective on rural ageing. In: N. Keating, ed., *Rural Ageing: A Good Place to Grow Old?*. Bristol: Policy Press, pp. 1–10.

Kendig, H. (2003). Directions in environmental gerontology: a multidisciplinary field. *The Gerontologist*, *43*(5), pp. 611–614.

Menec, V., Bell, S., Novek, S., Minnigaleeva, G.A., Morales, E., Ouma, T., Parodi, J.F. and Winterton, R. (2015). Making rural and remote communities more age-friendly: experts' perspectives on issues, challenges, and priorities. *Journal of Aging & Social Policy*, 27(2), pp. 173–191.

Milbourne, P. (2012). Growing old in rural places (special issue). *Journal of Rural Studies*, 28(4), pp. 315–317.

Naskali, P., Harbison, J. R., and Begum, S., (eds.). (2019). *New Challenges to Ageing in the Rural North: A Critical Interdisciplinary Perspective*. New York: Springer.

Peace, S., (ed.). (2019). *Environment and Ageing: Space, Place and Materiality*. Bristol: Policy Press.

Pelcastre-Villafuerte, B. E., Meneses-Navarro, S., Ruelas-González, M. G., Reyes-Morales, H., Amaya-Castellanos, A. and Taboada, A. (2017). Aging in rural, indigenous communities: an intercultural and participatory healthcare approach in Mexico. *Ethnicity & Health*, 22(6), pp. 610–630.

Rowles, G. D. (1988). What's rural about rural aging? An Appalachian perspective. *Journal of Rural Studies*, 4(2), pp. 155–124.

Rowles, G. D. and Bernard, M. A., (eds.). (2013). *Environmental Gerontology: Making Meaningful Places in Old Age*. Cham: Springer.

Scharf, T., Walsh, K. and O'Shea, E. (2016). Ageing in rural places. In: D. Shucksmith and D. L. Brown, eds., *Routledge International Handbook of Rural Studies*. London: Routledge, pp. 50–61.

Scheidt, R. J. and Schwarz, B., (eds.). (2013). *Environmental Gerontology: What Now?* London: Routledge.

Sidney, I. (2008). Towards a critical rural gerontology: a multi-disciplinary journey to a place where others age. *Generations Review*. British Society of Gerontology. Available at: www.britishgerontology.org/publications/generations-review-the-newsletter.

Skinner, M. and Hanlon, N., (eds.). (2016). *Ageing Resource Communities: New Frontiers of Rural Population Change, Community Development and Voluntarism*. London: Routledge.

Skinner, M. W., Andrews, G. J. and Cutchin, M. P., (eds.). (2018). *Geographical Gerontology: Perspectives, Concepts, Approaches*. London: Routledge.

Skinner, M. W. and Winterton, R. (2018a). Interrogating the contested spaces of rural ageing: implications for research, policy and practice. *The Gerontologist*, 58(1), pp. 15–25.

Skinner, M. W. and Winterton, R. (2018b). Rural ageing: dynamic places, contested spaces. In: M. W. Skinner, G. J. Andrews and M. P. Cutchin, eds., *Geographical Gerontology: Perspectives, Concepts, Approaches*. London: Routledge, pp 136–148.

Spelten, E.R. and Burmeister, O.K. (2019). Growing old gracefully in rural and remote Australia? Special issue: ageing in rural Australia. *Australian Journal of Rural Health*, 27(4), pp. 272–274.

Twigg, J. and Martin, W., (eds.). (2015). *The Handbook of Cultural Gerontology*. London: Routledge.

Wahl, H. W. and Weisman, G. D. (2003). Environmental gerontology at the beginning of the new millennium: reflections on its historical, empirical, and theoretical development. *The Gerontologist*, 43(5), pp. 616–627.

Walsh, K., O'Shea, E., Scharf, T. and Shucksmith, M. (2014). Exploring the impact of informal practices on social exclusion and age-friendliness for older people in rural communities. *Journal of Community & Applied Social Psychology*, 24(1), pp. 37–49.

Wenger, G. C. (1988) *Old People's Health and Experiences of the Caring Services: Accounts from Rural Communities in North Wales*. Liverpool: Liverpool University Press.

Wenger, C. G. (2002). Introduction: intergenerational relationships in rural areas. *Ageing & Society*, 21(5), pp. 537–545.

Wiles, J. L., Leibing, A., Guberman, N., Reeve, J. and Allen, R. E. S. (2012). The meaning of "aging in place" to older people. *The Gerontologist*, 52(3), pp. 357–366.

Warburton, J. and Winterton, R. (2017). A far greater sense of community: the impact of volunteer behaviour on the wellness of rural older Australians. *Health & Place*, 48, pp. 132–138.

Winterton, R. and Warburton, J. (2014). Healthy ageing in Australia's rural places: the contribution of older volunteers. *Voluntary Sector Review*, 5(2), pp. 181–201.

World Health Organization. (2018). *The global network for age-friendly cities and communities: Looking back over the last decade, looking forward to the next*. Geneva: World Health Organization.

Youmans, E. G. (1967). *Older Rural Americans: A Sociological Perspective*. Lexington: University of Kentucky Press.

Part II

Interdisciplinary foundations

2 Demographic ageing and rural population change

E. Helen Berry

Introduction

Ageing is studied using demography, which is a foundational discipline on the dynamics of human populations (e.g., see Kulcsár and Curtis, 2012). The purpose of this chapter is to describe the ageing of rural populations. Population ageing is the increase in the average age of a country's or region's population as a whole, which occurs when its birth rates decline, its life expectancy increases and elders and youth move into or out of the place (Gavrilov and Heuveline, 2003). Ageing is most acute in rural places due to their relatively smaller populations (Poston and Micklin, 2018).

While most of this chapter will focus on rural ageing, it is useful to realize that the definition of rurality is problematic. The term rural describes different types of places in that it varies from country-to-country, and even in the ways that the word rural is used. For example, India defines rural as places with a density of less than 400 persons per square kilometre with less than 5000 total population but in which more than a quarter of the male working population are engaged in agriculture, thus incorporating gender, population, occupation and geography in the definition (Dunn, 1993; Berry, 2015; John, 2019). In the United States, the US Census defines rural much less specifically, describing rural as any place not urban (US Census, 2020) although for some governmental purposes rural is classified as incorporating types of land use, density of population and distance to higher population areas (Ratcliffe et al., 2016). The effect is that rural varies dramatically from country to country and region to region with some countries like Afghanistan being all but entirely rural while others, notably Singapore, are completely urban.

Virtually every country is experiencing increases in the median age of the population which is the age at which half the population is older and half younger. Population ageing is producing dramatic social changes in the nature of family, intergenerational relations, workforce composition and governmental services (like social security/insurance systems, discussed later in this text). But readers should understand that rural places are becoming older more rapidly than are urban ones. An example of this can be found in the US: from 1920 to 1940 the median age of rural Americans was five years younger than that of urban Americans.

Between 1950 and 1970, the rural-urban age difference narrowed, leading to a reversal in 1980 when median ages in rural areas exceeded the median in the urban US, a pattern that continues into the 2020s (Kirschner, Berry and Glasgow, 2006). This chapter will describe the demographic processes behind ageing. Other chapters in this text describe the many other social changes associated with ageing (e.g., see Glasgow and Doebler, Chapter 10; Duvvury et al., Chapter 8; Hansen et al., Chapter 12).

Population ageing: how ageing happens – first, birth rates decline

The ageing of a population comes about as a function of the three fundamental demographic variables: births, deaths and migration, of which, interestingly, birth rates are the most important. As fewer children are born, the average age of a population, by default, becomes older. In much of the developed world the number of persons over age 40 outnumbers those in younger ages (Dunn, 1993). A contributing factor is that life expectancy throughout the world is increasing as mortality in the older age groups declines (Roser, Ortiz-Ospina and Ritchie, 2020). Worldwide, life expectancy has doubled since 1900, as well as increased at all ages (Riley, 2005).

To explain how low fertility, in combination with longer lives, results in a society that is ageing, please see Table 2.1. There is much international variation in the median age. In Japan, the median age is 47.3 while the United Kingdom's median age, at 40.5 is relatively younger than Japan's. When rural and urban areas are compared, differences also exist: England's most rural areas had an average age of 44.8 in 2016, five years older than its major urban centres (Small Area Population Estimates, 2016).

Demographers use the Total Fertility Rate (TFR) to summarize the number of children a women would have if, throughout her childbearing years, she bore children at age-specific rates (Shryock and Siegel, 1975). If the TFR is below 2.1, then fertility is below the level required to replace the population. Afghanistan's TFR of 5.0 indicates that the women of the country, on average, have five children (see Table 2.1). On the other hand, Thailand's TFR, at 1.5, indicates women have an average of only 1.5 children.

Take note that in Table 2.1, countries with the lowest fertility rates also have substantially older populations and higher proportions of persons over ages 65 and 85. The lower birth rates and smaller proportions in the younger ages thereby directly increase the median age of the population. A striking example of this can be found in the rural areas of Japan. Selena (2017) and Matanle (2008) report that Japan has become a 93% urban nation (also reported in Table 2.1). The urbanisation has occurred in large part because young adults have moved from rural prefectures to the large urban corridors for jobs and universities, taking their youth and childbearing potential with them. Their parents and grandparents continue to live in rural areas. Rural towns have few young people because job, university and economic opportunities are located in cities. Lack of opportunity leaves rural

Table 2.1 Population characteristics for 20 countries

Country	*Total population in 1000s (a)*	*Percent rural (b)*	*Median age (a)*	*Dependency ratio (a)*	*Aged dependency ratio (a)*	*Total fertility rate (TFR) (a)*	*Population percentage age 65+ (a)*	*Population percent age 85+(a)*	*Life expectancy at birth (a)*	*Infant mortality rate (a)*
Argentina	44694	8.1	31.7	58.8	18.5	2.2	11.8	1.3	77	10
Japan	126168	8.4	47.3	69.8	48.2	1.4	28.4	4.8	86	2
United Kingdom	65105	10.0	40.5	55.7	28.3	1.9	18.2	2.6	81	4
Sweden	10041	12.6	41.2	61.0	32.8	1.9	20.4	2.6	82	3
United States	329256	19.3	38.1	53.0	24.5	1.9	16.0	2.0	80	6
Russia	142143	25.6	39.6	46.8	21.5	1.6	14.7	1.7	71	7
Italy	62247	29.6	45.5	54.5	33.5	1.4	21.7	3.6	82	3
China	1384689	40.9	37.4	39.8	15.8	1.6	11.3	0.7	76	12
Guatemala	16581	49.0	22.1	64.0	7.3	2.9	4.5	0.2	72	23
Thailand	68616	50.5	37.7	38.3	15.2	1.5	11.0	0.8	75	9
India	1296834	66.0	27.9	50.1	9.6	2.4	6.4	0.3	69	38
Burma	55623	69.4	28.2	47.6	8.4	2.1	5.7	0.3	69	34
Afghanistan	34941	74.5	18.8	77.1	4.6	5.0	2.6	0.1	52	108
Eswatini	1087	76.2	21.7	61.8	6.1	2.6	3.8	0.2	57	47
Ethiopia	108386	87.0	17.9	85.8	5.5	4.9	3.0	0.1	63	48

Median Age = Age at which half the population is older and half younger; Dependency Ratio = ((0-14 + 65 +)/15-64)*100); Aged Dependency Ratio = (Ages 65+0Ages 15-64)*100); Total Fertility Rate = Number of children a woman would have if bearing children at age-specific rates through her lifetime; Infant Mortality Rate = Deaths during the first year of life per 1000 infants born alive;

Sources: (1) U.S. Census International Data Base, 2018 (2) https://tradingeconomics.com/japan/rural-population-percent-of-total-population-wb-data.html

and particularly agricultural areas with fewer and fewer residents after the youth leave. The effect is that large parts of the Japanese countryside are emptying out, with half of rural communities unable to support municipal services and projected to disappear, barring an influx of youth (Masuda, 2014 as quoted in Selena, 2017).

Increased longevity plays a role

Longer life expectancy is often assumed to be associated with ageing but has less influence than one would suppose. Life expectancy is essentially a sum of the individual's probability of dying at every age during the biological life course. Throughout most of human history, the likelihood of dying was highest in the earliest years of life. Wrigley's (1972) estimates of life expectancy in pre-industrial England were low – between 37 and 43 years of age. Durand reports life expectancy of 25–30 for the Roman Empire of 100–200 A.D. (1960). Estimates in 2018 are for substantially longer lives, ranging from the low of 52 and 57 for Afghanistan and Eswatini (formerly Swaziland), two countries that are about 75% rural. Japan, that is highly urban, has a projected life expectancy of age 86 (Table 2.1).

Because life expectancy is heavily influenced by deaths at younger ages, expectancy becomes longer when mortality at younger ages declines as well as when health care in older ages improves. One theory explaining the decline in birth rates is that births tend to decline when infant mortality declines (Kirk, 1996). Certainly, births also decline as countries urbanize (Kirk, 1996). Turning again to Table 2.1, it should be clear that the infant mortality rate (IMR) is lower among countries where birth rates are also low and life expectancy and median age are higher. There is, however, no magic number at which IMRs are so low that birth rates suddenly begin to decline. Rather, birth rates and IMRs tend to decline together because they are both related to the improved health of the population. Also related is the presumption that if couples have a child when the IMR is high, they should have another child as insurance so that they may be certain that at least one child will survive. Low infant mortality rates are thus indicators of the overall health of the population.

Life expectancy tends to be associated with rurality in that when the overall health of a population is good, persons live longer. Thus, increased expectancy is related to both declines in infant mortality and improved health for older persons, whether in rural or urban places. But because access to health care is less easily accessed in rural places, more rural places tend to have higher IMRs. Examples include Guatemala and India that are 49% and 66% rural, respectively. India, the more developed country, has a surprisingly high infant mortality rate of 38 deaths per 1000 live births, while Guatemala, somewhat less rural, although experiencing extensive internal unrest, has a lower rate at 18. Compare those rates to the rates in the much less rural Argentina (8.1% rural) with an IMR of 10 or Sweden (12.6% rural) at three. Rurality is not the cause but is closely correlated with the rates.

Migration is key to rural ageing

The third factor contributing to ageing, and the one most relevant to rural areas, is migration (see Cheng et al., Chapter 9). When persons in a particular age group either migrate in or migrate out of a place, the addition or subtraction of those persons changes the average age of the population. Rural areas in most of the world have tended to have higher birth rates than urban places and are prime sending regions for young adults who move away. Indeed, it is a rite of passage for young adults to depart rural places to join the military, head to college, move for marriage or move to the city for more plentiful employment opportunities (Garasky, 2002; Brooks et al., 2011).

Migration is also a factor for older persons nearing or at retirement ages (Pickering and Crooks, 2019). Places with natural amenities are attractive, especially those places with gentler climates and plentiful recreational activities. These areas often experience the in-migration of older, more affluent persons who have retired and have time on their hands to enjoy the recreational options available, thereby increasing the age of the destination place (Pickering and Crooks, 2019; Cromartie, 2018). This type of rural retirement migration, also called amenity-migration, is best documented in the US (Johnson and Lichter, 2013) and throughout Europe but also, increasingly, in vacation destinations worldwide (e.g., Williams and McIntyre, 2011).

As the Japanese case illustrates, when young people move out, elders tend to stay (Cromartie, 2018; Johnson, Field and Poston, 2015; Masuda, 2014 as reported in Selena, 2017; Matanle, 2008), an extensively documented worldwide (Bailey, Jensen and Ransom, 2014; Kulcsár and Curtis, 2012). The particularly evocative text by Carr and Kefalas, *Hollowing out the Middle: The Rural Brain Drain and What it Means for America* (2009) described the emptying out of the rural Midwestern US rural out-migration due to globalisation of world markets and agriculture is associated with fewer and fewer rural jobs, as well as an overall de-industrialisation of the economy as a whole.

While the migration of youth makes rural areas older, occasionally such migration may make a place younger. Specifically, Johnson and Lichter (2009) document that Hispanic in-migration to areas in the US Great Plains region, during the period 1990–2006 halted an overall population decline and led to an increase in births in several rural communities. The US non-Hispanic white population has long had a declining birth rate, whether in rural or urban regions. In 2018, the overall birth rate for the US was 1.73 but the non-Hispanic white TFR was 1.64 and non-Hispanic Asian population was at 1.5 children per woman (Martin et al., 2019). But for Hispanics, the largest US ethnic group, the TFR was 1.96 (Martin et al., 2019), thereby raising the overall fertility rate. International immigrants generally move to urban areas, but, as Massey (2008) and Kandel and Parrado (2005) illustrate, immigrants also move to new destinations in rural places. The in-migration of this group to the rural Plains and other so-called New Destinations areas has kept these populations younger and more diverse than might have otherwise been anticipated.

Elsewhere, both in and out migration are associated with ageing in both rural and urban contexts. For example, both Shanghai, China and Seoul, South Korea are experiencing the in-migration of young professionals and workers who are single or have no children (Kim, 2005; Merli and Morgan, 2011). The effect of their out-migration from rural areas is that the places they leave have fewer people of childbearing age making the sending places older (Kim, 2005; Merli and Morgan, 2011). But the receiving cities, Shanghai and Seoul, are also ageing because the childbearing of these young singles is quite low. Migration, in other words, can have the effect of ageing both the sending and the receiving populations.

Illustrating ageing with population pyramids

The ageing process is strikingly described using population pyramids. Three pyramids are shown in Figure 2.1 for the countries of Afghanistan, Burma and Japan. In the pyramids, the number of males in five-year age groups is on the left side while the number of females in the same age groups is on the right.

Afghanistan is three-quarters rural and its pyramid shows a very young population, with more males and females in the 0–4 age group than in the 5–9 group; more children ages 5–9 than 10–14, and so forth, with few individuals, either male or female over age 65, relative to younger age groups. Another way to describe the youthfulness of Afghanistan is to realize that, for a population of just under 35 million, 41% are under age 15 while only 3% are age 65 and older (Table 2.1).

The pyramid for Burma (aka Myanmar) illustrates a population that is less rural, at 69%, less youthful and more middle-aged. There are fewer children age 0–4 than 5–9, but its population age 20–24 is approximately the same size as the 15–19 year and 10–14 year groups. The size of each group, relative to the ages younger than themselves, is closer to the group below it on the pyramid, resulting in the relatively "fatter" pyramid for Burma than for Afghanistan. The under 15 population is only 27% of the total, compared to Afghanistan's 41% while Burma's over age 65 group is twice as large as is found in Afghanistan. Burma's slightly greater urbanism suggests an increasingly modernising country that is still heavily reliant on its rural population and economy.

Japan is one of the most urban (93%) and the oldest country, with 28% of its population over age 65 and only 13% under age 15. The pyramid for Japan looks radically different from the previous two, with larger proportions of the population in the middle age and older age groups. The group of women older than the childbearing ages of 15–44, is larger than those currently in childbearing ages. Women of childbearing age have not yet reproduced themselves by bearing as many children as their older counterparts. They are rapidly "ageing out" of childbearing and, even if the women who are earliest in their childbearing years begin to have many more infants, their ability to reproduce the older generation is limited by the smaller size of their own cohort.

The population pyramids, for Afghanistan, Burma and Japan show three very different fertility, ageing and rural regimes. Afghanis are quite young with lots of children and are primarily rural and agricultural. The Burmese still have a young

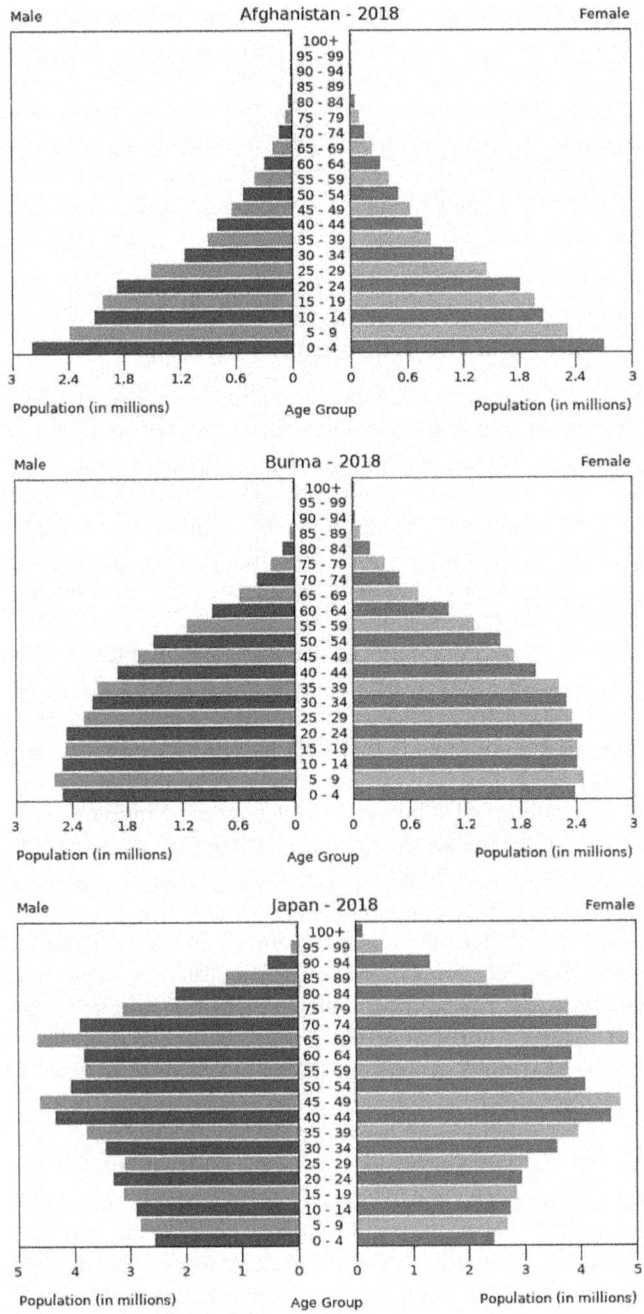

Figure 2.1 Population pyramids for Afghanistan, Burma and Japan, 2018

Source: Census Bureau (2018)

population but are having fewer children. The lowered birth rate for Burma means deaths are only just being replaced by births so that their population will continue to age although persons of childbearing age are still replacing themselves with new births. Japan, on the other hand, is a country where population is declining and has been since 2008, and where rural populations are disappearing. The older cohorts of Japanese had more children than those who are currently in the child-bearing years. Families are shrinking in size. There are many aunts and uncles, grandparents and great grandparents but few grandchildren and few nieces, nephews and cousins.

Why does ageing matter? Dependency ratios, aged dependency ratios and the demographic dividend

The dependency ratio and aged dependency ratio provide ways to understand the impact of older populations on a country's working populations. The dependency ratio is the sum of persons under age 15 plus persons over 65 divided by the working age population, then multiplied by 100. The ratio is a measure of how many people in the non-working age groups (old and young) there are relative to the working ages. The dependency ratio together with the aged dependency ratio, which compares those over age 65 to those ages 15–64, is another way to think about age structure.

Consider Italy when examining these two measures (Table 2.1). Italy's dependency ratio is 54.5 with an aged dependency ratio of 33.5. That means that for every 100 working age people there are nearly 55 (34 elders and 20 children) people not of working age. Implied is that Italians must spend more to support their elders since there are half-again more elders than children. By contrast, Eswatini has a dependency ratio of 61.8 but an aged dependency ratio of 6.1. Eswatini has few elders but ten times the school age population. One can similarly see this by looking at the percentage of the population over age 65 that is 3.77% in Eswatini. The percentage for Italy is just under 22% over age 65.

Implicit is that where there are lots of children, a country can look forward to an increasing workforce, but must also provide infrastructure to support the expanding workforce (schools, roads, jobs, housing) and must do so with a workforce that is smaller than the future workforce will be. The circumstance is called a demographic dividend and is, in large part, what fuelled the Industrial Revolution: birth rates had historically been high in England, France and the Netherlands in the 1700–1800s but infant mortality was also high (Wrigley, 1972). In that era, children provided the social security system for their parents because there was no other retirement system, or, as Caldwell (2005) puts it, the larger the number of surviving children, the more wealth parents have since many children mean plenty of security in one's older age. In a rural, agrarian society, children provide workers for parents so the family can produce more goods. With industrialisation and accompanying changes in farming, families increasingly leave the land, providing a surplus of workers for cities (Wrigley, 1972). Because younger persons are more likely to move, the migration results in older persons left behind and the resulting difference in age structure between rural and urban places. Ageing is, in essence, a reverse effect from the youthful fertility and migration that fuelled the Industrial Revolution. When there are more children surviving to adulthood, people have fewer children (Caldwell, 2005). The children live into and through their childbearing ages and into old age. As lives become easier, as birth rates fall from formerly high levels, and as life expectancies increase, societies age.

Ageing matters first, and foremost, because it changes the fundamental basis upon which societies focus. Societies are, in many ways, reproduction factories. By having children and devoting between 15 to 20 years to raise and socialize each child, societies reproduce themselves. Families, by reproducing the norms, traditions, behaviours of the parents, reproduce those norms, traditions and behaviours. Societies are, in essence, machines designed to reproduce the past. When birth rates decline, the society changes from being primarily focused on the children to caring for the older members of society. Care for the young will continue, but caring for older generations as well as younger ones means that there is a substantially greater focus on elder health care, on finding ways to deal with the increased disability of the aged and on finding ways to support the elderly, as opposed to supporting the young. Indeed, these are the topics described throughout the rest of this text (e.g., see Hanlon and Poulin, Chapter 4; Duvvury et al., Chapter 8; Herron and O'Shea, Chapter 26).

This is where a demographic dividend can come into play. A demographic dividend may create an economic benefit, a benefit produced by the increased proportion of the population that is working while the proportion of children declines and the percentage of retirees is not yet large. When this occurs, the society has the opportunity to build on the dividend of lots of working age people, while not requiring heavy investment in schools or health care. Examples of building on a dividend can be seen with the rapid growth of South Korea and Taiwan. Rwanda, too, has taken advantage of the dividend by stimulating family planning use and lowering child mortality, investing in the accelerated economic growth that comes with fewer children to support and an older, but not yet truly aged, population (Population Reference Bureau, 2012).

Conclusion

What makes ageing so important for rural areas is that it is associated with both (1) the departure of youth and (2) a change in the proportion of elderly. In the first case, schools close, towns lose enough population that their grocery stores and gasoline stations close and the taxable population declines, but there is still need for health care and services for a rising proportion of elders. In places that experience the in-migration of elders to a high amenity area, there is an influx of relatively higher income elders whose presence has resulted in a demand for new kinds of services, and hence an associated influx of younger migrants as well as the elders (e.g., Johnson and Lichter, 2013). Either way, rural places become older relative to urban ones, while urban places are becoming younger. The resultant challenges for rural places are topics that are addressed extensively in the several chapters in this text, particularly those in Part III (Contemporary Scope) that look at providing services in rural areas and in Part IV (Emerging Critical Perspectives).

Understanding that the reason for ageing is not simply longer lives is important. Ageing in rural places is influenced by low fertility combined with the out-migration of youth who leave elders behind, or, by the in-migration of relatively well-off elders to high amenity rural places. Caring for the needs of elders who are in rural places will depend on whether they are long-time residents who choose to age-in-place or on whether they are new in-migrants who have moved to rural residences. The requirements of each group are likely to be influenced by the economic characteristics of the rural setting itself, in that well-off aged populations are able to command more resources than are poorer, less mobile populations.

Acknowledgements

The author acknowledges the assistance of Corinna Knowles and the support of the Utah State University Agricultural Experiment Station Grant #1361 and WAAESD multistate project W4001: Social, Economic and Environmental Causes and Consequences of Demographic Change in Rural America.

References

Bailey, C., Jensen, L. and Ransom, E., (eds). (2014). *Rural America in a Globalizing World: Problems and Prospects for the 2010s*. Morgantown: West Virginia University Press.

Berry, E. H. (2015). *What is Rural?* (Presidential Address). Madison, WI: Rural Sociological Society.

Brooks, T., Lee, S. L., Berry, E. H. and Toney, M. B. (2011). The effects of occupational aspirations and other factors on the out-migration of rural youth. *Journal of Rural and Community Development*, 5, pp. 19–36.

Caldwell, J. C. (2005). On net intergenerational wealth flows: an update. *Population and Development Review*, 31(4), pp. 721–740.

Carr, P. J. and Kefalas, M. J., (eds). (2009). *Hollowing Out the Middle: The Rural Brain Drain and What It Means for America*. Boston: Beacon Press.

Cromartie, J. B. (2018). *Rural aging occurs in different places for very different reasons*. Research and Science, U.S. Department of Agriculture. Available at: www.usda.gov/media/blog/2018/12/20/rural-aging-occurs-different-places-very-different-reasons [Accessed 16 December 2019].

Dunn, M. G., (ed.). (1993). *Exploring Your World: The Adventure of Geography*. Washington, DC: National Geographic Society. Available at: www.nationalgeographic.org/encyclopedia/rural-area/ [Accessed 12 January 2020].

Durand, J. D. (1960). Mortality estimates from Roman tombstone inscriptions. *American Journal of Sociology*, 65(4), pp. 365–373.

Garasky, S. (2002). Where are they going? A comparison of urban and rural youths' locational choices after leaving the parental home. *Social Science Research*, 31(5), pp. 409–431.

Gavrilov, L. A. and Heuveline, P. (2003). Aging of population. In: P. Demeny and B. McNicoll, eds., *The Encyclopedia of Population*. New York: Macmillan, pp. 32–37.

John, P. O. (2019). *What is Rural?* Rural Information Center. USDA National Agricultural Library (Revised and updated by J. Flood). Available at: www.nal.usda.gov/ric/what-is-rural#INTRO [Accessed 12 January 2020].

Johnson, K. M., Field, L. M. and Poston, D. L. Jr. (2015). More deaths than births: subnational natural decrease in Europe and the United States. *Population and Development Review*, 41(4), pp. 651–680.

Johnson, K. M. and Lichter, D. T. (2009). Immigrant gateways and Hispanic migration to new destinations. *International Migration Review*, 43(3), pp. 496–518.

Johnson, K. M. and Lichter, D. T. (2013). Rural retirement destinations: natural decrease and the shared demographic destinies of elderly and Hispanics. In: N. Glasgow and E. H. Berry, eds., *Rural Aging in 21st Century America*. Dordrecht: Springer, pp. 275–294.

Kandel, W. and Parrado, E. A. (2005). Restructuring of the U.S. meat processing industry and new Hispanic destinations. *Population and Development Review*, 31(3), pp. 447–461.

Kirk, D. (1996). Demographic transition theory. *Population Studies: A Journal of Demography*, 50(30), pp. 361–387.

Kim, D. S. (2005). Theoretical explanations of rapid fertility decline in Korea. *The Japanese Journal of Population*, 3(1), pp. 1–25.

Kirschner, A., Berry, E. H. and Glasgow, N. (2006). The changing faces of rural America. In: W. A. Kandel and D. L. Brown, eds., *Population Change and Rural Society*. Amsterdam: Springer, pp. 53–74.

Kulcsár, L. J. and Curtis, K. J., (eds). (2012). *International Handbook of Rural Demography*. Dordrecht: Springer.

Martin, J. A., Hamilton, B. E., Osterman, M. J. K. and Driscoll, A. K. (2019). *Births: Final data for 2018*. National Vital Statistics Reports No. 68(13). Available at: https://www.cdc.gov/nchs/data/nvsr/nvsr68/nvsr68_13-508.pdf [Accessed 22 September 2020].

Massey, D. S., (ed.). (2008). *New Faces in New Places: The Changing Geography of American Immigration*. New York: Russell Sage Foundation.

Masuda, H. (2014). Extinct municipalities. In: A. Selena (2017). Can anything stop rural decline? *The Atlantic*. Available at: www.theatlantic.com/business/archive/2017/08/japan-rural-decline/537375/

Matanle, P. (2008). Shrinking Sado: education, employment and the decline of Japan's rural regions. In: P. Oswalt, ed., *Shrinking Cities: Complete Works 3 Japan*. Berlin: Project Office, pp. 42–53.

Merli, M. G. and Morgan, S. P. (2011). Below replacement fertility preferences in Shanghai. *Population*, 66(3–4), pp. 519–542.

Pickering, J. A. J. and Crooks, V. A. (2019). Retirement migration. In: D. Gu and M. E. Dupre, eds., *Encyclopedia of gerontology and population aging*. New York: Springer.

Population Reference Bureau. (2012). *Attaining the Demographic Dividend*. Available at: www.prb.org/demographic-dividend-factsheet/ [Accessed 18 July 2019].

Poston, D. L. and Micklin, M. (2018). Prologue. In: D. L. Poston and M. Micklin, eds., *Handbook of Population*. New York: Springer, pp. 1–15.

Ratcliffe, M., Burd, C., Holder, K. and Fields, A. (2016). *Defining Rural at the U.S. Census Bureau, ACSGEO-1*. Washington, DC: U.S. Census Bureau.

Riley, J. C. (2005). Estimates of regional and global life expectancy, 1800–2001. *Population and Development Review*, 31(3), pp. 537–543.

Roser, M., Ortiz-Ospina, E. and Ritchie, H. (2020). *Life Expectancy*. OurWorldInData.org. Available at: https://ourworldindata.org/life-expectancy

Selena, A. (2017). Can anything stop rural decline? *The Atlantic*, 23 August. Available at: www.theatlantic.com/business/archive/2017/08/japan-rural-decline/537375/

Shryock, H. and Siegel, J. (1975). *Methods and Materials of Demography*. Washington, DC: U.S. Department of Commerce, Bureau of the Census.

Small Area Population Estimates. (2016). Office for National Statistics. Available at: www.ons.gov.uk/peoplepopulationandcommunity/populationandmigration/populationestimates/bulletins/annualsmallareapopulationestimates/mid2018

US Census Bureau International Data Base Revised: September 18, 2018 Version: Data: 18.0822 Code:12.0321. Available at: www.census.gov/data-tools/demo/idb/informationGateway.php.

US Census Bureau. (2020). *Urban and Rural Classifications*. Available at: www.census.gov/programs-surveys/geography/guidance/geo-areas/urban-rural.html [Accessed 12 January 2020].

Williams, D. R. and McIntyre, N. (2011). Place affinities, lifestyle mobilities and quality-of-life. In: M. Yusal, R. Perdue and R. Sirgy, eds., *Handbook of Tourism and Quality-of-Life Research*. Dordrecht: Springer, pp. 209–231.

Wrigley, E. A. (1972). The process of modernization and the industrial revolution in England, *The Journal of Interdisciplinary History*, 3(2), pp. 225–259.

3 Rural studies of ageing

Jesse Heley and Michael Woods

Introduction

This volume is testament to the ongoing interest in the relationship between ageing and the constitution of rural space and rural communities. This is, at least in part, a reflection of current political concerns regarding long-standing demographic trends and the fact that, in many parts of the world, the population is getting older (see Berry, Chapter 2 in this volume). A regional trend long-noted in the global north, including the US (Glasgow and Brown, 2012), Canada (Wilson, Errasti-Ibarrondo and Low, 2019), Japan (Kavedžija, 2016) and the UK (Blake and Mayhew, 2006), recent scholarship has also pointed to this trend in China (Woo et al., 2002) and India (Bloom, Canning and Fink, 2010).

The transition from high birth and mortality rates to low fertility and low mortality rates is generally positioned one of the principal problems facing society as a whole. Tending to focus on health care provision, there are wider implications for employment and intergenerational relations, as well as living arrangements and shifts in attitudes (Tinker 2002: 729). In Europe, the combination of population ageing, and declining fertility is of considerable political concern, with many governments grappling with rising social security costs and a smaller working population from which to generate tax income. With higher levels of economic activity among older people (including deferred retirement) and a greater provision of private pensions going a limited way in addressing this issue, these pressures remain and will continue to vex policy makers (Banks and Smith, 2006).

In this chapter we discuss how ageing populations in rural communities and rural experiences of ageing have been studied in rural geography, rural sociology and related fields, and how conceptual and methodological approaches have evolved. The first section examines how analyses of urbanisation and counterurbanisation in the mid twentieth century onwards raised questions about the age structure of rural communities and how these were viewed through a political-economy lens; particularly in respect to retirement in-migration and gentrification. We then consider the influence of the "cultural turn" in social science in the 1990s in drawing attention to the experiences and characterisations of older people in rural communities and their attachment to place, as well as later research on the participation of older people in rural society. Finally, the chapter counters the primarily European and North American emphasis of this narrative by noting briefly the increasing study of rural ageing in other parts of the world, especially east Asia.

Rural change and shifting demographics

The underlying reasons as to why some rural communities have a greater relative proportion of ageing residents are varied and have changed markedly over the past century (Champion and Shepard, 2006). They defy simplistic characterisation and are very much dependent upon national and regional contexts. Nevertheless, in Europe and North America the phenomenon of the ageing rural has been very much driven by an exodus of young adults to urban centres as industrialisation took hold from the 1850s onwards (Stockdale, 2006).

Considering this process in detail, a tranche of sociological studies in United States and postwar Britain (e.g., Jenkins et al., 1962; Landis, 1940) noted that rural populations had become disproportionately older than urban populations. Connecting this to distinctive characteristics of rural communities, key themes included: the relative absence of age stratification in community life; the significance of informal care through extended families; and respect for older people and their knowledge, especially in farming. However, Landis (1940) also warned that "too often the conservative hand of the aged rules community affairs, blocking innovations and the forces of progress" (50–51).

By the 1970s, concerns regarding the loss of younger people from rural communities were overlain by an evident trend of counter-urbanisation from larger cities to non-metropolitan areas. Widely referred to as the population 'turnaround', this process was characterised as being driven predominately by an increasingly mobile, middle-aged middle class. This corresponded with the growing popularity of Marxist analyses and, noted Champion (1989), "fitted in very neatly with contemporary ideas of a shift from an industrial to a post-industrial society, providing a physical and easily measurable manifestation of more complex and deep-seated changes believed to be taking place in economic and social structures" (1). Within this framework "empty nesters" and return-migrants became a focus of attention within rural studies, as well as the increased presence of these cohorts in areas adjacent to large urban locations in American research (e.g., Schwarzweller, 1979).

The notion of the population "turnaround" proved compelling and a number of studies quickly emerged which sifted through the evidence for such a trend in other industrialised economies, including Australia and Canada (see Berry, 1976), and Western Europe (Fielding, 1982). Focusing on those "push" and "pull" factors that enable or constrain an individual's capacity to live, work and move between the city and the country, a typology of those contributing processes as identified in the literature include (after Kontuly, 1998): regional investment cycles; new spatial divisions of labour; environmental factors (particularly housing costs); government policies and investments; and, technological innovations (particularly communications).

It was apparent that each of these processes have implications in terms of rural ageing, and many studies considered ageing implicitly and in terms of occupation and labour market position. Characteristic of class analysis, this stemmed from a determined critique of postwar rural studies as lacking focus and a rigorous epistemological backbone (Buttel and Newby, 1980). With a critical Marxist perspective coming to dominate the discipline, protagonists effectively recast class as the most important lens on the shifting composition of rural communities (Newby, 1977; Pahl, 1966). Set against the declining role agricultural employment in many "developed" economies and the rise of "newer", innovative industries in non-metropolitan areas (Petrulis, 1979), a corresponding out-migration of the lower-class workforce was seen as creating an opportunity for middle-class commuters, second-home owners and retirees to settle in rural localities. The latter were identified as making up a significant proportion of "aspirational migrants", seeking environments "free from urban problems and rich in natural amenities" (Clay and Price, 1979: 4). Such research on amenity in-migration tended to identify a process of re-enforcement. As more of this cohort took up residence, they would tend to become involved in local politics and work to ensure the preservation of their new domain through planning apparatus and political forums. Within these spheres, the comparatively high standards of education, professional credentials and affordances of free time enjoyed by middle-class in-migrants – and particularly those in retirement – was widely viewed as key to their success in shaping rural localities (Murdoch and Marsden, 1994).

By the mid-1980s, the population turnaround thesis was subjected to increased criticism as an explanatory framework for understanding migration patterns for different age groups at regional and local scales (Woods, 2005). For example, the expansion of higher education and the concentration of graduate jobs in urban centres has undoubtedly influenced the movement of younger people out of many rural localities. A proportion of these will return to their rural "home" in later life, given opportunities and financial resources, or else settle in rural locations elsewhere. In either case, they are potentially able draw on their upbringing and "rural" credentials and better integrate into the community. While there continues to be a paucity of research on "returnees", that work which considers the influence of this cohort on the social fabric of rural communities highlights a strong commitment to place, a tendency to adopt leadership roles and the capacity to translate their "education and training into substantial economic and social benefits" (Cromartie et al., 2015: 22).

Retirement migration and rural communities

Studies of counter-urbanisation reveal the significant role of post- and pre-retirement migration, including former residents returning to their childhood localities at the end of their working lives and lifestyle-oriented migration by those seeking to spend their retirement in high amenity environments. Initial research from the 1990s onwards called attention to the substantial flows of people over the age of 60 into non-metropolitan areas (Fulton, Fuguitt and Gibson, 1997).

More recent studies, however, have indicated that this trend may have peaked, or has been over simplified. Von Reichert, Cromartie and Arthun (2013), for example, note that for couples in their mid-50s through early retirement years, there is only a marginally greater propensity to migrate as compared to the 30–50 age category. The key difference is that those in the mid-50s+ age bracket are significantly more likely to move to rural and small town settings. On this point they note that retiree destinations are highly selective, including sites in Florida, Arizona and Nevada (Longino and Bradley, 2003). This resonates with the identified growth of retirement "resort" locations in Australia – such as New South Wales (Gurran and Blakely, 2007) – and Europe, including the Costa del Sol in Spain (Rodríguez et al., 1998).

At a smaller scale, the growth of purpose-built retirement developments and villages in Australia and the US has also been reflected in research which emphasises the important role of environmental and recreational motifs in attracting new residents (Gardner, 1994). Charting the transition from "medicalized to a marketized model of service delivery" for retirement complexes in New Zealand, Simpson and Cheney (2007: 191) set out the growth of a "customer-centred" organisational model for aged-care services in many parts of the industrialised world. Recognising the ever-more diverse needs and expectations of those older people looking to move into these locations, this resonates with work on rural destinations of older migration (RDOMs) in the US by Brown and Glasgow (2008).

Defined as nonmetropolitan counties with 15% or higher net in-migration at ages 60 and older (Glasgow and Brown, 2012), RDOMs tend to develop in association with natural amenities (mountains, lakes, coastlines) and where long-term recreation and tourism translate into the in-migration of older persons to select communities. With successive cohorts being recruited via family and friendship networks, the role of private enterprise plays an important role in developing these sites, as does promotional activity by regional and local government. This is on the basis that this cohort is inclined to have more "positive" attributes than longer-term older residents. More likely to assume active roles in the community, more able to increase capacity to stimulate demand for local goods, services and facilities, and with a lower tendency to draw on public services, RDOMs are understandably seen as an effective strategy for economic development (Brown and Glasgow, 2008). Interest in RDOMs is, however, pointedly set in contrast to those other (much more numerous) rural regions where demographic ageing is characterised as a foremost challenge.

In their overview of the relationships between retirement, migration and rural communities in the UK, Stockdale (2011) observes the propensity for policy makers and academics to approach demographic ageing as a pensions and care issue. This inclination towards negative portrayals has created a clear need for a more balanced agenda which accounts for the benefits of older people in and for place. On this theme, a number of fertile avenues for research are apparent. These include; the impacts of ageing demographics at different points on the rural – urban continuum; the interrelationship between ageing and specific aspects of employment, consumption, health and care and the role of older volunteers (Stockdale, 2011: 216).

Ageing, community and place-attachment

A critical reappraisal of retirement migration resonates with a growing body of work which takes a more qualitative, experiential approach to relationships between ageing, community and attachment to place (see Wiles et al., Chapter 15). Burholt and Naylor (2005), for example, have explored the ways in which rural communities are more than just "settings" but rather active components of identity formation for older residents. Comparing empirical data gathered in Wales for "retirement" communities (featuring high levels of in-migration) and "native" communities (with low levels of in-migration), they argue that these types of settlements can be characterised by differing forms of place attachment. With specific reference to the Welsh language as a conduit of identity, they speculate that such a basis for place-attachment is likely in other rural areas where minority languages are prevalent. This tallies with a number of inter-generational studies of minority languages in other parts of Europe, including Brittany (Kneafsey, 2000) and Galicia (Bermingham, 2018).

Other processes that have been identified as having an important bearing on experiences of ageing in rural place include length of residency, historical attachments, existing social networks and shared life events (Stedman, 2006). This is particularly so among those older residents who have spent a significant proportion of their life in one location. For example, Burholt and Naylor (2005) call attention to the important role of shared memories as a source of emotional support and that, for older people in particular, proximity to family and friends often functions as an important buttress for independence. In contrast to established or "native" residents, later life- and retirement-migrants have been shown to alternatively emphasise aesthetic qualities, environmental attributes and outdoor recreation (Cuba, 1989).

In drawing comparisons between "ageing in" and "out" of place (Jackson, 2002) we should be mindful of differences in experience but also of (re)affirming problematic dichotomies. For example, there is an emergent body of work which highlights an increased risk of social isolation for rural in-migrants in later life. This has in turn been associated with increased levels of depression, morbidity and mortality (Curtin et al., 2017). However, we should also be cautious in connecting these processes to "culture clashes" and that conventional wisdom which pits "native" values against the attitudes of in-migrants. In their study of in-migration and cultural divisions in the South Appalachian Mountains, Brennan and Cooper observe that "while the distinction between natives and in-migrants is salient . . . it is different than conventional wisdom would have us believe" (Brennan and Cooper, 2008, p. 292). Here we are reminded that those choosing to move to rural regions might do so on the basis of sharing attitudes towards community, heritage and amenities.

The turn to research on rural ageing which emphasises place-based relations and the experiential geographies of rural ageing is reminiscent of the approach to social relations embedded in postwar community studies. It also symptomatic of the influence of the oft-cited "cultural turn" on rural studies during early 1990s. Infusing aspects of political-economy with poststructuralism, this combined an attentiveness to the role of class as a manifestation of uneven access to capital alongside a wider range of processes connected to identity and cultural difference (Shucksmith, 2012). Here interventions by Philo (1992), among others, highlighted the multiplicity of experiences of the countryside, drawing attention to historically side-lined or "othered" groups, including older people.

This critical repositioning also involved embracing theories of performance and "embodied rural geographies" (Little and Leyshon, 2003). Making a comprehensive case for studies which explore the role of the body in actively reproducing of rural spaces and constructing rural identities, as well articulating uneven power relations in given settings, this has set the groundwork for a wide range of studies which explore the relationship between performance, routines and rurality and of "more-than-representational" understandings of rural living (Carolan, 2008). Although clearly apparent in those studies of rural ageing and attachment to place as previously discussed, few studies have directly sought to connect theories of performance to studies of ageing in rural communities. Where embodiment is discussed, it tends to be considered in respect to declining personal mobility, isolation and – again – as part of the medicalised agenda (e.g., Dwyer and Hardill, 2011). Such work is clearly pertinent, but there remains much scope for studies which explore the relationship between ageing, dwelling and socialisation in more varied and upbeat terms within a "productive rural ageing" agenda (Davis et al., 2012).

Participation of older people in rural communities

Underscoring the need for more research on the positive aspects of the "greying countryside" Le Mesurier has described older people as the "most vital source of social capital in the countryside" (2006: 133), with older volunteers playing a significant role in maintaining socio-cultural institutions (see Winterton and Warburton, Chapter 23; Colibaba et al., Chapter 24). Facilitated by the increased level of free time post-retirement, as well as a degree of financial flexibility often enjoyed by many older citizens, this sits alongside the wide range of skills and experiences retained by older people in general.

In their study of northern Victoria, for example, Davis et al. (2012) demonstrate the often-crucial role older people play in maintaining civic life through their involvements in a wide range of groups and committees. Often articulated in terms of obligation, these activities merge with and overlap leisure pursuits. This echoes a shift in recreation studies from focusing almost solely upon the physical and mental benefits of leisure pursuits to the individual, to the benefits of leisure bestowed upon communities as a whole (Arai and Pedlar, 2003).

That work on volunteering among the rural old in institutional settings has been accompanied by a parallel interest in the role that the over 60s and retirees play as part of the more informal "social glue" of rural community life. There now exists is a wealth of literature signifying the importance of older people as informal care providers for family and friends, and in carrying out a range of day-to-day tasks such as collecting groceries and checking on elderly neighbours (see Heley and Jones, 2013). The reliance of many families in rural areas on grandparents for childcare duties is also well-established, often connected to the lack of local pre-school facilities (see King et al., 2009), Reflecting on these informal acts of time-giving by older rural residents, we would, however, repeat warnings made elsewhere regarding rural policy formation centred on ageing volunteers, and further work could usefully be undertaken on the "darker side" of social capital under this guise (Jones and Heley, 2016).

International perspectives on ageing in rural communities

The narrative outlined earlier has been dominated by studies from Europe and North America, reflecting demographic trends the intellectual trajectories of rural studies in these settings. Nevertheless, recent years have seen increased research on rural ageing in other countries, notably in Asia and South America, as extended life expectancy and migration dynamics have combined to inflate the prominence of older people in rural areas. Countries such as Argentina, Brazil, China, India, Indonesia and South Korea are all forecast to experience rapid growth in the proportion of their populations aged over 65, located disproportionately in rural communities (Wen, 2014). In China, for example, the proportion of the population aged over 60 is projected to reach 16% in 2020, but is already higher in most rural communities as an effect of working-age men moving to cities for work leaving "left-behind" populations of children, women and the elderly (Wen, 2014). As such, there were estimated to be 45 million "left-behind" older residents in rural China in 2013 (Ye, 2017).

As Ye (2017) observes, "Chinese scholars have paid more attention to the economic contribution of migration to both the sending areas and receiving cities than to the impacts on the families and people left behind in the countryside" (973), and it is only in the last decade that research has begun to study the rural "left-behind" elderly. The evidence from these studies to date is furthermore mixed, with some emphasising positive impacts of remittances, yet others highlighting issues of isolation and loneliness as well as increased responsibilities on older people to undertake family chores, childcare and farming. A common finding is the disruption of household structures and cultures of care, where elderly parents have traditionally lived with their children and received care through extended family networks. The erosion of such structures has led to an expansion of formal care for the elderly in rural communities, such as communal housing, yet as Liu, Eggleston and Min (2016) note, "these facilities cannot be equated with nursing homes or even Western concepts of assisted living facilities" (2046). Accordingly, the situation of older people in rural China, as in many developing nations, is complex and contingent on geographical context.

Conclusion

Research on older people has become increasingly prominent in rural studies. This is in part a reflection of changing demographics in the countryside, and in part due to a growing emphasis on the cultural and experiential dimensions of different rural social groups. This shift in focus is arguably overdue given that rural areas have long had greater proportions of older people in their populations than cities. This skew has been reinforced in recent decades by the dual processes of out-migration by working-age residents and – in some localities – in-migration by older retirees and returnees.

Demographic ageing due to out-migration can be associated with problems of social isolation and loss of productivity, while retirement in-migration may lead to competition for property and social conflicts. In both situations, however, the presence of a substantial older population creates demands for care services that are not always readily met in rural districts, contributing to sometimes "hidden" problems of ill-health and deprivation. At the same time, more recent studies have recognised older people as a social resource within rural communities, whose voluntary activity commonly underpins many aspects of community life.

In terms of more specific avenues for future research, we have called attention to a number of areas which warrant further exploration in this chapter. First, we would argue that there is a need for research which address the ongoing social and economic dynamics of RDOMs and how the characteristics of these locations has variously transformed as residents move out of comparatively active Third Age, and into Fourth Age. In the context of Anglo-European research we also feel that there is a need for more research which unpacks the seductive local-incomer dichotomy and the tendency for uncritical reproductions of the "culture clash" narrative. This reflects growing levels of mobility and more complex residential histories involving both urban-to-rural and rural-to-rural migration flows. Last, there continues to be a requirement which considers the relationship between ageing, caregiving and (the dark side) of social capital. This includes the shifting role of families and the state in caring for older people but also a growing dependency on older people as care providers. This is an issue that cuts across developed and developing world contexts, and there are significant opportunities for a critical dialogue on ageing in rural place which connects and transcends these spheres (see Urbaniak, Chapter 18).

References

Arai, S. and Pedlar, A. (2003). Moving beyond individualism in leisure theory: a critical analysis of concepts of community and social engagement. *Leisure Studies* 22, pp.185–202.

Banks, J. and Smith, S. (2006). Retirement in the UK. *Oxford Review of Economic Policy* 22(1), pp. 40–56.

Bermingham N. (2018). Double new speakers? Language ideologies of immigrant students in Galicia. In: C. Smith-Christmas, N. Ó Murchadha, M. Hornsby and M. Moriarty, eds., *New Speakers of Minority Languages*. London: Palgrave Macmillan, pp. 111–130.

Berry, B., (ed.). (1976). *Urbanization and Counterurbanization*. Beverly Hills: SAGE Publications.

Blake, D. and Mayhew, L. (2006). On the sustainability of the UK state pension system in the light of population ageing and declining fertility. *The Economic Journal*, 116(512), pp. 286–305.

Bloom, D., Canning, D and Fink, G. (2010). Implications of population ageing for economic growth. *Oxford Review of Economic Policy*, 26(4), pp. 583–612.

Brown, D. and Glasgow, N., (eds). (2008). *Rural Retirement Migration*. Dordrecht: Springer.

Burholt, V. and Naylor, D. (2005). The relationship between rural community type and attachment to place for older people living in North Wales, UK. *European Journal of Ageing*, 2, pp. 109–126.

Buttel, F. and Newby, H. (1980). Toward a critical rural sociology. In: F. Buttel and H. Newby, eds., *The Rural Sociology of the Advanced Societies*. London: Croom Helm, pp. 1–39.

Brennan, K. and Cooper, A. (2008). Rural mountain natives, in-migrants, and the cultural divide. *The Social Science Journal*, 45(2), pp. 279–295.

Carolan, M. (2008). More-than-representational knowledge's of the countryside: how we think as bodies. *Sociologia Ruralis*, 48, pp. 408–22.

Champion, T. (1989). The counterurbanization experience. In: A. Champion, ed.,. *Counterurbanization: The Changing Pace and Nature of Population Deconcentration*. London: Routledge, pp. 1–18.

Champion, T. and Shepard, J. (2006). Demographic change in rural England. In: P. Lowe and L. Speakman, eds., *The Ageing Countryside: The Growing Older Population of Rural England*. London: Age Concern Books, pp. 29–50.

Clay, D. and Price, M. (1979). *Structural Disturbances in Rural Communities: Some Repercussions of Migration Turnaround in Michigan* (Report for Department of Agriculture). Washington, DC: USDA.

Cuba, L. (1989). Retiring from vacationland: from visitor to resident. *Generations*, 13, pp. 63–67.

Curtin, A., Martins, D., Gillsjö, C. and Schwartz-Barcott, D. (2017). Ageing out of place: The meaning of home among Hispanic older persons living in the United States. *International Journal of Older People Nursing*, 12(3), p. 12150.

Cromartie, J., von Reichert, C. and Arthun, R. (2015). *Factors affecting former residents' returning to rural communities*. ERR 185. United States Department of Agriculture Economic Research Service Research Report. Available at: https://www.ers.usda.gov/webdocs/publications/45361/52906_err185.pdf?v=645.3

Davis, S., Crothers, N., Grant, J., Young, S. and Smith, K. (2012). Being involved in the country: productive ageing in different types of rural communities. *Journal of Rural Studies*, 28(4), pp. 338–346.

Dwyer, P. and Hardill, I. (2011). Promoting social inclusion? The impact of village services on the lives of older people living in rural England. *Ageing & Society*, 31(2), pp. 243–264.

Fielding, A. (1982). Counterurbanisation in Western Europe. *Progress in Planning*, 17(1), pp. 1–52.

Fulton, J., Fuguitt, G and Gibson, R. (1997). Recent changes in metropolitan to non-metropolitan migration streams. *Rural Sociology*, 62, pp. 363–384.

Gardner, I. (1994). Why people move to retirement villages: homeowners and non-homeowners. *Australian Journal on Ageing*, 13(1), pp. 36–40.

Glasgow, N. and Brown, N. (2012). Rural ageing in the United States: trends and contexts, *Journal of Rural Studies*, 28(2), pp. 422–431.

Gurran, N. amd Blakely, E. (2007). Suffer a sea change? Contrasting perspectives towards urban policy and migration in coastal Australia, *Australian Geographer*, 38(1), pp. 113–131.

Heley, J. and Jones, L. (2013). Growing older and social sustainability: considering the 'serious leisure' practices of the over 60s in rural communities. *Social and Cultural Geography* 14(3), pp. 276–299.

Jackson, J. (2002). Conceptual and methodological linkages in cross- cultural groups and cross- national ageing research. *Journal of Social Issues*, 58(4), pp. 825–835.

Jenkins, D., Jones, E., Jones Hughes, T. and Owen, T., (eds.). (1962). *Welsh Rural Communities*. Cardiff: University of Wales Press.

Jones, L. and Heley, J. (2016). Practices of participation and voluntarism among older people in rural Wales: choice, obligation and constraints to active ageing. *Sociologia Ruralis*, 56(1), pp. 176–196.

Kavedžija, I. (2016). The age of decline? Anxieties about ageing in Japan. *Ethnos*, 81(2), pp. 214–237.

King, S., Kropf, N., Perkins, M., Sessley, L., Burt, C. and Lepore, M. (2009). Kinship care in rural Georgia communities: responding to needs and challenges of grandparent caregivers. *Journal of Intergenerational Relationships*, 7, pp. 225–242.

Kneafsey, M. (2000). Tourism, place identities and social relations in the European rural periphery. *European Urban and Regional Studies*, 7(1), pp. 35–50.

Kontuly, T. (1998) Contrasting the counter-urbanisation experience in European nations. In: P. Boyle and K. Halfacree, eds., *Migration into Rural Areas*. Chichester: Wiley, pp. 61–78.

Landis, P., (ed.). (1940). *Rural Life in Process*. New York: McGraw-Hill.

Le Mesurier, N. (2006). The contributions of older people to rural community and citizenship. In: P. Lowe and L. Speakman. eds., *The Ageing Countryside, the Growing Older Population of Rural England*. London: Age Concern Books.

Little, J and Leyshon, M. (2003). Embodied rural geographies: developing research agendas. *Progress in Human Geography*, 27(2), pp. 257–72.

Liu, H., Eggleston, K. and Min, Y. (2016). Village senior centres and the living arrangements of older people in rural China: considerations of health, land, migration and intergenerational support, *Ageing & Society*, 37(6), pp. 2044–2073.

Longino, C. and Bradley, D. (2003). The first look at retirement migration trends in 2000. *The Gerontologist*, 43(4), pp. 904–907.

Murdoch, J. and Marsden, T., (eds). (1994). *Reconstituted Rurality*. London: UCL Press.

Newby, H., (ed.). (1977). *The Deferential Worker: A Study of Farm Workers in East Anglia*. London: Allen Lane.

Pahl, R. (1966). Rural-urban continuum. *Sociologia Ruralis*, 6(1), pp. 299–329.

Petrulis, M. (1979). *Growth Patterns in Nonmetro-Metro Manufacturing Employment* (Development Research Report 7). Washington, DC: U.S. Department Agriculture.

Philo, C. (1992). Neglected rural geographies: a review. *Journal of Rural Studies*, 8(1), pp. 193–207.

Rodríguez, V., Fernández-Mayoralas, G. and Rojo, F. (1998). European retirees on the Costa del Sol: a cross-national comparison. *International Journal of Population Geography*, 4(2), pp. 183–200.

Schwarzweller, H. (1979). Migration and the changing rural scene. *Rural Sociology*, 44(1), pp. 7–23.

Shucksmith, M. (2012). Class, power and inequality in rural areas: Beyond social exclusion? *Sociologia Ruralis*, 52(4), pp. 377–397.

Simpson, M. and Cheney, G. (2007). Marketization, participation, and communication within New Zealand retirement villages: a critical – rhetorical and discursive analysis. *Discourse and Communication*, 1(2), pp. 191–222.

Stedman, R. (2006). Understanding place attachment among second home owners. *American Behavioral Scientist*, 50(2), pp. 187–205.

Stockdale, A. (2006). Migration: pre-requisite for rural economic regeneration? *Journal of Rural Studies*, 22(3), pp. 354–366.

Stockdale, A. (2011). A review of demographic ageing in the UK: opportunities for rural research population. *Space Place*, 17, pp. 204–221.

Tinker, A. (2002). The social implications of an ageing population. *Mechanisms of Ageing and Development*, 123(7), pp. 729–735.

Von Reichert, C., Cromartie, J. and Arthun, R. (2014). Impacts of return migration. *Rural Sociology*, 79(2), pp. 200–226.

Wen, L., (ed.). (2014). *China's Contemporary Society*. Beijing: China Intercontinental Press.

Wilson, D., Errasti-Ibarrondo, B. and Low, G. (2019). Where are we now in relation to determining the prevalence of ageism in this era of escalating population ageing? *Ageing Research Reviews*, 51(1), pp. 78–84.

Woo, J., Kwok, T., Sze, F. and Yuan, H. (2002). Ageing in China: health and social consequences and responses, *International Journal of Epidemiology*, 31(4), pp. 772–775.

Woods, M., (ed.). (2005). *Rural Geography*. London: SAGE Publications.

Ye, J. (2017). Left-behind elderly: Shouldering a disproportionate share of production and reproduction in supporting China's industrial development, *Journal of Peasant Studies* 44(7), pp. 971–999.

4 Rural health and ageing

Making way for a critical gerontology of rural health

Neil Hanlon and Laura Poulin

Introduction

Rural older adult health, care and well-being have emerged as major foci of research, policy and practice in the twenty-first century. This is due to the worldwide trend of increasingly disproportionate numbers of older adults ageing in place in and/or migrating to non-metropolitan (rural) environments (WHO, 2015) and the concurrent recognition of health status vulnerabilities of older rural adults and the particular challenges they face in accessing limited health and social care in underserviced rural settings (Hanlon and Kearns, 2016). The latter are longstanding interests in rural health scholarship integrating health studies and medical sciences, as evident in the proliferation of university schools, research centres and international journals dedicated to rural health and medicine over the last half century (e.g., Australia's system of university-based Schools of Rural Health; Centre for Rural and Northern Health Research, Laurentian University, Canada; and *Journal of Rural Health* and *Canadian Journal of Rural Medicine*) (Humphreys and Gregory, 2012; Pong, 2000; Pong, DesMeules and Lagacé, 2009). Notwithstanding decades of scholarship that emphasises the health disparities faced by rural older adults (e.g., Davis and Bartlett, 2008; Gesler et al., 1998; Glasgow and Berry, 2013; Jensen et al., 2020; Keating, Swindle and Fletcher, 2011; Kulig and Williams, 2012), there is still a need to better understand the relationship between the interconnected yet surprisingly disparate fields of rural health and rural gerontology that each inform current scholarship on rural older adult health (Poulin, Skinner and Hanlon, 2020). Towards this end, this chapter explores the connection between these fields and asks what can be learned from an increasingly critical turn in rural gerontology to inform a more holistic approach to understanding "rural gerontological health".

As introduced by Skinner et al. (Chapter 1), rural gerontology has engaged with critical turns in the gerontological literature as it continues to illuminate the many ways in which mainstream discourses of rural ageing reflect and serve dominant social and political interests at the expense of less privileged groups (see Burholt and Scharf, Chapter 6). Critical gerontology is not so much a unified approach as it is an assemblage of perspectives on the mechanisms that perpetuate socially structured processes of discrimination and inequality of the older population

(Estes, 2001). Its proponents urge an ongoing commitment to reflexivity; that is, an approach to scholarship that continually challenges mainstream (and, hence, dominant) narratives of ageing as a necessary means to illuminate and improve the conditions in which people are ageing (Katz, 1996). As Holstein and Minkler (2003) outline, critical perspectives seek to uncover whose interests are best served by prevailing gerontological paradigms, and to offer guidance about what it means to pursue research and policy that enables everyone to age well. Drawing on political economy, feminism, disablement studies and various strands of cultural theory, critical gerontology continues to offer a powerful counterbalance to, and deconstruction of, dominant discourses (e.g., austerity, resilience, autonomy, "successful ageing") shaping ageing policy and research on a worldwide scale (Van Dyk, 2014).

For all of its contributions in speaking truth to power about the many guises and instances of age discrimination, however, it is surprising the dearth of critical gerontology engagement with matters of rural health. To redress this gap in the literature and align ourselves with Heley and Woods (Chapter 3 in this volume), we begin by taking into account relational and political economic ways of understanding rural ageing and health (see also Heley and Jones, 2012). We seek to integrate these considerations with calls for embodied and intersectional approaches to rural ageing and health research. We then make a case for an emplaced critical approach to older adult health that is reflective of the framework outlined by Thomas et al. (2011) concerning the wider political economic processes, social spatial structures, and socio-cultural practices operating in rural settings. In these ways, we hope to encourage a critical gerontology of rural health that challenges the binary and reductionist tendencies of mainstream policy and thinking around "health" and "rural ageing" and enables more contextually nuanced understandings of the experiences of older populations in rural areas.

Taking a closer look at ageing in the countryside

In many ways, the lack of attention to rural health in critical gerontology can be traced to ongoing debates about what "rural" means and how one goes about studying rural ageing (Hanlon and Kearns, 2016). In its crudest sense, we can think of rural as a territorial container of people, resources and activities that is arranged and manipulated over time by political economic processes operating at a variety of scales to look, feel and function differently than urban community settings. It is important, however, not to put too much stock in purely topological categorisations of rural territory, as these tend to reinforce thinking about rural ageing as a uniform or isolated process in an increasingly interconnected and urban-dominated world (McCarthy, 2008). Instead, place-based and critically derived accounts of rural health and ageing are needed to challenge the urban-centrism and binary thinking that pervades gerontological theory, policy and practice (Heley and Jones, 2012).

Since it is unreasonable to expect that a single conceptualisation would ever fully capture the diversity and dynamism of rural ageing experiences (Rowles, 1988), we make no attempt here to propose such a universalising account. Instead, we feel there is much to be gained by adopting an approach to theory that recognises degrees of difference rather than discrete categories. Rural spaces are increasingly influenced by widening circuits of trade, production, marketisation, neoliberalism and technological change that influence how people age (McCarthy, 2008; Poulin and Skinner, 2019; Skinner and Hanlon, 2016). These political economic forces operate in and through rural territories that are more typically valued by, and transformed to serve, the commercial interests of primary economic extraction and production, or else to provide utilities and ecosystem services for nearby urban and metropolitan centres (Skinner and Hanlon, 2016). In regions dominated by primary resource economies, this results in political economic processes and socio-cultural relations that ensure a continued source of resource procurement and production rather than attention to local community development and infrastructure (Hanlon et al., 2007). Thus, investments in health care, education, and other forms of social infrastructure are less generous in rural centres and are rarely framed expressly as a response to population ageing. These distributional inequities, in combination with the social and economic challenges that accompany extractive economies (e.g., boom and bust economic cycles, fewer formal services), tend to encourage individual and household resilience as well as more communitarian sensibilities (e.g., a greater affinity for informal social relations, stronger attachments to place and community) (Doheney and Milbourne, 2017; Scott, 2013; Wells, 2009). In these ways, the socio-cultural elements of rural ageing are influenced by the dynamic interplay of space and the political economic structure of the rural environment.

It is important to note that the processes of socio-spatial sorting are historically and geographically contingent, and that, consequently, the particular features of rural places, populations and societies should be understood as such rather than as essential qualities (Smith and Easterlow, 2005). It is incumbent on critical scholarship, therefore, to account for increasingly globalised political economic forces that shape the contexts of rural ageing (McCarthy, 2008). In keeping with developments in the field of critical gerontology and rural studies, such as those in this collection, we aim to emphasize the need for relational approaches that reveal the diverse, dynamic and interactive conceptions of ageing and health in rural communities.

Alternating narratives of older persons' health and well-being in the countryside

Much of the literature on older persons' health and well-being in the countryside tends to convey very disparate narratives about place (Skinner and Winterton, 2018). Rural landscapes have long been represented in popular media as bucolic settings that embody ideals of successful ageing, retirement aspirations and the rewards of a life of hard work and sacrifice (McHugh, 2003). The idea that idyllic rural landscapes innately possess regenerative powers has also been around for some time. A more recent instance of this may be found in applications of the concept of therapeutic landscape and the purported capacity of "green" and "blue" spaces to promote well-being (Andrews, 2004; Bell et al., 2018; Coleman and Kearns, 2015). Some critical scholars suggest, however, that this more recent revival of interest in the healing powers of natural landscapes is best understood as the work of elite imagination and idyllic fantasy (Keating, 2008; Shucksmith, 2018).

In contrast, an alternative narrative of rural disparity (as reflected in a wide body of work in fields such as epidemiology and community health) paints rural populations and communities as typically beset by stagnation and decline (Skinner and Winterton, 2018; Thomas et al., 2011). In fact, much of the academic literature on rural health and ageing tends to invoke negative scenarios whereby persons ageing in rural places are more likely to suffer poverty and isolation (Milne, Hatzidimitriadou and Wiseman, 2007). To age in such settings, according to this narrative, is to put oneself at elevated risk of chronic disease, co-morbidity and episodic injury (Baernholdt et al., 2012; Goeres et al., 2015). This "deficit view" of rurality pervades much of the gerontological literature, creating and reinforcing a view that ageing in rural communities is inherently fraught with personal and societal risk (Skinner and Rosenberg, 2006). Yet, in spite of its more convincing appeal to evidence, this deficit narrative can be overplayed to the point that critical scrutiny is warranted.

A bifurcated narrative is also evident in portrayals of rural care and support. On one hand, much is made of the seemingly inherent challenges of delivering care in rural settings. That is, small population bases and a lack of infrastructure are said to inhibit efforts to attract and retain health professionals and are assumed to make it more difficult and expensive to deliver formal health care services (Gesler et al., 1998). On the other hand, rural settings have long been regarded as places of abundant informal care and support for older people (Estes, 2001; Keating, 2008; Kelly and Yarwood, 2018). Such representations often presume the presence of numerous and nearby family members and tight-knit community bonds that can be counted on to help out in times of need. Recent demographic and social trends, however increasingly call these assumptions into question (Hanlon et al., 2007; Scharf et al., 2016). Others go so far as to suggest that the persistence of "rural community values" is a narrative best understood as fiction propagated by urban-centric policy makers in order to evade their responsibilities to address shortcomings in rural welfare systems and supports (Doheney and Milbourne, 2017; Leipert and Reutter, 2005; Scott, 2013; Wells, 2009).

In reality, most rural communities are unable to live up to idyllic communitarian expectations (Keating, 2008; Scharf et al., 2016). Instead, social relations in many rural areas have been fundamentally transformed by commercial exploitation of natural resources and the "resource-dependency" that so often accompanies this (McCarthy, 2008; Thomas et al., 2011). That is, the nature of working conditions in the primary sector, exposure to environmental hazards (e.g., floods, wildfires, dangerous roads) and diminished access to cultural goods are factors that often lead to poorer social development in rural areas (e.g., poorer health outcomes, lower levels of income and education attainment). In spite of decades of industrial and service sector de-agglomeration, most rural economies remain heavily tied to primary extractive activities (i.e., agriculture, forestry, mining), and thus suffer cyclical economic downturns that prompt many residents, especially young adults, to leave in search of more lucrative opportunities (Hanlon et al., 2007; Skinner and Hanlon, 2016). Population out-migration, especially of younger adults, tends to disrupt the care and support networks within rural areas (Hanlon et al., 2007). Compounding this is a set of spatial configurations (i.e., small populations, geographic remoteness) that disadvantage rural areas as places to attract health care providers and organize and deliver services (e.g., diseconomies of scale). These factors tend to work in combination to reproduce and normalize the notion that rural spaces are "beyond the pale" of most models of age-friendly communities (Herron and Skinner, 2018).

Yet, for many rural residents, there is a strong preference to age-in-place despite these challenges and shortcomings (Gesler, Rabiner and DeFriese, 1998; Keating, 2008). Some suggest that concepts such as community (Doheney and Milbourne, 2017) and resilience (Scott, 2013) are useful lenses through which to make sense of these rural place attachments. As others point out, however, concepts such as resilience tend to reinforce a discourse of diminished expectation for welfare entitlement, especially in the realm of publicly funded health care (Skinner and Winterton, 2018; Wild, Wiles and Allen, 2013). The work of Milbourne (2016) is especially illuminating in this regard, revealing the presence of an embedded culture of diminished expectations in a rural Welsh community; a cultural trait that muted the impacts of, and opposition to, welfare austerity measures imposed by central government.

These alternating narratives tend to suggest that rural environments are either very good or very bad places to grow old, with not much in the way of scenarios residing somewhere between these two extremes. For our purposes, the question of whether or not rural settings are good places to grow old is predicated on a degree of ecological or topological determinacy that tends to overlook the interconnectivity and mutuality of health/place experiences (Andrews, 2004). Since rural places are fully capable of harming and healing simultaneously, more research is needed to explore the connections between wellness, care and the dynamics of rural places (Bell et al., 2018; Herron and Skinner, 2018).

Thinking beyond binary alternatives

As a starting point to elicit critical reflection, we must challenge binary forms of thinking about rural gerontological health in favour of relational modes of inquiry about these issues. The challenge is to move beyond binary characterisations, such as idyllic or deprived, supportive or inhospitable, inclusive or exclusive (Skinner and Winterton, 2018). As Rowles and Cutchin discuss more thoroughly in Chapter 21, thinking relationally involves more than merely abandoning discrete polarising categorisations. A critical account should also recognize the fluid and dynamic interrelationships of rural ageing, as well as the variability of experiences that contribute to spatial and temporal understandings of health.

A critical gerontology of rural health thus challenges taken for granted representations of rural realities and seeks to move both policy and research beyond a tendency to treat rural as a discrete and disconnected "other". More importantly, a critical gerontology approach to rural health should insist that policy and research employ relational thinking to recognize that rural places are complex and fluid settings affected by wider socio-political forces and capable of being very different things to different groups of rural older adults. By way of illustration, Katz and Calasanti (2015) argue that dominant gerontological paradigms such as "successful ageing" reflect and reinforce a heteronormative elitism that valorise neoliberal concepts such as agency and choice (achieved through acts of self-reliance, individualism and private responsibility), but negate the social inequities that systematically prevent disadvantaged groups from ageing well. This pervasive heteronormativity is especially evident in the lack of attention to the issues and experiences of members of the older lesbian, gay, bisexual, transgender, intersex or queer (LGBTIQ) communities in age-friendly research (Van Wagenen, Driskell and Bradford, 2013).

It is important to take into account the myriad ways in which aspects of identity-based discrimination intersect with the "rural-ness" of older adult health experiences (Van Wagenen, Driskell and Bradford, 2013). In particular, considerations of gender, class, ethnicity, sexuality and able-ness come into play in the construction of older adults' health that is further influenced by the rural community in which they live. In so doing, critical scholarship challenges the urban-normativity of successful ageing scholarship by looking more closely at intersectional identities across the full spectrum of community settings, including, and especially, in rural places (Van Dyk, 2014).

A critical gerontology of rural health should also reveal how disease and various conditions are powerful metaphors in shaping and re-shaping the relationship between individuals and their lived environments. Such an inquiry eschews an emphasis on the biological and physiological aspects of disease and disability to focus instead on how medicalised conditions are an important means by which processes of social domination and control are enacted on older persons. This approach turns the more mainstream study of rural older adult health and care on its head by identifying the social processes of stigma, discrimination, and marginalisation that should be the primary focus of attention rather than any supposed deficiencies of older rural populations. Critical scholarship then seeks to uncover the ways in which dominant discourses about health and care are inscribed on ageing individuals. These inscriptions are particularly germane to people living in rural settings, calling into question idyllic assumptions such as those about social and environmental connectedness, or the inevitability of service scarcity (Doheney and Milbourne, 2017; Scharf, Walsh and O'Shea, 2016).

The embeddedness of rural older adult health in rural contexts

An older person's relation to place is likely to be deeply affected by their experiences of health and well-being. Questions of rural ageing and health are important as they tend to bring to the surface otherwise dormant or hidden dimensions of rural experience, such as place attachments, values and expectations. Matters of health and care are thus embedded in each of the critical dimensions of rural space (i.e., political economy, spatial structure, socio-cultural relations) outlined by Thomas and colleagues (2011), and provide a powerful impetus by which these different dimensions of rurality are mutually constituted.

Considering rural political economies

Rural delivery of care and support for older persons is closely bound up with considerations of livelihood and political economy, or what Thomas et al. (2011) refer to as the structural dimension of rural identity. Applying this construct to discussions regarding accountabilities for health and care reveals situated negotiations around what is deemed to be "appropriate" levels of collective response. As critical scholarship has revealed, these negotiations are deeply embedded in political economic processes that tend to sort older populations between productive-extractive spaces with typically poorer health outcomes and health care deficiencies and spaces of elite consumption with typically better health and care profiles (Hanlon et al., 2007; Skinner and Winterton, 2018; Smith and Easterlow, 2005).

Relational and political economic framings offer a powerful means to challenge dominant initiatives that most often originate in urban and metropolitan seats of power (e.g., age-friendly strategies directed by senior levels of government) and that seek to place responsibility for older adult health and care on local "community" and personal resources (Bulow and Soderqvist, 2014; Katz and Calasanti, 2015). The dominant narratives in question seek to formalize an ethic of (diminished) care as an immutable feature of enacted rural citizenship (Kelly and Yarwood, 2018). Such narratives tend to reinforce a muted expectation for welfare entitlement, especially in the realm of public goods such as health care (Skinner and Winterton, 2018; Wild, Wiles and Allen, 2013). In this sense, efforts to mobilize "community" align with Van Dyk's (2014) critique of the "successful ageing" paradigm whereby "community" is invoked as a means to promote the political objectives of marketisation and welfare state retrenchment rather than to align with the priorities and preferences of rural residents. It is imperative, therefore, to consider the ideological means by which an ethic of care is negotiated and to question the tendency to represent rural communities as more innately charitable, resilient, self-reliant and stoic in regard to older adult care (Shucksmith, 2018). Such an approach would yield more relational understandings of care provision in rural communities that challenge prevailing (neo)liberal understandings of support.

Considering rural spatial structures

A critical gerontology of rural health is incomplete without careful consideration of the ways in which spatial structures support (or do not support) older adults living in rural community settings. Such considerations require that care be theorised by including the ways in which institutional processes produce persistent socio-spatial inequalities and, thus, foster a geographically differentiated need for care (Smith and Easterlow, 2005). Alternatively, the adoption of relational and networked approaches to care and support may challenge the prevailing narratives of rural care gaps and deficiencies. Relational approaches would thus express how community health organisations may re-negotiate the support of local ageing populations in the context of regional and multi-faceted systems of health governance (Heley and Jones, 2012). This may help unlock potential to leverage new technologies and models of service delivery to enable rural communities to better attend to the needs of its ageing residents.

In concentrating on the efforts of rural community members to challenge externally driven development strategies that reflect hegemonic conceptions of "rural" space and place, we encourage active resistance to urban dominance (Thomas et al., 2011). This would help counteract the urban normativity of older adult care provision, thus targeting the injustices long faced by rural ageing populations when urban oriented health strategies have been adopted uncritically rather than confronted. Such efforts hold much promise in offering a means to advocate for local systems of care, and to work towards changes that are needed to funding, power, and resource allocation so as to more appropriately align with the spatial configurations of rural areas.

Considering rural socio-cultural relations

Finally, efforts to articulate socio-cultural relations embedded in rural communities are key considerations towards making critical sense of the experiences and impressions of people living with various forms of health conditions in rural settings. Critical scholarship in fields such as geography and sociology alert us to the extent to which matters of health and care are implicated in the full spectrum of social relations operating in rural places (Heley and Jones, 2012). This line of inquiry is especially useful in highlighting the extent to which negotiations of care in rural settings are just as capable of generating social division and exclusion as they are of producing social cohesion and support (Walsh et al., 2014).

As a starting point, we propose that a more relational account of health is needed to encompass the dynamic and interacting factors (e.g., values, biases, understandings) that influence the experience of well-being in rural settings. This account would recognize that health and care are embedded in the socio-cultural relations that construct rural identities, and would reflect on how local values, expectations, and community relations have important implications for the personal health experiences of older residents (Hanlon et al., 2007; Wiles et al., 2009). Highlighting both spatial and temporal dimensions, the health experiences of older persons cannot simply be reduced to rates of medical diagnoses but rather must be understood as fluid and dynamic experiences that have both subjective and objective elements. Taking account of the enabling and constraining socio-cultural factors that shape older adult health experiences, the health of older persons in rural places is an avenue of inquiry that can no longer be left uncritically examined.

Conclusion: towards an emplaced critical approach

Much of the literature on the health of rural older adults, scarce as it is, suffers from undertheorised sociological and geographical categorisations of place. As will be explored more thoroughly by our co-authors later in this book, uncritical scholarship assumes that rural places are structurally deficient as places of formal care delivery (e.g., Urbaniak, Chapter 18), and/or are innately supportive communities populated by resilient residents (e.g., Hennessy and Innes, Chapter 17). Such characterisations tend to condition expectations about welfare entitlement and the limits of formal care delivery (e.g., Glasgow and Doebler, Chapter 10). This, together with the persistent representation of "rural" as a distinct and separate geographical entity, acts further to stifle the range of possibilities for service delivery and policy response in rural places. These narratives of deficiency tend to be reproduced and reinforced over time in spite of innovations in organisation and technology that make it possible to overcome the "disadvantages" of distance and smaller populations.

To the extent that disparities are a common feature of rural settings, a critical gerontology of rural health must attend to Smith and Easterlow's (2005) call to illuminate geographies of health discrimination, or, in their words:

> [the] socially structured, institutionalized processes which accumulate across the lifecourse, subtly but systematically differentiating populations according to health histories and prospects.
>
> (174)

For our purposes, this is a call to explore the socio-geographic construction of differentiated life chances, opportunities, and barriers of older adult health, while not merely reducing the experiences of rural ageing to inherent disadvantage, or else idyllic community living. This more critical and relational approach instead recognises the diversity in older adult health and care and takes into account the full range of personal experiences and the influences that are possible in particular rural environments.

The health of rural older adults, and the negotiation of care and social support networks, are important instances in shaping and re-shaping relations between people and place. Rather than reproducing uncritical and crude characterisations of rural, we align ourselves with the co-authors of this book to advocate for a critical approach that recognises the tensions, contradictions and negotiations which are ongoing features of emplaced experiences of health and care. In this way, we hope to promote a greater engagement of critical gerontology in matters of rural health and to encourage more contextually nuanced and emancipatory approaches such as those exhibited and advocated for throughout this book.

References

Andrews, G. (2004). (Re)thinking the dynamics between healthcare and place: therapeutic geographies in treatment and care practices. *Area*, 36(3), pp. 307–318.

Baernholdt, M., Yan, G., Hinton, I., Rose, K. and Mattos, M. (2012). Quality of life in rural and urban adults 65 years and older: Findings from the National Health and Nutrition Examination Survey. *The Journal of Rural Health*, 28(4), pp. 339–347.

Bell, S. L., Foley, R., Houghton, F., Maddrell, A. and Williams, A. M. (2018). From therapeutic landscapes to healthy spaces, places and practices: a scoping review. *Social Science & Medicine*, 196, pp. 123–130.

Bulow, M. H. and Sodergvist, T. (2014). Successful ageing: a historical overview and critical analysis of a successful concept. *Journal of Aging Studies*, 31, pp. 139–149.

Coleman, T. and Kearns, R. (2015). The role of bluespaces in experiencing place, aging and wellbeing: insights from Waiheke Island, New Zealand. *Health and Place*, 35, pp. 206–217.

Davis, S. and Bartlett, H. (2008). Healthy ageing in rural Australia: issues and challenges. *Australian Journal of Ageing*, 27(2), pp. 56–60.

Doheney, S. and Milbourne, P. (2017). Community, rurality and older people: critically comparing older people's experiences across different rural communities. *Journal of Rural Studies*. 50, pp. 129–138.

Estes, C. L., (ed.). (2001). *Social Policy and Aging: A Critical Perspective*. Thousand Oaks; London; New Delhi: SAGE Publications.

Gesler, W. M., Rabiner, D. J. and DeFriese, G. H., (eds). (1998). *Rural Health and Aging Research: Theory, Methods and Practical Applications*. New York: Baywood.

Glasgow, N. and Berry, E., (eds.). (2013). *Rural Aging in 21st Century America*. Dordrecht: Springer.

Goeres, L., Gille, A., Furuno, J., Erten-Lyons, D., Hartung, D., Calvert, J., Ahmed, S. and Lee, D. (2015). Rural-urban differences in chronic disease and drug utilization in older Oregonians. *The Journal of Rural Health*, 32(3), pp. 269–279.

Hanlon, N., Halseth, G., Clasby, R. and Pow, R. (2007). The place embededness of social care: restructuring work and welfare in Mackenzie, BC. *Health and Place*, 13, pp. 466–481.

Hanlon, N. and Kearns, R. (2016). Health and rural places. In: M. Shucksmith and D. L. Brown, eds., *Routledge International Handbook of Rural Studies*. London; New York: Routledge, pp. 62–70.

Heley, J. and Jones, L. (2012). Relational rurals: some thoughts on relating things and theory in rural studies. *Journal of Rural Studies*, 28, pp. 208–217.

Herron, R. and Skinner, M. (2018). Rural places and spaces of rural health and health care. In: V. Crooks, G. Andrews and J. Pearce, eds., *Routledge Handbook of Health Geography*. London; New York: Routledge, pp. 267–272.

Holstein, M. S. and Minkler, M. (2003). Self, society and the "new gerontology". *The Gerontologist*, 43(6), pp. 787–796.

Humphreys, J. S. and Gregory, G. (2012). Celebrating another decade of progress in rural health: what is the current state of play? *The Australian Journal of Rural Health*, 20(3), pp. 156–163.

Jensen, L., Monnat, S. M., Green, J. J., Hunter, L. M. and Sliwinski, M. J. (2020). Rural population health and aging: toward a multilevel and multidimensional research agenda for the 2020s. *American Journal of Public Health*, 110(9), pp. 1328–1331.

Katz, S., (ed.). (1996). *Disciplining Old Age: The Formation of Gerontological Knowledge*. Charlottesville; London: University Press of Virginia.

Katz, S. and Calasanti, T. (2015). Critical perspectives on successful aging: does it "appeal more than it illuminates"? *The Gerontologist*, 55(1), pp. 26–33.

Keating, N., (ed.). (2008). *Rural Ageing: A Good Place to Grow Old?* Bristol: Policy Press.

Keating, N., Swindle, J. and Fletcher, S. (2011). Ageing in rural Canada: A retrospective and review. *Canadian Journal on Aging*, 30(3), pp. 323–338.

Kelly, C. and Yarwood, R. (2018). From rural citizenship to the rural citizen: farming, dementia and networks of care. *Journal of Rural Studies*, 63, pp. 96–104.

Kulig, J. and Williams, A. M., (eds). (2012). *Health in Rural Canada*. Vancouver, BC: University of British Columbia Press.

Leipert, B. and Reutter, L. (2005). Developing resilience: how women maintain their health in northern geographically isolated settings. *Qualitative Health Research*, 15(1), pp. 49–65.

McCarthy, J. (2008). Rural geography: globalizing the countryside. *Progress in Human Geography*, 32(1), pp. 129–137.

McHugh, K. E. (2003). Three faces of ageism: society, image and place. *Ageing & Society*, 23(2), pp. 165–187.

Milbourne, P. (2016). Austerity, welfare reform, and older persons in rural places: competing discourses of voluntarism and community? In: M. W. Skinner and N. Hanlon, eds., *Ageing Resource Communities: New Frontiers of Rural Population Change, Community Development and Voluntarism*. New York; London: Routledge, pp. 74–88.

Milne, A., Hatzidimitriadou, E. and Wiseman, J. (2007). Health and quality of life among older people in rural England: exploring the impact and efficacy of policy. *Journal of Social Policy*, 36(3), pp. 477–495.

Pong, R. W. (2000). Rural health research in Canada: at the crossroads. *The Australian Journal of Rural Health*, 8(5), pp. 261–265.

Pong, R. W., DesMeules, M. and Lagacé, C. (2009). Rural-urban disparities in health: how does Canada fare and how does Canada compare with Australia? *The Australian Journal of Rural Health*, 17(1), pp. 58–64.

Poulin, L. and Skinner, M. (2019). Leveraging a contextually sensitive approach to rural geriatric interprofessional education. *Healthcare Management Forum*, 33(2), pp. 70–74.

Poulin, L. I., Skinner, M. W. and Hanlon, N. (2020). Rural gerontological health: emergent questions for research, policy and practice. *Social Science & Medicine*, 258, p. 113065.

Rowles, G. (1988) What's rural about rural aging? An Appalachian perspective. *Journal of Rural Studies*, 4(2), pp. 115–124.

Scharf, T., Walsh, K. and O'Shea, E. (2016). Ageing in rural places. In: M. Shucksmith and D. L. Brown, eds., *Routledge International Handbook of Rural Studies*. London and New York: Routledge, pp. 80–91.

Scott, M. (2013). Resilience: a conceptual lens for rural studies? *Geographical Compass*, 7(9), pp. 597–610.

Shucksmith, M. (2018) Re-imagining the rural: from rural idyll to good countryside. *Journal of Rural Studies*, 59, pp. 163–172.

Skinner, M. and Hanlon, N., (eds). (2016). *Ageing Resource Communities: New Frontiers of Rural Population Change, Community Development and Voluntarism*. New York: Routledge.

Skinner, M. W. and Rosenberg, M. W. (2006). Managing competition in the countryside: non-profit and for-profit perceptions of long-term care in rural Ontario. *Social Science & Medicine*, 63(11), pp. 2864–2876.

Skinner, M. W. and Winterton, R. (2018). Rural ageing: contested spaces, dynamic places. In: M. W. Skinner, G. J. Andrews and M. P. Cutchin., eds., *Geographical Gerontology: Perspectives, Concepts, Approaches*. London; New York: Routledge, pp. 136–148.
Smith, S. and Easterlow, D. (2005). The strange geography of health inequalities. *Transaction of the Institute of British Geographers NS*, 30(2), pp. 173–190.
Thomas, A. R., Lowe, B. M., Fulkerson, G. M. and Smith, P. J., (eds). (2011). *Critical Rural Theory: Structure, Space, Culture*. Lanham, MD: Lexington Books.
Van Dyk, S. (2014). The appraisal of difference: critical gerontology and the active-ageing-paradigm. *Journal of Aging Studies*, 31, pp. 93–103.
Van Wagenen, A., Driskell, H. and Bradford, J. (2013). "I'm still raring to go": successful aging among lesbian, gay, bisexual, and transgender older adults. *Journal of Aging Studies*, 27, pp. 1–14.
Walsh, K., O'Shea, E., Scharf, T. and Shucksmith, M. (2014). Exploring the impact of informal practices on social exclusion and age-friendliness for older people in rural communities. *Journal of Community and Applied Psychology*, 24, pp. 37–49.
Wells, M. (2009). Resilience in rural community-dwelling older adults. *The Journal of Rural Health*, 25(4), pp. 415–419.
World Health Organization (WHO). (2015). *WHO Global Strategy on People-Centred Integrated Health Services*. Geneva: WHO Document Production Services.
Wild, K., Wiles, J. and Allen, R. E. S. (2013). Resilience: thoughts on the value of the concept for critical gerontology. *Ageing & Society*, 33, pp. 137–158.
Wiles, J. L., Allen, R. E. S., Palmer, A. J., Hayman, K. J., Keelings, S. and Kerse, N. (2009) Older people and their social spaces: a study of well-being and attachment to place in Aotearoa New Zealand. *Social Science and Medicine*, 68, pp. 664–671.

5 Critical human ecology and global contexts of rural ageing

Norah Keating, Jacquie Eales, Judith Phillips, Liat Ayalon, Mayeso Lazaro, Verónica Montes de Oca, Patricia Rea and Prakash Tyagi

Introduction

The mission of the UN Sustainable Development Goals is bold and inclusive. In endorsing its principles, UN Member States pledged to ensure "no one will be left behind" and to "endeavour to reach the furthest behind first" (UNDP 2018: 3). The goals have a social justice agenda, aimed at reducing disparities within and across world regions (see Burholt and Scharf, Chapter 6 in this volume for further articulation of the social justice agenda).

Older persons are one of the groups that has received global attention. Contexts that address their status and potential vulnerabilities have been the subject of major reports by UN agencies that emphasise that risks of older people being left behind are influenced by their social, physical and policy settings (World Health Organization, 2015). Rural residents are seen as especially vulnerable as a result of the places in which they live that may leave them "isolated, without access to their immediate families or to social and other types of infrastructure" (UNECE, 2017).

Given their international remit, it seems surprising that global NGOs have not highlighted the powerful influence of global contexts such as climate change (see Carroll and Walker, Chapter 28), political instability and pandemics that go beyond political borders and may shape the experiences of older people in ways that we do not yet fully understand. It is time to take up the challenge of understanding "how diverse rural older adults become empowered or disempowered within certain contexts" (Skinner and Winterton, 2018: 16). In our view, despite considerable theorising and empirical examination of rural ageing, we have not adequately considered the contemporary relevance of global contexts in shaping experiences and processes of rural ageing.

The purpose of this chapter is to address this theoretical gap through:

1 Incorporating global contexts into Critical Human Ecology Theory, considering how macro contexts create constraints and opportunities in the lives of older people.
2 Grounding this global theorising with case examples from Malawi, India, Mexico and Israel, that illustrate the immense diversity in rural communities and their older residents.

3 Reconsidering key assumptions of Critical Human Ecology Theory in light of the incorporation of global contexts of rural ageing.
4 Proposing theoretical challenges to frame the next decade of rural ageing studies.

By incorporating global contexts into Critical Human Ecology theory, we illustrate how such contexts provide constraints (and a few opportunities) for rural people and settings. We believe that the increasing global perspective of ageing and engagement with international contexts requires such theoretical work as a basis for providing evidence of the importance of contexts in framing the lives of older people.

Critical human ecology theory: incorporating the global

Critical Human Ecology (Keating and Phillips, 2008) is a contextual framework that theorises near environments as the key contexts of rural ageing. Informed by Human Ecology and Environmental Gerontology traditions, the most relevant elements of near environments for older rural residents are physical (including built environments of home and community) and interpersonal (networks of family members, friends and neighbours). Interactions with physical and social aspects of peoples' immediate surroundings are emphasised (Keating and Phillips, 2008). The framework draws on critical theory, interrogating assumptions about rural areas as good places to grow old.

Global statements about contexts of rural ageing, suggest that we must broaden our ecological theorising, turning to macro environments as we have done so well and for so long with the near environments of rural communities and people. We can no longer ignore global contexts of uncertainty and unpredictability, climate change, political instability and mass migration. Rather, we must better understand these macro environments and how they intersect with and influence rural places and people. Examining global contexts of rural ageing will indeed help us better understand ageing in rural places (Skinner and Winterton, 2018) and the extent to which contemporary global contexts may marginalise and exclude.

Case examples of rural ageing

Theorising global contexts of rural ageing is in its infancy, in part because our theoretical understandings and empirical knowledge are based largely in high income countries in Europe, North America and Asia-Oceania. Yet, a key assumption of Critical Human Ecology is that understanding diverse experiences of ageing requires explicit consideration of ageing in various contexts (Keating and Phillips 2008). We address the admonition to broaden our contextual remit, grounding this discussion of global contexts within case examples from Malawi, India, Mexico and Israel. They represent different world regions. They also reflect that range of income groups designated by the World Bank List of Economies (World Bank Data Team, 2018) in which gross national yearly income (GNI) per capita is calculated. These are: low income ($995 or less, Malawi); lower middle income ($996 to $3,895, India); upper middle income ($3,896 to $12,055, Mexico); and high income ($12,056 or more, Israel). Importantly, they illustrate how different components of global contexts shape lives, often, but not always, in ways that leave older people behind. The case examples, presented in the next section of the chapter, are written by people who live in the countries they describe and who are social science and health professionals. Each was invited to write a short statement about older rural people in their country that illustrates broad contexts of rural ageing.

Rural Malawi: Grandfathers raising orphaned grandchildren

Malawi is one of the least developed countries in the world. It is largely rural, with an economy based in agriculture (World Population Review, 2019). It is heavily reliant on international aid. Mayeso Lazaro writes of a group of older rural residents of Malawi who have been rendered invisible by global and national contexts that contribute to their marginalisation. They are older men

In rural Malawi, the state plus local and international NGOs provide cash and food transfers to mitigate livelihood challenges for vulnerable households. Among the most vulnerable are households with children orphaned by HIV/AIDs. Orphans account for 12% of children in Malawi. Over 80% are being raised by grandparents, including grandfathers. Yet support programs have limited coverage, marred by widespread fraud and corruption, nepotism and local politics. They are often gendered due to the feminisation of care in which grandmothers are credited with orphan care. Grandfathers are systematically excluded from welfare support that targets orphans raised by grandparents. This exclusion is justified based on the assumption that grandfathers (like other men) cannot be trusted because they will use the money for beer or other personal needs instead of benefiting the children. Further, grandmothers are viewed as more vulnerable – frailer than grandfathers and with more limited economic opportunities – and are deliberately targeted in welfare safety nets. Subsequently, gendered conceptions of care may influence who (grandfathers versus grandmothers) gets access to social support in rural Malawi. Despite performing important care responsibilities for orphans, many grandfathers remain invisible in plain sight, as the care they provide is discounted and overlooked.

Figure 5.1 Rural Malawi: Grandfathers raising orphaned grandchildren in sub-Saharan Africa

who are raising grandchildren in the wake of the continuing HIV/AIDS epidemic that has left 12% of children in the country without parents. They are taking on unexpected family care in a setting in which families are the assumed source of support, but grandfathers are not seen as primary or even suitable caregivers. They are marginalised because of gender stereotypes that position them as undeserving of the limited amount of financial support for grandparent care. They are further marginalised because of the precariousness of international aid and widespread corruption in its distribution. Few receive any financial assistance.

The difficulties experienced by the Malawi grandfathers illustrate influences of global contexts on older rural residents. Hyde and Higgs (2016: 40) describe these influences as different "spatial logics" whose relations may continually shift but that must both be taken into account. The term "glocal" has been coined to describe these contextual relationships in which global trends "pass through the local like light passing through glass", changing the local and reflecting back on the global (Roudometof, 2015: 403).

The rather poetic use of refraction of light seems in stark contrast to the devastating influence of the HIV/AIDS pandemic on local communities and on the family lives of those who survive. There has been much research on the HIV/AIDS epidemic in response to a global public health crisis. Much less is known about how the lives of older persons are changed. On the global stage the plight of grandmothers raising grandchildren has been reflected back through NGOs (see, for example: https://stephenlewisfoundation.org/get-involved/grandmothers-campaign). Grandfathers remain invisible. Do such global reflections and invisibilities provide opportunities for rural researchers to see "the local" in new ways and to advance theorising of global-local interfaces? Do we need to incorporate gender more explicitly into our rural ageing research?

Thar Desert, India: climate and deep exclusion

In his description of the Thar Desert in India, Prakash Tyagi provides a picture of the powerful impact of climate on older people. We see a harsh rural setting in which frequent crop failures due to drought have a cascading effect, resulting in livelihood failures and food and water insecurity. These in turn are associated with a "double burden" of disease, exacerbated by the inability to access health services. Lack of literacy and family neglect mean that older persons, especially women, are vulnerable and have little agency. In this place where the marker of entry into old age is 55 (P Tyagi 2019, pers. comm. 23 September), there is deep exclusion (Walsh, Scharf and Keating, 2017). We know little about how climate change might be creating an even more dire situation for older people in the desert.

In this case example, little of the global impact of climate on local communities is reflected back. Researchers have begun to document the disproportionate impact of natural disasters such as earthquakes and tsunamis on the lives of

The Thar Desert of India is the most densely populated desert eco-system in the world. Over 27 million people in the Thar Desert within the Rajasthan State live under extreme poverty and chronic droughts. There are about 2 million older people who live under distress, in poor health and with very little support. Most rural areas are overlooked in India, this this one in particular because of being a desert.

Older people of Thar do not have regular sources of income. Agriculture is the only viable occupation, often failing due to droughts. Water insecurity and food insecurity are rampant causing severe malnutrition. Prevalence rates of diseases are very high with a double burden of communicable and non-communicable diseases (NCDs). Older women have even more serious challenges caused by isolation and social norms that give women no rights or resources. Widowhood is common in older women and is a serious social challenge. Older women always live within families where there is isolation and neglect. Health facilities are available at significant distances, are not affordable and are not of satisfactory quality. Lack of literacy among older people is a barrier to self-care education and in prevention and management of diseases.

Figure 5.2 Thar Desert, India: Climate and deep exclusion

older persons (Burholt and Dobbs, 2012; Labra, Maltais and Gingras-Lacroix, 2018). Global climate disasters are similar to pandemics – they are dramatic events that capture world attention. The difference with the example of the Thar Desert is the stable and pervasive exclusion of older people. There is nothing dramatic and so we do not see – an invisibility that surely is the hallmark of deep exclusion.

As rural researchers and ecological theorists, engaging with unfamiliar contexts can prompt us to challenge our critical thinking about the near environments that have long been the focus of our work. If we "see" the Thar desert, might we rethink ageing in place when food and water insecurity are endemic; or family support in settings of widespread abuse and neglect of older women within their families. Might we then really, critically, undertake the challenge raised by the editors of this volume to better understanding "how diverse rural older adults become empowered or disempowered within certain contexts" (Skinner and Winterton 2018: 16)?

Rural Mexico: International migration and well-being of older people

In their discussion of rural ageing in Mexico, Patricia Rea Angeles and Veronica Montes de Oca also describe a setting of poverty among older rural people where livelihoods based in agriculture are increasingly uncertain. What makes Mexico unique among the case countries illustrated in this chapter is that global contexts have long been a part of the national consciousness.

Mexico is one of the countries in Latin American with the highest rates of poverty and marginalisation; 41% of its population is poor. Rural communities which comprise 23% of the country's areas, are particularly affected by immigration agreements with North America. Changes in cropping patterns, declining crop yields due to global warming, low prices for agricultural products, drug trafficking and armed violence, have led to a depopulation of the countryside and a process of migration to the United States that began in the 1940s. Older people who stay in the communities, live alone or care for grandchildren, counting on remittances from absent migrants. Now migration has stopped and mass return and deportation processes have begun, threatening the well-being of older people who are in poor health and live without pensions, in homes without basic services, with lack of transport to the cities and limited access to health and social services. They often are isolated with limited family networks and are subject to violation of their human rights. Despite this, the elderly continue to confirm their support role for family members residing in Mexico or abroad. They have an important role as transmitters of cultural traditions and ancestral knowledge to younger generations.

Figure 5.3 Rural Mexico: International migration and well-being of older people

For more than 70 years, people have left rural areas of Mexico and migrated to the United States. They leave for economic opportunities and to escape the violence and instability associated with drug trafficking. The remittances they send home serve as a buffer against low and uncertain income as they do in many receiving countries (Adams and Page, 2005). Yet just at a time when the stability provided by remittances is increasingly important to older rural residents, the flow of people and of resources has been halted. Abrupt changes in US immigration policy mean that Mexican migrants are being barred or expelled from US residence. Global contexts of migration policy and international movement of illegal drugs frustrate governments and disrupt lives (Durán-Martínez, 2017; Slack, 2019). We know little about how the return of migrants to their home communities affects those communities or the social connections across generations in their families. Global politics have built a wall against global opportunities.

Israel Kibbutz: global ideological shifts

Liat Ayalon's narrative of the transformation of philosophy of the kibbutzim in Israel bears striking resemblance to global ideological shifts towards neoliberalism. Kibbutzim were established in the early twentieth century based on a philosophy of equality of production and consumption, values that are often associated with rural communities (Keating, Swindle and Fletcher, 2011).

Since the early twentieth century, the kibbutz was an important type of rural settlement in Israel. Kibbutzim were established based on communal living and equality of consumption and production. Members were afforded high social status and often assumed national leadership roles. In the past, older adults in such settlements reported higher well-being and more family support than those living in other settings.

Israel now is predominantly urban with only 8% of the population living in rural settings and less than 2% of the population living in kibbutzim. In turn, kibbutzim have moved away from a focus on agriculture and communal life towards privatisation and individual goals. Those in desirable locations have turned the land into their number one commodity. This evolution is associated with some challenges to well-being of older adults, although even following the privatisation of the kibbutz, well-being remains high. A recent study (Dahan and Schwartz, 2014) shows that older people in the kibbutz still feel more connected to their community, have better access to services, report better financial status and are more socially engaged compared with older adults who live in other small communities or in the city.

Figure 5.4 Israel Kibbutz: Global ideological shifts

However, as Ayalon notes, both rural living and such rural values have all but disappeared. Of the few remaining kibbutzim, only about 25% are still communal. Most have been privatised, emphasising markets as the organising feature of political, economic and social arrangements (Springer, Birch and MacLeavy, 2016).

How have older people fared in this transition? Political theorists and gerontologists argue that a feature of neoliberalism is the individualisation of risk (Hamilton, 2014), often associated with older people who have little agency in the face of such global changes. However, there is an indication that older people who have remained in the kibbutz fare well as owners of desirable properties with a strong sense of belonging. They may be exemplars of those who are the advantaged in what Rubinstein and de Medeiros (2015) have described as a two-class system of those who age "successfully" and those who do not.

The dominant global narrative about those who remain in rural communities that are experiencing out-migration is that they are "left behind", a phrase that is often associated with deprivation, isolation and advanced age (Antman, 2010; Zhong et al., 2017). Yet for these older people it seems plausible that rather than being "left behind", they have "stayed behind" enjoying the benefits of the culture of sharing and supportiveness that were fundamental to the establishment of the kibbutzim and those of the market economy that have contributed to their economic well-being.

Updating key assumptions from critical human ecology theory

Critical Human Ecology Theory was developed to draw together theoretical work from Environmental Gerontology and Human Ecology on important contexts in the lives of older persons and from Critical Theory that stresses the need to challenge discourses that foster inequities (Keating and Phillips, 2008). In our view, theories must be developed, applied and revised in an ongoing process that keeps them fresh, relevant and useful. In this final section of the chapter, we revisit four assumptions of Critical Human Ecology Theory that need to be reconsidered in light of the addition of global contexts (Keating and Phillips, 2008) and global settings, discussing revisions and challenges for rural ageing research.

Boundaries between environments are permeable so that characteristics of one environment interact with and influence others

The lens of rural ageing research has been on physical and social environments that comprise rural communities. Thus our theorising has been about the ideal interfaces between the resources of older rural residents and those of their communities (Keating, Eales and Phillips, 2013). The addition of global contexts requires that we rethink the influences of contexts on each other. Our case examples suggest that for the most part, global contexts have a powerful and often negative influence on local contexts. Pandemics, climate change and isolationist policies of other governments restrict lives.

We propose the following revised assumption: *Environments interact with and influence each other. Macro global contexts have a differential impact on other contexts, shaping local places and local lives.*

People have varying capacities to make choices and to act upon or adapt their environment

"Varying capacities to act" seems in hindsight a rather naive first-world assumption given the immensely difficult lives of older people described in most of the case examples. Do we risk deepening the exclusion of those who have little agency by assuming capability in the face of the profound inequities they experience? Does thinking globally create a conceptual imperative to think about where responsibility lies for supporting those who have limited ability to adapt their environments?

We hold different actors responsible for providing such support. UN reports place the onus on governments to leave no one behind by reducing widening inequalities in income, health status and access to services. Yet governments increasingly demur, placing the onus on communities to foster supportive environments. Winterton et al. (2016) provide an example. They state that "policy discourses promote the economic and health-related benefits of older adults ageing in community settings rather than in residential care, place considerable responsibility on communities to foster supportive environments" (321). Neoliberal discourses of individualisation of risk further shift responsibility to older persons themselves through narratives of empowerment and successful and unsuccessful ageing. We must be vigilant in the face of what Mégret (2011) describes as the power/vulnerability paradox which positions older people as affluent and powerful as well as isolated and vulnerable. He argues that this paradox "structures society's very ambivalent rapport to its elderly class" (47). Thus, we propose the following revised assumption: *People have varying capacities to make choices and to act upon or adapt their environments. There must be shared responsibility for supporting those with limited agency.*

Critical Human Ecology challenges the homogeneity of older persons

We have not undertaken a systematic consideration of elements of homogeneity that we need to challenge. In the case examples we are reminded of age diversity (Malawi grandfathers ranged from 54 to 92); and of age disadvantage (in the harsh Thar Desert, the marker of entry into old age is 50). Gender diversity is evident though not always so. Women's suffering in the Thar Desert is disproportionate to that of men; in contrast, men are rendered invisible in Malawi. Are there similarly gendered experiences in who benefits more from land as commodity on kibbutzim or from remittances sent home to older people who remain behind in rural Mexico?

Importantly the immense diversity in rural settings also challenges us to address the question of whether rural itself remains distinctive. If low population density is a rural hallmark, how do we think of the Thar Desert that has millions of people but is designated rural because it is more sparsely populated than other parts of India? What is a rural community or rural people where boundaries are blurred by in and out migration and by global connections (Mexico), or where rural is economically advantaged (Israel)? Can we reconcile ideas of rural communities as supportive when we see evidence of a sense of belonging (Israel) against that of suffering and family neglect (Thar Desert)? We do not propose changing the assumption of homogeneity but call for further consideration of which "heterogeneities" we wish to highlight and address.

Critical Human Ecology attempts to make the voices of marginalised groups heard

For the most part, researchers come from a place of careful, dispassionate creation of the evidence. We position ourselves as informers of action agendas rather than as leaders of them. Yet if we take seriously the incorporation of a global lens on environments of rural ageing, then we must see those who are most left behind. There is a place for a kind of theoretical activism here in making the invisible visible by creating evidence of the contexts that exclude at both global and local levels (Walsh, Scharf and Keating, 2017). It is only then that we can address the broad sweep of the UN assumption that [all] rural residents are at risk because of the places in which they live or that access to their immediate families reduces the risk. We propose the following revised assumption: *Critical Human Ecology makes the voices of marginalised groups heard by identifying and creating evidence of those who are rendered invisible by contexts that exclude.*

Framing the next decade of rural ageing studies

In this chapter we have incorporated global contexts into Critical Human Ecology theory, grounding our theorising within four cases from different world regions. Revised theoretical assumptions point to the next phases in our theory, research and action agendas on rural ageing studies. We present three main challenges:

1 To further develop components of global environments/contexts, parallel to the physical and social components of near environments that should be articulated and investigated for their impact on older (rural) people. As a starting point, we have identified: climate change, political instability, national values and ideologies and transnational migration. These require further articulation across diverse settings.
2 To further develop critical perspectives on marginalisation, including recognising inequities, determining which can be addressed and where agency lies in addressing them.
3 To be self-reflective as rural researchers about whether "rural ageing studies" is a useful disciplinary label given the theoretical and empirical challenges we have raised.

Disciplinary labels establish theoretical and substantive lenses on specific populations or topics. In the social sciences, they often arise from a wish to address aspects of the human condition that create vulnerabilities. Global discourses of older rural residents at risk of being "left behind" place rural gerontology at the nexus of marginalised people and places. Yet our critical theorising and evidence belie a unidimensional experience of exclusion. We must avoid reifying either rural or age, focusing on the intersectionalities of diverse environments and people across the global north and south; and about the ways in which these may create exclusion for some.

References

Adams, R. H. and Page, J. (2005). Do international migration and remittances reduce poverty in developing countries?, *World Development*, 33(10), pp. 1645–1669.

Antman, F. M. (2010). Adult child migration and the health of elderly parents left behind in Mexico. *American Economic Review: Papers and Proceedings*. 100(2), pp. 205–208.

Burholt, V. and Dobbs, C. (2012). Research on rural ageing: where have we got to and where are we going in Europe? *Journal of Rural Studies*, 28(4), pp. 432–446.

Dahan, D. and Schwartz, E. (2014). *Veteran Citizens in Small Settlements: Main Characteristics in Comparison to Veteran Citizens in the City: A Research Report*. Jerusalem: The Hebrew University (in Hebrew).

Durán-Martínez, A. (2017). Drug trafficking and drug policies in the Americas: change, continuity, and challenges. *Latin American Politics and Society*, 59(2), pp. 145–153.

Hamilton, M. (2014). The 'new social contract' and the individualisation of risk in policy. *Journal of Risk Research*, 17(4), pp. 453–467.

Hyde, M. and Higgs, P., (eds). (2016). *Ageing and Globalization*. Bristol: Policy Press

Keating, N., Eales, J. and Phillips, J. (2013). Age-friendly rural communities: conceptualising best-fit. *Canadian Journal on Aging*, 32(4), pp. 319–332.

Keating, N. and Phillips, J. E. (2008). A critical human ecology perspective on rural ageing. In: N. Keating, ed., *Rural Ageing: A Good Place to Grow Old?* Bristol: Policy Press, pp. 1–10.

Keating, N., Swindle, J. and Fletcher S. (2011). Aging in rural Canada: a retrospective and review. *Canadian Journal on Aging*, 30(3), pp. 323–338.

Labra, O., Maltais, D. and Gingras-Lacroix, G. (2018). Medium-term health of seniors following exposure to a natural disaster. *INQUIRY: The Journal of Health Care Organization, Provision, and Financing*, 55, pp. 1–11.

Mégret, F. (2011). The human rights of older persons: a growing challenge. *Human Rights Law Review*, 11(1), pp. 37–66.

Roudometof, V. (2015). Theorizing glocalization: three interpretations. *European Journal of Social Theory*. 19(3), pp. 391–408.

Rubinstein, R. and de Medeiros, K. (2015). Successful aging, gerontological theory and neoliberalism: a qualitative critique. *The Gerontologist*, 55(1), pp. 34–42.

Skinner, M. and Winterton, R. (2018), Interrogating the contested spaces of rural ageing: implications for research, policy and practice. *The Gerontologist*, 58(1), pp. 15–25.

Slack, J., (ed.). (2019). *Deported to Death: How Drug Violence is Changing Migration on the US – Mexico Border*. Oakland, CA: University of California Press.

Springer, S., Birch, K. and MacLeavy, J. (2016). An introduction to neoliberalism. In: S. Springer, K. Birch and J. MacLeavy, eds., *Handbook of Neoliberalism*. New York: Routledge, pp. 1–14.

UNECE (2017). Older persons in rural and remote areas. *UNECE Policy Brief on Ageing No. 18*. Available at: www.unece.org/population/ageing/policybriefs.html [Accessed 9 October 2019].

UNDP (2018). What does it mean to leave no one behind? *A UNDP discussion paper and framework for implementation*, United Nations Development Programme. Available at: www.undp.org/content/undp/en/home/librarypage/poverty-reduction/what-does-it-mean-to-leave-no-one-behind-.html [Accessed 9 October 2019].

Walsh, K., Scharf, T. and Keating, N. (2017). Social exclusion of older persons: a scoping review and conceptual framework. *European Journal on Aging*, 14(1), pp. 81–98.

Winterton, R., Warburton, J., Keating, N., Petersen, M., Berg, T. and Wilson, J. (2016). Understanding the influence of community characteristics on wellness for rural older adults: a metasynthesis. *Journal of Rural Studies*, 45, pp. 320–327.

World Bank Data Team. (2018). *New Country Classifications by Income Level: 2018–2019*. Available at: https://blogs.worldbank.org/opendata/new-country-classifications-income-level-2018-2019 [Accessed 21 September 2019].

World Health Organization. (2015). *World Report on Ageing and Health*. Available at: http://apps.who.int/bookorders/anglais/detart1.jsp?codlan=1andcodcol=15andcodcch=903 [Accessed 9 October 2019].

World Population Review. (2019). *Malawi*. Available at: http://worldpopulationreview.com/countries/malawi-population/ [Accessed 9 October 2019].

Zhong, Y., Schön, P., Burström, B. and Burström, K. (2017). Association between social capital and health-related quality of life among left behind and not left behind older people in rural China. *BMC Geriatrics*, 17, p. 287.

6 Critical social gerontology and rural ageing

Vanessa Burholt and Thomas Scharf

Introduction

This chapter focuses on rural places as the context for critical social gerontology. Critical gerontology is a meta-framework for challenging power and social injustice in ageing societies. The framework considers ageing to be a social construct that is shaped by societal forces and characteristics, which is an important expansion on the social gerontology literature (Dannefer and Phillipson, 2010). We illustrate the role of critical rural gerontology by drawing on Fraser's (1998) social justice framework and consider the distribution of resources (such as, material and social resources or community resources including access to services), recognition (such as social status, cultural visibility through social participation and cultural worth through valued social roles) and representation (in research and policy making processes) of older people in rural areas.

Our approach questions taken-for-granted assumptions about ageing. For example, critical gerontologists reject, scrutinise or critique commodification through the market economy, rationalisation of resources through ageing policies and the construction of ageist assumptions through culture or media imagery that result in instrumental domination of (some) older people and limit opportunities (access to roles or resources), choices and experience in later life. As critical gerontology assumes power, equity and social justice are related to the political and cultural economy of particular countries and contexts, the approach draws on political economy, moral economy and humanistic theories, to explain the experiences of older populations. Age could be regarded as an organising principle of the economic structure of society and is related to the distribution of resources (Fraser, 1998). Critical gerontology is concerned with seeking egalitarian redistribution of goods that does not discriminate against or exclude older people.

With rural areas often portrayed as traditional and conservative, they are frequently viewed as a historical snapshot or repository of a bygone culture, or "backward areas" in need of modernisation (Shucksmith, 2018). Some authors argue that globalisation will mean that "there is no culture in the world that can be said to be fixed and bounded, separate from other cultures" (Gillis, 2004: 118). From this perspective, it is assumed that rural communities will succumb to the homogenisation of cultures and eventually conform to family, religious, social and economic structures found elsewhere. However, this approach glosses over the agency of rural inhabitants and the "politics of recognition" (Fraser, 1998).

As an alternative to global homogenisation, critical gerontology sheds light on how spaces, places, environments and landscapes shape older inhabitants' health and well-being (see Hanlon and Poulin, Chapter 4) and experiences, while simultaneously individuals shape or adapt their environments in both the physical and socio-cultural milieu in which they are situated (Skinner and Winterton, 2018). A critical gerontological approach to social justice moves away from assimilation to a set of global norms and instead focuses on respecting the intersection of social and environmental differences.

In using research and theory to influence policy, practice and social action, critical gerontology is also underpinned by an anti-oppressive emancipatory philosophy (Stephens, Burholt and Keating, 2018). It attempts to strengthen connections between academics, older people, practitioners and policy makers to challenge and reshape ways in which societies construct ageing and impact on older people's lives. In terms of social justice, it focuses on parity of participation (Fraser, 2007). Critical gerontology seeks ways to ensure older people have a "voice" in society, can interact with researchers as peers and are able to influence outcomes (e.g., political recognition of distinctiveness, social inclusion and redistribution of resources) that impact on their lives (see Winterton and Warburton, Chapter 23).

The critical lens acknowledges the influence of macro-level policies, norms, values and attitudes, meso-level social networks, organisations and communities and micro-level social contexts. It examines the interplay between these and access to resources and roles at the individual level as explanations for ageing experiences in rural areas.

In this chapter, we consider whether critical gerontology has raised awareness of the heterogeneity of rural ageing in High Income Countries (HICs) in Europe, North America and Australasia and compare this to our knowledge of the issues that are associated with rural ageing in Low- and Middle-Income Countries (LMICs) (see Amoah and Phillips, Chapter 7). By doing so, we provide a reflexive interpretation of the state of critical rural gerontology that takes into account any disjuncture between HIC theories and policies emphasising productivity/participation and the rise of a global political economy promoting risk/instability that may be especially prevalent in rural LMICs. The chapter is organised into three sections focusing on aspects of social justice: resources, recognition and representation.

Demography of rural areas and the distribution of resources

LMICs are at an earlier stage of "demographic transition" than HICs, with the former experiencing either stable or declining fertility and decreasing mortality. These features contribute to increasing numbers of older people in expanding populations. By contrast, in HICs, low fertility and stable mortality contribute to an increase in the proportion of older people in stable or declining populations. In both scenarios there are more older people, but in LMICs there are *proportionally* fewer older people in the population than in HICs (Rishworth and Elliott, 2019).

The age structure of society is important because it is used as a political tool to "justify" the distribution of resources to older people in rural areas. For example, in LMICs with a younger population age structure, the focus of government expenditure may be on reproductive health, communicable diseases, education and training for children and young adults. Legislation in LMICs often mandates families to "care for" older persons. On the other hand, in HICs with an older demographic structure, the State may focus on health and social care provision in rural areas that addresses chronic conditions that are more common in later life (e.g., cardiovascular disease, dementia), appropriate accommodation and accessible communities and social protection through pension provision and/or extending working lives.

Many academic publications on rural ageing in LMICs focus on demographic trends such as migration, urbanisation and counter-urbanisation, socio-economic and health features and the consumption of health and welfare services by rural older populations (Hosseinpoor et al., 2016). However, most comprise an uncritical collection of data that do not challenge the unequal distribution of disease, poverty and access to services.

The ways in which demographic change is used to represent a "drain on resources" is crucial. A typical example would focus on projected burdens of non-communicable diseases prevalent in rural older populations in terms of predicted health care use and cast this as putting "enormous pressure on health, welfare and social systems" (Tollman, Norris and Berkman, 2016: 998) and as a factor that will limit future economic growth (Brinda et al., 2014). Portraying older populations in apocalyptic scenarios contributes to a moral panic about how resources are distributed to older rural cohorts (Gee, 2002). This contrasts with a critical gerontological perspective, which regards understanding diversity and inequalities *within* cohorts in LMICs as being more important in shaping the distribution of resources.

A focus on age-based "burden of disease" typically homogenises needs. This approach fails to distinguish between specific requirements of certain groups of older people (defined by such characteristics as gender, socio-economic status, level of disability) and between different locations. Some rural places are characterised by weak local economies and poor provision of services (Milbourne, 2012) and/or have an inadequate transport infrastructure that limits social contacts outside the community or region (Shergold and Parkhurst, 2012). By contrast, other rural areas have experienced gentrification (Stockdale, 2010). While this may lead to investment in and restructuring and regeneration of rural communities (Burholt et al., 2018), counter-urbanisation is also associated with rising property prices and issues of housing affordability (Winter and Burholt, 2019).

The age structure of rural areas may differ from country-level indicators and is often more skewed towards ageing populations. For example, in many European countries out-migration of younger people and in-migration of retirees is anticipated by 2030 to lead to higher proportions of older people in rural areas than urban areas (Burholt et al., 2018). In LMICs, rural out-migration is also age-selective comprising mostly young adults seeking to escape rural poverty by migrating to cities for employment (Anríquez and Stloukal, 2008). With outmigration of younger people potentially leading to "shortages" of family care in rural areas, provision of age-related resources may be more imperative in some rural locations (see Berry, Chapter 2 and Cheng et al., Chapter 9 in this volume).

Critical gerontology challenges the use of crude measures, such as "old-age dependency ratios, as the basis of resource distribution, considering instead intersectionality and inequality to develop a deeper understandings of the lives of rural older people and their caregivers, which both shape and are shaped by their geographic location. In HICs, scholars have applied the concept "precarity" to explore future uncertainties, insecurity and exclusion in older people's lives (Grenier, Phillipson and Settersten, 2020). For example, in many rural areas social protection offered through the voluntary sector has been eroded by austerity measures. Lack of resources, in terms of formal services and support, places greater pressures on caregivers, or leaves older rural people without support altogether (Power, 2018). The impacts are amplified for people who have accumulated disadvantage across the life course.

Research in LMICs has highlighted precarity in rural livelihoods of older people (Burholt, Maruthakutti and Maddock, 2020). Livelihoods are only sustainable if they can recover from a shock (e.g., poor health) and this is dependent on an older person's access to resources. For example, in Cambodia catastrophic health expenditure was associated with living in a rural area with an older person in the household (Jacobs, de Groot and Fernandes Antunes, 2016). In China, there is significantly less access to formal support services for older people living in rural areas and the impact of the out-migration of children is felt more acutely (Bai, Lai and Guo, 2016).

Globally, it is estimated that 1.2 billion people live in poverty in rural areas (75% of those in extreme poverty), and there is an "overwhelming concentration of the poor and hungry in the rural areas" of LMICs (Rodríguez-Pose and Hardy, 2015: 11). From a critical gerontology perspective, it is imperative that more research is conducted on the experiences of rural older people who may have accumulated disadvantage throughout the life course (see Milbourne, Chapter 25).

Recognition

Recognition "designates an ideal reciprocal relationships between subjects in which each sees the other as equal" (Fraser and Honneth, 2003:10). Recognition is a prerequisite for the fair distribution of resources and vice versa: maldistribution impinges on misrecognition (Fraser and Honneth, 2003).

Critical gerontology is a useful lens to examine the recognition of older people in rural areas. The dynamic processes of being excluded from key systems and institutions can shape the economic and social integration of people within society (Burholt et al., 2018). The combined effects of exclusion from representation and characteristics of geographical location, age *and* gender, disability, race and sexuality influences social status and access to valued social roles and social participation and institutionalises "misrecognition" (Fraser, 2000). Misrecognition can contribute to negative experiences for older people who may be disrespected and subjected to humiliation, social exclusion or abuse (Honneth, 2007). Examining the social construction of "old age" through the experiences and perceptions of normative social values, ageism, prejudice and discrimination provides one means of exploring the misrecognition of older people in rural areas.

In HICs, few research studies have explored older people's experiences of discrimination in rural areas. These studies have tended to focus on specific communities of older people, such as lesbian, gay and trans people (Westwood et al., 2019), gypsies and travellers (Hennessy et al., 2014), people from ethnic minority groups (Westwood et al., 2019), or indigenous populations (Svensson, 2019). The scarcity of research is, in itself, a mark of the lack of recognition of the heterogeneity of older people in rural areas. For example, while the social acceptance of older lesbian, gay, bisexual, transgender, intersex or queer (LGBTIQ) people in HICs has grown markedly in recent years, this has not been reflected in an increase in research with LGBTIQ older adults living in rural areas: their existence is frequently unrecognised (Adams, 2019).

In LMICs, there is a body of research focusing on transformations in social structures that influence the recognition of older people. Frequently, in these contexts theorisation links changes in social status to "modernisation". In a gross over-simplification of trends, social mores and social values, modernisation theory suggests that traditional societies are bound together by territorial tribalism, economic interdependence and family solidarity (filial piety and intergenerational co-residence) whereas modern societies are characterised by diffused ties, geographic separation and independence of nuclear units across generations. In economically advanced societies, characterised by rapid increases in specialised knowledge and family dispersion for educational and employment benefits, it is assumed that older people have a lower status than in more traditional societies. In this respect, rural areas in LMICs and some HICs have been portrayed as "traditional" societies undergoing ideational change in which values and beliefs about older adults are transformed (e.g., Bai, Lai and Guo, 2016; Silverstein, Burholt and Wenger, 1998).

The preoccupation of research with ideational transformation in rural areas of LMICs has led to a body of knowledge concerning the recognition of older adults. While some studies note that respect and recognition of older people in rural areas is strong (Latif et al., 2019), others point to changing social-cultural values in which adult children aspire to autonomy (individualism), there is a normalisation of disrespect, neglectfulness or verbal abuse (e.g., Bai, Lai and Guo, 2016). Older women living alone appear to be particularly prone to discrimination in many LMICs and are excluded from full participation in rural communities (Nguyen et al., 2019). Despite contradictory evidence on the status of older people, it is notable that the volume of studies illustrating the endurance of "traditional" values and recognition of older people has been declining since the late 1990s.

In HICs, the rural bucolic myth has influenced the misrecognition of older people. Visual portrayals of imagined rural places are popularly duplicated and mythologised in the media, with normative supportive rural practices of close-knit communities reproduced through film, television, literature, art, music and poetry. The dominant policy focus in many HICs on "active" and "healthy" ageing also contributes to the dominant ideology governing supportive behaviours by attempting to harness the power of local communities to provide mutual support. In general, rural citizens are expected simultaneously to take responsibility for their own well-being by adopting lifestyle practices that minimise the likelihood for need for care and accept their public duty to help support others in the community (Sevenhuijsen, 2003). For example, the voluntary sector is expected to provide services to older people, while simultaneously being reliant on older volunteers to deliver services (Colibaba et al., Chapter 24; Skinner et al., 2014). The "rights and responsibilities" approach assumes that people (and places) have both the means and ability to engage, thus misrecognising those who lack these attributes.

Media and policy rhetoric that (mis)recognises older people in rural areas risks becoming self-fulfilling, as dominant ideologies of "self-help" influence the way that people behave. However, the discourse is constructed by powerful and political voices in society and excludes the voices of marginalised older people (discussed later). While it has been argued that "devaluing" older people is challenged by positive portrayals of older people (e.g., as supportive volunteers), this is problematic when disadvantaged rural communities lack the resources to fill service gaps for older people, or when individuals with different modes of power relating, for example, to age, gender, disability and class, fail to meet social expectations for voluntarism (Burholt et al., 2018).

The disjuncture between theory and policy that emphasises productivity and/or participation is especially stark in rural areas of LMICs where the global political economy has contributed to risk and precarity. Despite the growing trend for "radical individualism, narcissism, consumer acquisitiveness and the erosion of collective (social) responsibility" (Sharma, 2016: 267), concepts such as volunteerism and the consumption of leisure or cultural activities are largely irrelevant to older people engaged in subsistence living with no other means of external support.

Misrecognition of older people in rural areas is important because self-awareness of social (in)justice impacts on older people's quality of life, social cohesion and sense of belonging (Honneth, 2007). Internalisation of negative stereotypes or the inability to live up to normative expectations can lead to poor self-esteem, low self-respect, unhappiness and depression (e.g., Tran, Nguyen and Van Vu, 2018). For example, in rural China, where older people are likely to be perceived as a burden to family and society (Bai, Lai and Guo, 2016), suicide rates for rural older people are around six times the average (Cao, 2019). By contrast, in rural China when older people's status is recognised through filial piety and multi-generational co-residence, they achieve better psychological outcomes (Silverstein, Cong and Li, 2006).

Applying a critical gerontological approach to rural research on (mis)recognition may help with operationalisation of concepts such as "social status". While significant differences and inequalities exist between and within rural areas, there is little scientific evidence of how features of rural communities might structure rural life. Although evidence summarises the impact of social changes on distribution of resources, such as rural social support (Keating and Dosman, 2009), there is limited evidence about impact of misrecognition (e.g., the inability to live up to policy rhetoric) on other outcomes. For example, increasing population turnover may influence local neighbourliness or social cohesion even in rural communities which are presumed to have strong support networks that compensate for poor formal infrastructure (Winterton et al., 2016). Moreover, within areas typified by "ageing in place" (older people staying in the communities of origin) or as "ageing places" (communities that have a growing population of older migrants), there may be different degrees of marginalisation (ageing with few financial or health resources) (Burholt and Sardani, 2017) that can impact on recognition and social status variously for different groups of older people.

Representation

A particular strength of critical gerontology is its focus on a "moral economy" which is concerned with power in relation to the construction of old age. Here, we consider how critical gerontology has influenced research in HICs and LMICs and promoted empowerment of older people in rural areas. Empowerment is intended to enable older people and communities to transform their lives and communities through civic action (Minkler, 1996). Thus, participatory research and its resulting evidence can be used to challenge social and political assumptions and disempowerment that often accompanies ageing. Simultaneously, the representation of rural older people in research and policy making opens opportunities to challenge the misrecognition of older people as burdensome and "unproductive".

Critical gerontologists use participatory methods to enable older people to assume control of the research process and the production of knowledge. However, biomedical perspectives on rural ageing dominate the literature (Burholt and Dobbs, 2012). Arising from an approach where ageing is viewed as pathologised as a "problem" linked to decline and dependency, very few rural studies actively involve older people in the research process. In HICs, some positive examples of rural participatory research have either empowered older people (Nomura et al., 2009) or influenced policies (e.g., John and Gunter, 2015). However, there are fewer rural than urban examples of participatory research. While examples of good practice involving older people in the development of public services exist in HICs, these also predominantly focus on urban areas with the voices of rural older people rarely taken into consideration (Le Mesurier, 2006). From a critical gerontological perspective, it is imperative that rural older people's voices contribute to policy making to ensure that "structured dependency" is avoided and that policies developed in urban settings are not assumed to map unproblematically onto rural areas.

The critical gerontology gaze is firmly focused on macro-level institutional processes (e.g., social protection and health systems) to understand how these impact on individuals' lives in rural settings. Consequently, the outcomes of a few participatory rural research projects have been used by stakeholders, for example, to leverage the provision of community centres in the USA (John and Gunter, 2015). In LMICs, participatory research has tended to focus on developing appropriate health, care (e.g., Pelcastre-Villafuerte et al., 2017) and transport services (e.g., Porter et al., 2015). However, few rural studies in either HICs or LMICs have challenged policies that generate dependency.

India is an example of a country that depends on informal social protection and mandates families to take care of their older members. The Maintenance and Welfare of Parents and Senior Citizens Act (Ministry of Social Justice and Empowerment, 2007), states that parents, grandparents and "childless" older people who are unable to maintain themselves from their own income and property are entitled to demand and receive income from children, grandchildren and other relatives who have sufficient resources. Cases (where support is not forthcoming) can be taken to Tribunal and can result in the issue of maintenance orders with penalties for non-compliance including fines and imprisonment. Presently, 70% of the older population in rural areas of India are fully or partially financially dependent on others (Alam et al., 2012).

The International Labour Organization (ILO) (2011) reports that three-quarters of the world's population are not covered by "*adequate* social security" and, as noted earlier, the poor are disproportionately located in rural areas. In rural areas of many LMICs there is a reliance on *informal* social protection. However, changes in rural family structures and values, migration of family members and increasing numbers of women working outside of the home place increasing strain on families to provide support. Traditional forms of solidarity and collectivism are eroded by market economies: increasing monetisation impacts on forms of reciprocity, and requirement for a responsive mobile labour force has impacted on availability. Therefore, it is important to challenge the "realities" of family support systems in rural areas which may not be as robust as portrayed by policy makers (Burholt, Maruthakutti and Maddock, 2020).

Representation of rural-dwelling older people in research is important in both HICs and LMICS, in both cases this could facilitate contribution to policy making and/or challenge the dominant discourses. While community action is not impossible in LMICs, it is difficult as participants have limited resources and fewer options for "free time" to devote to civic participation. Similarly, in HICs people with the least resources to participate in civic action in rural areas are those who are most likely to be disempowered and unrepresented (e.g., older carers of people living with dementia; disadvantaged older women). Canada provides an illustrative example, whereby women are rarely represented in rural governance (Bock and Derkzen, 2008). The conceptualisation of highly gendered rural connectivity that dictates the "correct and proper" roles for women (i.e., domestic responsibilities and caring tasks) is unlikely to be challenged substantially and continues to be perpetuated in the dominant policy discourses of "active ageing" and volunteerism.

Conclusions

While critical gerontology has been used as a meta-framework in research in HICs and LMICs for challenging power and injustice in an ageing society, there are gaps in the main topics of attention. Resource distribution could be improved by further research addressing such questions as "What are the trajectories and vulnerabilities that affect later life?" Critical gerontology can contribute to the recognition of rural older people and help to determine (1) under what circumstances traditional social values persist and communities recognise and value older people's contributions of older people or (2) cultural change leads to the misrecognition of older people and poor outcomes. Structured dependency is overwhelmingly an issue for rural-dwelling older people (especially in LMICs) but has yet to be tackled adequately by research on (mis)representation. Critical gerontology could facilitate the involvement of rural older people in decision-making processes that determine resource distribution, access to services and define support arrangements while simultaneously contributing to the recognition of older people. However, HICs with strong clusters of critical gerontologists have a clear role to play in developing opportunities to work with others and extending modes of practice and insights to rural areas around the world.

Above all, the critical gerontology lens reminds us of our moral and ethical obligations to older people in rural areas. The tools we adopt to undertake research and the language we use to articulate research findings can either challenge hegemony, power and injustices or reinforce exclusion, patterns of oppression (silencing) and undermine the social status of older people in rural areas. Echoing Phillipson and Walker's (1987) argument for "a more value-committed approach to social gerontology", our view is that critical gerontologists should not only conduct research that explores the characteristics and impacts of inequality in rural areas, but that they also pick up the challenge to translate their findings into social action aimed at improving the lives of older people in rural areas.

References

Adams, M. (2019). LGBT older adults in Latin America: an emerging movement. *ReVista (Cambridge)*, 19, pp. 1–7.

Alam, M., James, K. S., Giridhar G., Sathyanarayana, K. M., Kumar, S., Siva Raju, S. Syamala, T. S., Subaiya, L. and Bansod, D. W. (2012). *Report on the Status of the Elderly in Select States of India, 2011*. New Delhi, India: UNFPA.

Anríquez, G. and Stloukal, L. (2008). Rural population change in developing countries: lessons for policymaking. *European View*, 7, pp. 309–317.

Bai, X., Lai, D. W. L. and Guo, A. (2016). Ageism and depression: perceptions of older people as a burden in China. *Journal of Social Issues*, 72, pp. 26–46.

Bock, B. B. and Derkzen, P. (2008). Barriers to women's participation in rural policy making. In: T. Marsden, ed., *Gender Regimes, Citizen Participation and Rural Restructuring*. Oxford: JAI Press, pp. 263–281.

Brinda, E. M., Rajkumar, A. P., Enemark, U., Attermann, J. and Jacob, K. S. (2014). Cost and burden of informal caregiving of dependent older people in a rural Indian community. *BMC Health Services Research*, 14, p. 207.

Burholt, V. and Dobbs, C. (2012). Research on rural ageing: where have we got to and where are we going in Europe? *Journal of Rural Studies*, 28, pp. 432–446.

Burholt, V., Foscarini-Craggs, P. and Winter, B. (2018). Rural ageing and equality. In: S. Westwood, ed., *Ageing, Diversity and Equality: Social Justice Perspectives*. London: Routledge, pp. 311–328.

Burholt, V., Maruthakutti, R. and Maddock, C. A. (2020). A cultural framework of care and social protection for older people in India. *GeroPsych*.

Burholt, V. and Sardani, A. V. (2017). The impact of residential immobility and population turnover on the support networks of older people living in rural areas: Evidence from CFAS Wales. *Population, Space and Place*, 24, p. e2132.

Cao, F. (2019). The marginalisation of the rural elders. In: F. Cao. ed., *Elderly Care, Intergenerational Relationships and Social Change in Rural China*. Singapore: Palgrave Macmillan, pp. 111–132.

Dannefer, D and Phillipson, C., (eds.). (2010). *The SAGE Handbook of Social Gerontology*. London: SAGE Publications.

Fraser, N. (1998). *Social Justice in the Age of Identity Politics: Redistribution, Recognition, Participation* (WZB Discussion Paper). Berlin: WZB Berlin Social Science Center.

Fraser, N. (2000). Rethinking recognition. *New Left Review*, 2000, pp. 107–120.

Fraser, N. (2007). Feminist politics in the age of recognition: A two-dimensional approach to gender justice. *Studies in Social Justice*, 1, pp. 23–35.

Fraser, N. and Honneth, A., (eds.). (2003). *Redistribution or Recognition? A Political-Philosophical Exchange*. London; New York: Verso Books.

Gee E. M. (2002). Misconceptions and misapprehensions about population ageing. *International Journal of Epidemiology*, 31, pp. 750–753.

Gillis, J., (ed.). (2004). *Islands of the Mind: How the Human Imagination Created the Atlantic World*. New York: Palgrave Macmillan.

Grenier, A., Phillipson, C. and Settersten, R., (eds.). (2020). *Precarity and Ageing: Understanding Insecurity and Risk in Later Life*. Bristol: Policy Press.

Hennessy, C. H., Staelens, Y., Lankshear, G., Phippen, A., Silk, A. and Zahra, D. (2014). Rural connectivity and older people's leisure participation. In: C. H. Hennessy, R. Means and V. Burholt, eds., *Countryside Connections: Older People, Community and Place in Rural Britain*. Bristol: Policy Press, pp. 63–94.

Honneth, A., (ed.). (2007). *Disrespect: The Normative Foundations of Critical Theory*. Cambridge: Polity Press.

Hosseinpoor, A. R, Bergen, N., Kostanjsek, N., Kowal, P., Officer, A. and Somnath, C. (2016). Socio-demographic patterns of disability among older adult populations of low-income and middle-income countries: results from World Health Survey. *International Journal of Public Health*, 61, pp. 337–345.

International Labour Organization (ILO). (2011). *Social Protection Floor for a Fair and Inclusive Globalization*. Geneva: International Labour Organization.

Jacobs, B., de Groot, R. and Fernandes Antunes, A. (2016). Financial access to health care for older people in Cambodia: 10-year trends (2004–14) and determinants of catastrophic health expenses. *International Journal for Equity in Health*, 15, p. 94.

John, D. H. and Gunter, K. (2015). engAGE in community: using mixed methods to mobilize older people to elucidate the age-friendly attributes of urban and rural places. *Journal of Applied Gerontology*, 35, pp. 1095–1120.

Keating, N. and Dosman, D. (2009). Social capital and the care networks of frail seniors. *Canadian Review of Sociology/Revue Canadienne de Sociologie*, 46, pp. 301–318.

Latif, A., Ali, S. and Zafar, Z. (2019). Modernization and status of the aged people in South Asia: a mixed methods investigation from Pakistan. *Journal of Indian Studies*, 5, pp. 77–90.

Le Mesurier, N. (2006). The contribution of older people to rural community and citizenship. In: P. Lowe and L. Speakman, eds., *The Ageing Countryside*. London: Age Concern. pp. 133–146.

Milbourne, P. (2012). Growing old in rural places. *Journal of Rural Studies*, 28, pp. 315–317.

Ministry of Social Justice and Empowerment. (2007). Maintenance and Welfare of Parents and Senior Citizens Act, 2007. New Delhi, India.

Minkler, M. (1996). Critical perspectives on ageing: New challenges for gerontology. *Ageing & Society*, 16, pp. 467–487.

Nguyen, T. T., Le, N. B., Vu, L. H., et al. (2019). Quality of life and its association among older people in rural Vietnam. *Quality & Quantity*, 53, pp. 131–141.

Nomura, M., Makimoto, K., Kato, M., et al. (2009). Empowering older people with early dementia and family caregivers: a participatory action research study. *International Journal of Nursing Studies*, 46, pp. 431–441.

Pelcastre-Villafuerte, B. E., Meneses-Navarro, S., Ruelas-González, M. G., Reyes-Morales, H., Amaya-Castellanos, A. and Taboada, A. (2017). Aging in rural, indigenous communities: an intercultural and participatory healthcare approach in Mexico. *Ethnicity & Health*, 22, pp. 610–630.

Phillipson, C. and Walker, A. (1987). The case for a critical gerontology. In: S. DeGregorio, ed., *Social Gerontology: New Directions*. London: Croom Helm, pp. 1–15.

Porter, G., Tewodros, A., Bifandimu, F., Heslop, A. and Gorman, M. (2015). Qualitative methods for investigating transport and mobility issues among commonly socially excluded populations: a case study of co-investigation with older people in rural Tanzania. *Transportation Research Procedia*, 11, pp. 492–503.

Power, A. (2018). Informal caregivers: people, place and identity. In: V. A. Crooks, G. J. Andrews and J. Pearce, eds., *Routledge Handbook of Health Geography*. Abingdon: Routledge, pp. 166–171.

Rishworth, A. and Elliott, S. J. (2019). Global environmental change in an aging world: the role of space, place and scale. *Social Science & Medicine*, 227, pp. 128–136.

Rodríguez-Pose, A. and Hardy, D. (2015). Addressing poverty and inequality in the rural economy from a global perspective. *Applied Geography*, 61, pp. 11–23.

Sevenhuijsen, S. (2003). The place of care: the relevance of the feminist ethic of care for social policy. *Feminist Theory*, 4, pp. 179–197.

Sharma, S. (2016). Impact of globalisation on mental health in low- and middle-income countries. *Psychology and Developing Societies*, 28, pp. 251–279.

Shergold, I. and Parkhurst, G. (2012). Transport-related social exclusion amongst older people in rural Southwest England and Wales. *Journal of Rural Studies*, 28, pp. 412–421.

Shucksmith, M. (2018). Re-imagining the rural: from rural idyll to good countryside. *Journal of Rural Studies*, 59, pp. 163–172.

Silverstein, M., Burholt, V. and Wenger, C. (1998). Parent-child relations among very older parents in Wales and the United States: a test of modernization theory. *Journal of Aging Studies*, 12, pp. 387–409.

Silverstein, M., Cong, Z. and Li, S. (2006). Intergenerational transfers and living arrangements of older people in rural China: Consequences for psychological well-being. *The Journals of Gerontology: Series B*, 61, pp. S256-S266.

Skinner, M. W, Joseph, A. E., Hanlon, N., Halseth, G. and Ryser, L. (2014). Growing old in resource communities: exploring the links among voluntarism, aging, and community development. *The Canadian Geographer*, 58, pp. 418–428.

Skinner, M. W. and Winterton, R. (2018). Interrogating the contested spaces of rural aging: implications for research, policy, and practice. *The Gerontologist*, 58, pp. 15–25.

Stephens, C., Burholt, V. and Keating, N. (2018). Collecting qualitative data with older people. In: U. Flick, ed., *The SAGE Handbook of Qualitative Data Collection*. London: SAGE Publications, pp. 632–651.

Stockdale A. (2010). The diverse geographies of rural gentrification in Scotland. *Journal of Rural Studies*, 26, pp. 31–40.

Svensson, E. M. (2019). Older women, the capabilities approach and CEDAW: normative foundations and instruments for evaluation of the governance of the Nordic Arctic. In: P. Naskali, J. Harbison and S. Begum, eds., *New Challenges to Ageing in the Rural North: International Perspectives on Aging*. Cham: Springer, pp. 13–30.

Tollman, S. M., Norris, S. A. and Berkman, L. F. (2016). Commentary: the value of life course epidemiology in low- and middle-income countries: an ageing perspective. *International Journal of Epidemiology*, 45, pp. 997–999.

Tran, T. Q., Nguyen, C. V. and Van Vu, H. (2018). Does economic inequality affect the quality of life of older people in rural Vietnam? *Journal of Happiness Studies*, 19, pp. 781–799.

Westwood, S., Suen, Y-T. and Scott, V. (2019). Gaps within gaps: intersecting marginalisations of older Black, Asian and minority ethnic LGBT* people. In: A. King, K. Almack, Y-T. Suen, et al., eds., *Older Lesbian, Gay, Bisexual and Trans People: Minding the Knowledge Gaps*. London; New York: Routledge.

Winter, B. and Burholt, V. (2019). The welsh Welsh – Y Cymry cymreig: a study of cultural exclusion among rural-dwelling older people using a critical human ecological framework. *International Journal of Aging and Later Life*, 12, pp. 1–33.

Winterton, R., Warburton, J., Keating, N., Petersen, M., Berg, T. and Wilson, J. (2016). Understanding the influence of community characteristics on wellness for rural older adults: a meta-synthesis. *Journal of Rural Studies*, 45, pp. 320–327.

Part III

Contemporary scope

7 Rural ageing in low- and middle-income countries

Padmore Adusei Amoah and David R. Phillips

Introduction

Demographic ageing is now clearly evident not only in high-income nations but also in most low-and-middle-income countries (LMICs) in Latin America, the Caribbean, Asia and Africa (Harper, 2014; McCracken and Phillips, 2017; Rishworth and Elliott, 2018; UNDESA, 2017). In strictly monetary terms, LMICs are defined as countries with Gross National Incomes (GNI) per capita ranging from low income (< US$1,025) to high income (> $12,375), further subdivided as lower-middle-income ($1,026 to $3,995) and upper-middle-income countries (World Bank, 2020). Monetary classifications inevitably mean that individual countries can change positions, for example, between 2018–2019, Sri Lanka, Georgia and Kosovo moved to the upper-middle-income group from a lower-middle-income classification (World Bank, 2020). Many LMICs have similar demographic characteristics, though modified by region. Many have higher fertility rates and relatively low life expectancy (Table 7.1) (PRB, 2020).

Ageing is a relatively recent phenomenon in most LMICs and, while exact causes vary among countries, it generally stems from combined effects of rising standards of living, increasing longevity and falling, though several still have high fertility. Ageing populations have for some years been increasing more rapidly in developing regions than developed regions (Maharaj, 2013). This trend will very probably increase; by 2050 nearly 80% of the world's older persons (60+ years), about 1.7 billion people, are likely to be in current LMICs although the poorest among them will likely have the lowest growth in ageing. This exacerbates current global-regional ageing patterns: by 2050, Asia (with many LMICs) will have 61% of the world's older persons (approximately 1.3 billion people), up from 57% (549 million) in 2017. While numbers are somewhat lower, the rate of increase in ageing will be fastest in Africa, especially sub-Saharan Africa (SSA), with an increase of 229% (totalling some 226 million people in 2050), up from 69 million in 2017, or 10.9% of older persons globally. Countries in Latin America and the Caribbean are likely to see a similar trend, with their current 76 million older population increasing to 198 million in 2050, 9.5% of older persons worldwide (UNDESA, 2017). Correspondingly, issues surrounding ageing in LMICs will intensify and demand far greater research and development of appropriate policies to address current and emerging challenges, especially those relating to health and social care.

Crucially, the effects of ageing in LMICs and particularly those in Africa are further complicated as a significant proportion of older persons live in rural areas (Maharaj, 2013). Recent trends indicate rapid urbanisation across the globe, and about 58% of older persons live in urban areas globally, and urbanisation has been fastest among some LMICs (UNDESA, 2017). Despite this, the proportions of older persons living in rural areas in LMICs remain relatively higher than in wealthier countries. In Africa, 63% of older persons lived in rural areas in 2015 (UNDESA, 2017) and similar patterns exist in the developing parts of Asia. Figures 7.1 and 7.2 show that a significant proportion of populations in developing countries will still reside in rural areas in 2050 despite the anticipated ongoing rise in urbanisation and older persons will remain mostly in rural areas (Ritchie and Roser, 2019). Conservative estimates suggest the absolute numbers of older persons and people approaching older age in rural areas in developed countries will reduce by 2050, while those in less developed and least developed countries are likely to remain stable or increase (Figure 7.3). Therefore, rural living is likely to remain an essential feature for older persons in LMICs (Table 7.1), and suitable policies and services need to reflect this.

Today, rural lives in LMICs are often ones of extraordinary hardship. Despite decades of aid and local assistance programmes, ageing in rural LMICs, particularly in SSA, often remains accompanied by poverty, poor housing, minimal health and welfare services and poor infrastructure (including transport and sanitation) and older persons especially face numerous challenges (Amoah et al., 2018; Maharaj, 2013). These are exacerbated because age-related related frailty in LMICs generally occurs earlier than in higher-income countries (United Nations, 2015; WHO, 2015). Moreover, many face threats from infectious diseases and enormous growth in and risks from non-communicable diseases (NCDs) have been highlighted (Gyasi and Phillips, 2020). Unsurprisingly, older persons in LMICs, especially in SSA, tend to rank low on well-being (HelpAge International, 2015), and a significant proportion are known to rely on informal arrangements for health (Amoah et al., 2018; Maharaj, 2013). This chapter examines the patterns and challenges of ageing in LMICs, often especially evident in rural communities. It then analyses empirical data from Ghana to demonstrate the complex associations between social capital (SC) and health literacy (HL) and their implications for health-related quality of life (HRQoL) of older persons in rural LMICs.

Rural health and ageing in LMICs

The maintenance of health is often a challenge for many older persons but, for those in rural LMICs, it can be enormous. Most LMICs now battle the "double burden" of infectious and non-communicable disease (Maharaj, 2013; World Health Organization (WHO), 2015) although, in many sub-Saharan African countries, infectious diseases such as malaria and tuberculosis continue to be leading

Table 7.1 Demographic and ageing characteristics and rural percentages, selected LMICs

Country	*Pop. mid-2019 (millions)*	*Pop. mid-2050 (millions)*	*Pop. ages 65+, 2019 (%)*	*Pop. ages 65+, 2050 (%)*	*Life expectancy at birth (years), males, 2019*	*Life expectancy at birth (years), females, 2019*	*Life expectancy at age 65 (years), males, 2018*	*Life expectancy at age 65 (years), females, 2018*	*Total fertility rate, 2019*	*GNI per capita PPP 2018 (US$)**	*Percentage of population living in rural areas (2019)*
	3	2.4	12	23	72	79	15	17	1.6	4,230	36
Armenia											
Azerbaijan	10	11.1	7	17	73	78	13	16	1.8	4,050	47
Cambodia	16.5	22	4	12	69	73	13	15	2.5	1,390	77
China	1,398	1,343.90	12	26	75	79	15	17	1.6	9,460	40
Ecuador	17.3	23.2	7	16	74	80	19	20	2.5	6,110	36
Egypt	99.1	166.5	4	11	71	74	13	15	3.3	2,800	57
Ghana	30.3	51.3	3	6	62	65	12	13	3.9	2,130	44
Honduras	9.7	11.3	5	13	73	77	17	20	2.5	2,350	45
Indonesia	268.4	319.7	6	14	69	73	12	14	2.3	3,840	46
Kenya	52.6	95.5	2	7	64	68	14	16	3.6	1,620	68
Pakistan	216.6	306.6	4	8	66	68	14	14	3.6	1,590	63
Paraguay	7.2	8.9	6	13	72	76	17	19	2.5	5,670	39
Philippines	108.1	152	5	10	67	75	13	15	2.7	3,830	53
Sri Lanka	21.9	21.5	8	23	73	80	16	18	2.1	4,060	81
Zimbabwe	14.6	25.6	5	6	58	61	14	15	3.7	1,790	68

Sources: PRB (2020) and *World Bank (2020)

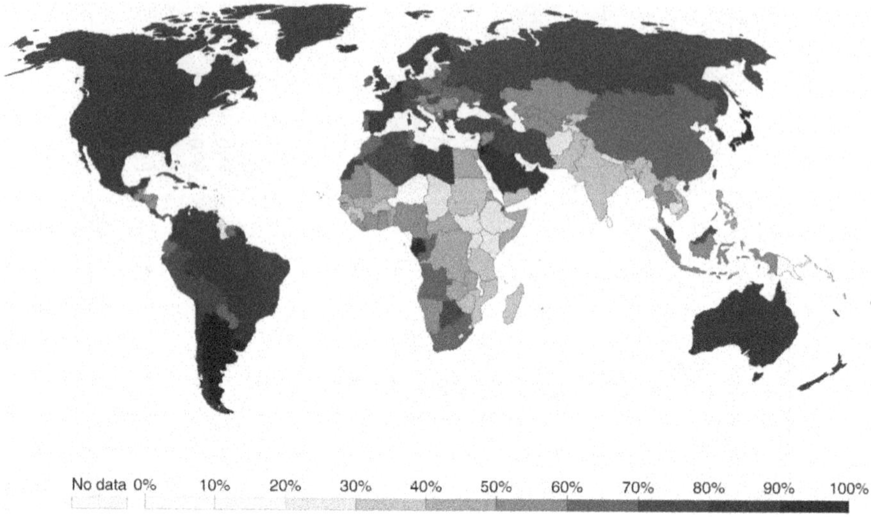

Figure 7.1 Share of the overall population in urban areas globally, 2020

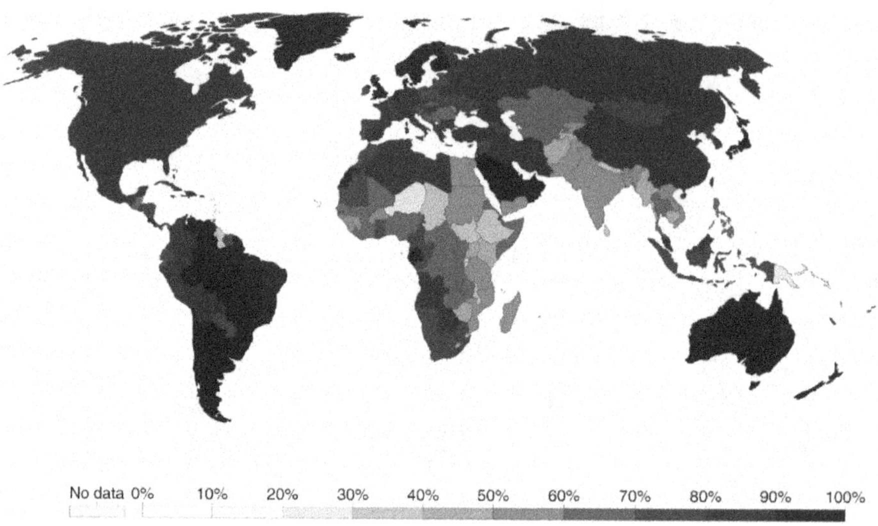

Figure 7.2 Share of the overall population in urban areas globally, 2050

causes of mortality and, importantly, morbidity, as well as HIV/AIDs which has bedevilled many lives in east and southern Africa. Yet in many LMICs in Africa, Asia, Eastern Europe and Latin America, chronic diseases relating to hearing, hypertension, mental illness, malnutrition, cancer and numerous organic diseases are becoming more prevalent than infectious diseases (Aboderin, 2010; Bigna and Noubiap, 2019). A study in rural Mexico found the most prevalent health

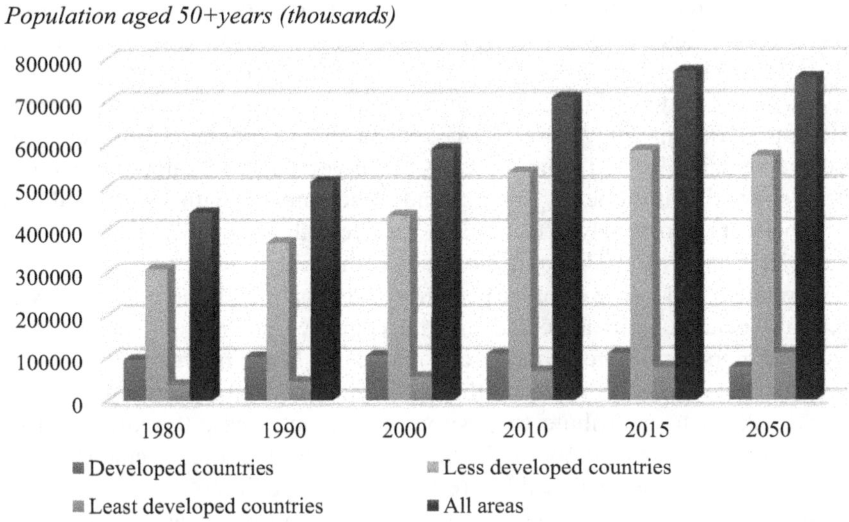

Figure 7.3 Proportions of older persons in developed, less developed and least developed countries, 1980–2050

Source: UNDESA (2014)

problems of older persons were chronic diseases with early diagnosis and continuous treatment presenting many challenges (Pelcastre-Villafuerte et al., 2017).

Rural health problems are frequently exacerbated by poor access to health facilities. Geriatric services are often scarce in urban areas and particularly in rural areas of LMICs (Maharaj, 2013). Moreover, ageing often increases health care expenditure at both household and national levels (Westerhout, 2014). The capacity of LMICs to address associated infrastructure and financial burdens in the health sector is often weak, with severe consequences for the well-being and equality of life chances of older persons and their families (Maharaj, 2013). Furthermore, even if health services are available, they are often inefficient and unaffordable in most LMICs, so health care use has several indirect costs (such as time and transportation) which deter rural dwellers from accessing needed care (Amoah et al., 2018). Complications widely recognised as fundamental causes of poor health outcomes are limited health knowledge and low access to health information among older persons, particularly those in deprived areas (Nutbeam, 2008). In rural LMIC areas, these especially reduce chances of achieving better health and active ageing among many older persons (Maharaj, 2013).

Social support, health and rural ageing in LMICs

Health problems and the corresponding poor health services emphasise the need to reconfigure formal and informal support and care systems for rural older persons in LMICs (Harper, 2014; Maharaj, 2013). Older persons in LMICs are more likely to require instrumental help for everyday chores and financial support, plus emotional and informational support due to physical and cognitive frailty (Maharaj, 2013). In LMICs and particularly their rural areas, the role of social support systems in the HRQoL of older persons is invaluable (Amoah, 2019; Gyasi, Phillips and Abass, 2018a), but an understanding of the mechanisms involved remains limited. Many factors are involved; for example, a study in Ghana showed that monetary support and close contacts with family and other social networks positively affect health service utilisation (Gyasi, Phillips and Abass, 2018b). In rural Thailand, close ties with children were found to improve health-related well-being of older persons (Abas et al., 2009) and, in rural Mexico, high social support appeared to limit cognitive decline among urban and rural older persons (Zamora-Macorra et al., 2017). Many related factors such as knowledge and remedies imparted through various forms of support from social networks seem to contribute to positive health outcomes. Research in Latin America and elsewhere supports the long-held belief that older persons in rural areas may prefer traditional health care providers due to the cultural acceptability, effective communication in the native languages and easier access (Pelcastre-Villafuerte et al., 2017).

However, the provision of social support in the well-being and health of older persons in rural LMICs is becoming more problematic given the rise in population ageing alongside rapid urbanisation across many LMICs. While urbanisation is associated with improved socio-economic conditions, its occurrence can limit access to social support systems for rural older persons because of urban migration of youthful populations (Maharaj, 2013; UNDESA, 2017). This has been clear in China where terms such as "left-behind older people" describe many who remain in rural areas after out-migration of younger populations (Liu, 2014; Phillips and Feng, 2018); it is also observed in Indonesia and elsewhere (Utomo et al., 2019). Indeed, longstanding norms in Asian societies such as filial piety (reciprocal intergenerational duties) are often disrupted by urbanisation because families may not live closely in modern societies (Liu, 2014). An important emerging issue is the problem of social exclusion and isolation, which can be considerable among older persons in LMICs (Phillips and Cheng, 2012). Nonetheless, the HRQoL of older persons in LMICs is strongly associated with the wealth of social networks (Amoah, 2019; Nilsson, Rana and Kabir, 2006). This role of social networks has been conceptualised as social capital (SC).

SC refers to resources (social support) embedded in different kinds of social relationships, facilitated by norms of reciprocity, civic engagements and trust (Putnam, 2000). Among older persons in LMICs, SC in the form of instrumental, emotional and informational support involves significant sources of money, daily chores assistance and companionship. Such support is significant for social capital at micro, meso and macro levels of societies and tends to have a favourable effect on HRQoL of individuals by triggering the immune systems to fight ill-health and improve psychological well-being (Putnam,

2000). SC, therefore, affects both needs for and access to health care, health-related decisions and health information (Amoah, 2018; Lee et al., 2009). It assists fundamental determinants of health such as health literacy (HL) which is increasingly acknowledged globally although research in LMICs relatively new (Amoah, 2019).

HL is linked to literacy and entails the motivation, knowledge and competencies to access, understand, appraise and apply health information in order to make judgements and take decisions in everyday life concerning healthcare, disease prevention and health promotion to maintain or improve quality of life throughout the course of life (Sorensen et al., 2012) Persons with sufficient HL are more likely to understand prevailing health systems, avoid deleterious health behaviours, have better health care access and positive health outcomes (Amoah and Phillips, 2018, 2019; Nutbeam, 2008). Some postulate a causal relation between HL and health (Paasche-Orlow and Wolf, 2007). Older persons are often likely to have lower HL than younger populations and, given lower education in rural areas, HL may even be worse among older persons in rural LMICs (Amoah, 2019). It appears how HL affects health can be affected by external factors such as one's SC. Support systems such as information and instrumental support can influence someone's HL. For instance, people with low HL but have trusting relationships with others in their household or communities can exploit such relationships to make healthy choices, effectively engage with the health system and follow medical instructions (Amoah et al., 2018; Lee et al., 2009). However, little is known about how different forms of SC affect the relations between HL and HRQoL among older persons in rural LMICs, especially in sub-Saharan Africa which has received relatively little gerontological research despite its rapid ageing (Rishworth and Elliott, 2018).

Social capital, health literacy and health-related quality of life of older persons in rural Ghana

Here, we examine the moderating role of three of SC proxies on the relations between HL and HRQoL of older persons in rural Ghana. Ghana is a lower-middle-income country with a relatively youthful population although some 7.1% of the population is aged 60+ plus and Ghana is projected to have the highest percentage of the older population in sub-Saharan Africa by 2030 (United Nations, 2015; World Bank, 2020). Approximately 44% of Ghana's population lives in rural areas and a significant proportion are older persons (Apt, 2013). Therefore, research-based evidence is essential to inform health promotion and service efforts in rural areas.

A cross-sectional household survey on Social Epidemiology, Universal Health Coverage, Health and Well-being in Ghana was conducted from July to December 2018 using a multi-stage cluster sampling approach. This substantial case study includes data from 51 rural communities located in four of the former ten administrative regions (Ashanti, Eastern, Brong Ahafo and Northern regions). The analyses include 244 participants aged 50 and older and those who were approaching old age (the relatively younger age of 50 is commonly used in LMIC research) (Gyasi, Phillips and Abass, 2018b).

Measures

The dependent variable, HRQoL, was measured using a validated five-item HRQoL scale (ISSP Research Group, 2015), with respondents rating their physical and mental health status on a five-point Likert scale. HL (independent variable) was measured on a short version of the European Health Literacy Survey Questionnaire (HLS-EU-Q16). In this, 16 items measure the extent to which people obtain, understand, appraise and apply health information in three domains: health care, disease prevention and health promotion on a four-point Likert scale (Pelikan et al., 2014). The moderating variable, SC, was represented by three proxies; bonding SC (resources from close relationships such as family and close friends), civic participation and trust to represent both the cognitive and structural components of the concept (Harpham, 2008). For bonding SC, participants selected those relationships that had provided any form of emotional, instrumental or informational support in the past year. Selections were summed to give an index for bonding SC. For civic participation, participants were asked whether they joined other community members to address a problem or common issue in the past year. Trust was measured by participant indicating whether they thought most people in their community or neighbourhood could be trusted, questions based on the Adapted SC Assessment Tool (S-ASCAT) (Harpham, 2008). Fundamental demographic variables were recorded. The analysis comprised six ordinal logistic regression models (Tables 7.2 and 7.3) testing the relations between HL and HRQoL and the moderating role of the three SC proxies.

Results

On average, participants had received support from at least two sources classified as bonding SC. Trust was prevalent, with 61.1% trusting their neighbours. However, only 37.3% had participated in organised community activities, perhaps due to frailty. HL was generally low, with only 19.7% identified as having sufficient levels. Most (50.8%) participants described their HRQoL in moderate terms, and 32% had high HRQoL. HL was consistently associated with HRQoL across all five models assessing the relations (Tables 7.2 and 7.3). All SC variables also showed positive relations with HRQoL independently (though significance was low for some). In the final model (Model 6), bonding SC negatively moderated the association between HL and HRQoL a consistent finding when the independent and interaction effects of this relation were examined. Importantly, trust positively modified (and even enhanced) the effect of HL on HRQoL in Model 5 and final Model 6. Although civic participation moderated HL and HRQoL in earlier models, its final effect was not significant when all other variables were considered.

Table 7.2 The role of SC in the relations between HL and HRQoL by ordinal logistics regression

	Model 1		*Model 2*		*Model 3*	
	B (95% CI)	*Stand. error*	*B (95% CI)*	*Stand error*	*B (95% CI)*	*Stand error*
Household size	0.007 (-0.061, 0.076)	0.035	0.011 (-0.062, 0.084)	0.037	0.009 (-0.064, 0.081)	0.037
Religion						
Christian	1.822 (0.404, 3.241)**	0.624	1.435 (0.428, 4.998)*	0.697	1.572 (0.244, 4.100)*	0.580
Islam	0.395 (-0.884, 1.674)	0.653	-0.445 (-1.795, 0.905)	0.689	-0.418 (-1.744, 0.909)	0.677
Traditional religion (ref)						
Income (log)	0.960 (0.154, 1.767)**	0.311	0.710 (-1.718, 0.297)	0.514	0.541 (-1.420, 0.337)	0.448
Marital status						
Married	0.787 (0.226, 1.600,)*	0.315	0.684 (-1.526, 0.157)	0.429	0.765 (-0.627, 2.156)	0.710
Not married (ref)						
Health literacy			0.813 (0.371, 1.316)**	0.305	1.661 (0.466, 2.855)**	0.610
Bonding SC			0.262 (-.206, 0.730)	0.169	1.730 (0.311, 3.149)**	0.724
Civic participation						
Yes			0.686 (0.219, 0.937)*	0.236		
No (ref)						
Trust						
Yes			0.514 (0.158, 0.818)*	0.247		
No (ref)						
Health literacy*Bonding SC					-0.557 (-1.032,-0.082)*	0.242
Health literacy *Civic Participation						
Health literacy *Trust						
Nagelkerke Pseudo R-Square	*0.146*		*0.247*		*0.264*	

p < .05. **p < .01. ***p < .001
Model 1: socio-economic correlates of HRQoL
Model 2: Model 1 and HL and moderators (SC proxies)
Model 3: Model 1, HL, bonding SC and interaction between bonding SC and HL.

Table 7.2 (Continued)

	Model 4		*Model 5*		*Model 6 (final model)*	
	B (95% CI)	*Stand. error*	*B (95% CI)*	*Stand error*	*B (95% CI)*	*Stand error*
Household size	0.005 (-0.077, 0.067)	0.037	0.007 (-0.064, 0.078)	0.036	0.013 (-.063 .089)	0.039
Religion						
Christian	2.151 (0.543, 3.760)**	0.821	1.441 (-2.987 .104)	0.789	1.174 (-2.970 .623)	.917
Islam	0.767 (-2.102, 0.568)	0.681	0.656 (-1.976 .665)	0.674	0.458 (-1.955 1.039)	.764
Traditional religion (ref)						
Income (log)	1.187 (0.496, 3.077)***	0.455	1.279 (0.344, 3.902)**	0.518	1.162 (0.397, 2.228)*	0.544
Marital status						
Married	0.358 (-0.980, 1.697)	0.683	0.474 (-0.829, 1.776)	0.665	0.779 (-0.634, 2.193)	0.721
Not married (ref)						
Health literacy	0.988 (0.393, 3.170)*	0.403	1.482 (0.271, 2.693)**	0.618	2.158 (0.613, 3.702)**	0.788
Bonding SC					2.055 (0.406, 3.704)**	0.841
Civic participation						
Yes	5.616 (0.898, 10.334)*	2.407			5.486 (0.742, 10.714)*	2.178
No (ref)						
Trust						
Yes			5.729 (1.312, 10.145)*	2.253	8.837 (3.334, 14.340)***	2.808
No (ref)						
Health literacy *Bonding SC					-0.677 (-1.210, -0.144)**	0.272
Health literacy *Civic Participation	1.768 (0.112, 3.424)*	0.745			2.268 (-0.012, 4.548)	1.163
Health literacy *Trust			1.721 (0.242, 3.200)*	0.804	2.936 (0.955, 4.517)**	1.011
Nagelkerke Pseudo R-Square	0.284		0.273		0.334	

*$p < .05$. **$p < .01$. ***$p < .001$

Model 4: Model 1, HL, civic participation and interaction between civic participation and HL.
Model 5: Model 1, HL, trust and interaction between trust and health literacy.
Model 6: Model 1 and all variables and interaction terms.

Rural gerontology in LMICs: the future

Rural ageing in LMICs presents an additional layer of complexity in efforts to promote active and productive ageing across the globe as the empirical evidence here indicates. The case study shows that, while HL is a critical determinant of HRQoL of older persons in rural LMICs, and should be promoted, the *kind* of SC can positively or negatively influence its effects. These findings have critical implications for rural gerontology in LMICs.

The role of HL in HRQoL highlights the need to make it a central approach in health promotion, disease prevention and management, especially among older persons in rural LMICs, where health services are inadequate (Maharaj, 2013). However, the case study shows that the role of HL in HRQoL is not unidimensional because SC can reduce or enhance its influence. For example, while bonding SC had positive relations with HRQoL, its influence can be detrimental to the extent to which HL affects health. This is a recognised manifestation of the inward-looking nature of bonding SC, which is often generated among people of the same or similar socio-economic status. It is unlikely to have a positive impact in terms of offering new information and resources (Putnam, 2000). This implies that rural gerontological approaches aiming to boost HL in LMICs must adopt community-level strategies as it can propagate appropriate health information through more resourceful persons/groups within broader communities instead of the close social networks of target groups (here, older persons). Such an approach can be orchestrated through the promotion of civic participation, which the case study and elsewhere shows has a positive effect on HRQoL (Putnam, 2000). Civic and other social participatory activities can be made possible through the establishment of community centres where activities for older persons can be promoted (Utomo et al., 2019). Furthermore, this approach is important in rural gerontology considering weakening filial piety in many LMIC societies from rising rural-urban migration (Liu, 2014).

Trust was another aspect of SC which affected HL's role in HRQoL of older persons. It seems that HL works best for health improvements when health information or instruction is received from individuals and even institutions considered trustworthy. However, access to trusting and resourceful people who can offer proper health information can be limited for rural older persons and rural localities in general (Amoah et al., 2018). This is partly why some older persons in China are embarking on rural-urban migration (see Cheng et al., Chapter 9 in this volume). The arguments earlier suggest that building rural communities of trust will be critical to effective health promotion as apart of rural gerontological efforts in LMICs. Furthermore, it reinforces calls for rural ageing in the right place, and age-friendly rural communities (Menec and Novek, Chapter 14; Wiles et al., Chapter 15). The data shows that the absence of distrust eases doubts about health information and instructions. Indeed, associated research in Ghana shows that trust plays a vital function in health-related decisions among rural residents (Amoah, Edusei and Amuzu, 2018), also observed in rural Mexico (Pelcastre-Villafuerte et al., 2017).

In conclusion, we see the huge importance of rural ageing in LMICs. The critical question remains, how can health and service use be enhanced in difficult circumstances? Potentially, enhancing HL could empower older rural adults in LMICs to better engage with their environment, including prevailing health systems. However, demographic ageing and its associated gerontological challenges position SC as a vital factor. Hence, designing and implementing effective strategies to integrate family and communal support systems in health promotion and services for older persons in rural LMICs are essential, although they are currently complex even in advanced economies (see Glasgow and Doebler, Chapter 10).

Acknowledgements

The University Grants Committee of Hong Kong (Lingnan University Faculty Research Grant, Grant 102159) funded the research on which this chapter's case study is based. However, the funder played no role in the study design, data gathering and analyses, manuscript preparation and the decision to publish the manuscript.

References

Abas, M. A., Punpuing, S., Jirapramupitak, T., Tangchonlatip, K. and Leese, M. (2009). Psychological wellbeing, physical impairments and rural aging in a developing country setting. *Health and Quality of Life Outcomes*, 7(1), p. 66.

Aboderin, I. (2010). Understanding and advancing the health of older populations in sub-Saharan Africa: policy perspectives and evidence needs. *Public Health Reviews*, 32(2), pp. 357–376.

Amoah, P. A. (2018). Social participation, health literacy, and health and well-being: a cross-sectional study in Ghana. *SSM – Population Health*, 4, pp. 263–270.

Amoah, P. A. (2019). The relationship between functional health literacy, self-rated health, and social support between younger and older adults in Ghana. *International of Environmental Research and Public Health*, 16(3188), pp. 1–14.

Amoah, P. A. Edusei, J. and Amuzu, D. (2018). Social networks and health: understanding the nuances of healthcare access between urban and rural populations. *International Journal of Environmental Research and Public Health*, 15(5), p. 973.

Amoah, P. A. and Phillips, D. R. (2018). Health literacy and health: rethinking the strategies for universal health coverage in Ghana. *Public Health*, 159, pp. 40–49.

Amoah, P. A. and Phillips, D. R. (2019). Socio-demographic and behavioral correlates of health literacy: a gender perspective in Ghana. *Women and Health*, 59(6), 1–17.

Apt, N. (2013). Older people in rural Ghana: health and health seeking behaviours. In: P. Maharaj., ed., *Aging and Health in Africa*. Boston, MA: Springer, pp. 103–119.

Bigna, J. J. and Noubiap, J. J. (2019). The rising burden of non-communicable diseases in sub-Saharan Africa. *The Lancet Global Health*, 7(10), pp. e1295–e1296.

Gyasi, R. M. and Phillips, D. R. (2020). Aging and the Rising burden of noncommunicable diseases in sub-Saharan Africa and other low- and middle-income countries: a call for holistic action. *The Gerontologist*, 60, pp. 806–811.

Gyasi, R. M., Phillips, D. R. and Abass, K. (2018a). Social support networks and psychological wellbeing in community-dwelling older Ghanaian cohorts. *International Psychogeriatrics*, 2018, pp. 1–11.

Gyasi, R. M., Phillips, D. R. and Amoah, P. A. (2018b). Multidimensional social support and health services utilization among noninstitutionalized older persons in Ghana. *Journal of Aging and Health*, 32(3–4), pp. 227–239.

Harper, S. (2014). Introduction: conceptualizing social policy for the twenty-first-century demography. In: S. Harper and K. Hamblin, eds., *International Handbook on Ageing and Public Policy*. Cheltenham: Edward Elgar, pp. 1–9.

Harpham, T. (2008). The measurement of community social capital through surveys. In: I. Kawachi, S. V. Subramanian and D. Kim, eds., *Social Capital and Health*. New York: Springer, pp. 51–62.

HelpAge International. (2015). *Global Rankings Table*. Available at: www.helpage.org/global-agewatch/population-ageing-data/global-rankings-table/

ISSP Research Group. (2015). *International Social Survey Programme: Health and Health Care – ISSP 2011*. https://zacat.gesis.org/webview/index/en/ZACAT/ZACAT.c.ZACAT/ISSP.d.58/by-Year.d.69/International-Social-Survey-Programme-Health-and-Health-Care-ISSP-2011/fStudy/ZA5800

Lee, S. Y. D., Arozullah, A. M., Cho, Y. I., Crittenden, K. and Vicencio, D. (2009). Health literacy, social support, and health status among older adults. *Educational Gerontology*, 35(3), pp. 191–201.

Liu, J. (2014). Ageing, migration and familial support in rural China. *Geoforum*, 51, pp. 305–312.

Maharaj, P., (ed.). (2013). *Aging and Health in Africa*. New York: Springer.

McCracken, K. and Phillips, D. R., (eds.). (2017). *Global Health: An Introduction to Current and Future Trends*, 2nd ed. New York: Routledge.

Nilsson, J., Rana, A. K. M. M. and Kabir, Z. N. (2006). Social capital and quality of life in old age: results from a cross-sectional study in rural Bangladesh. *Journal of Aging and Health*, 18(3), pp. 419–434.

Nutbeam, D. (2008). The evolving concept of health literacy. *Social Science and Medicine*, 67, pp. 2072–2078.

Paasche-Orlow, M. K., and Wolf, M. S. (2007). The causal pathways linking health literacy to health outcomes. *American Journal of Health Behavior*, 31, pp. 19–26.

Pelcastre-Villafuerte, B. E., Meneses-Navarro, S., Ruelas-González, M. G., Reyes-Morales, H., Amaya-Castellanos, A. and Taboada, A. (2017). Aging in rural, indigenous communities: an intercultural and participatory healthcare approach in Mexico. *Ethnicity and Health*, 22(6), pp. 610–630.

Pelikan, J., Röthlin, F., Ganahl, K. and Peer, S. (2014). *Measuring comprehensive health literacy in general populations: the HLS-EU instruments*. Paper presented at the Second International Conference of Health Literacy and Health Promotion, Taipei, Taiwan, 6–8 October.

Phillips, D. R. and Cheng, K. (2012). The impact of changing value systems on social inclusion: an Asia-Pacific perspective. In: T. Scharf and N. Keating, eds., *From Exclusion to Inclusion in Old Age*. Bristol: Policy Press, pp. 109–124.

Phillips, D. R. and Feng, Z. (2018). Demographics and aging. In: W. Wu and M. W. Frazier (eds), *The SAGE Handbook of Contemporary China*, vol. 2. New York; London: SAGE Publications, pp. 1049–1071.

PRB. (2020). *2019 World Population Data Sheet*. Washington, DC: PRB.

Putnam, R. D., (ed.). (2000). *Bowling Alone: The Collapse and Revival of American Community*. New York: Simon & Schuster.

Rishworth, A. and Elliott, S. J. (2018). Ageing in low- and middle-income countries: ageing against all odds. In: M. W. Skinner, G. J. Andrews and M. P. Cutchin, eds., *Geographical Gerontology: Perspectives, Concepts, Approaches*. London: Routledge, pp. 110–122.

Ritchie, H. and Roser, M. (2019). *Urbanization*. Available at: https://ourworldindata.org/urbanization

Sorensen, K., Broucke, S., Fullam, J., Doyle, G., Pelikan, J. and Slonska, Z. (2012). Health literacy and public health: a systematic review and integration of definitions and models. *BMC Public Health*, 12, p. 80.

UNDESA. (2014). *World Urbanization Prospects: 2014 Revision*. Geneva: UNDESA.

UNDESA. (2017). *World Population Ageing 2017*. New York: UNDESEA.

United Nations. (2015). *World Population Ageing 2015*. Available at: www.un.org/en/development/desa/population/publications/pdf/ageing/WPA2015_Report.pdf

Utomo, A., McDonald, P., Utomo, I., Cahyadi, N. and Sparrow, R. (2019). Social engagement and the elderly in rural Indonesia. *Social Science and Medicine*, 229, pp. 22–31.

Westerhout, E. (2014). Population ageing and health care expenditure growth. In: S. Harper and K. Hamblin (eds), *International Handbook on Ageing and Public Policy*. Cheltenham, UK: Edward Elgar, pp. 178–190.

World Health Organization (WHO). (2015). *Health in 2015: From MDGs, Millienium Development Goals (MDGs) to SDGs, Sustainable Development Goals*. Geneva: WHO.

World Bank. (2020). *World Bank Country and Lending Groups*. Available at: https://datahelpdesk.worldbank.org/knowledgebase/articles/906519-world-bank-country-and-lending-groups

Zamora-Macorra, M., de Castro, E. F. A., Ávila-Funes, J. A., Manrique-Espinoza, B. S., López-Ridaura, R., Sosa-Ortiz, A. L., . . . del Campo, D. S. M. (2017). The association between social support and cognitive function in Mexican adults aged 50 and older. *Archives of Gerontology and Geriatrics*, 68, pp. 113–118.

8 Rural women, ageing and retirement

Nata Duvvury, Áine Ní Léime and Tanya Watson

Introduction

Globally the discourse on ageing focuses on the impact of an ageing population for the overall economy, and in particular for sustaining entitlements and welfare benefits for those in retirement (OECD, 2019). While the economic dimensions are crucial for governments, there is a need to explore the lived experience of those ageing and the constraints faced by them to realise well-being, economic security and social stability. An aspect given less attention in the literature is unpacking the impacts of ageing for different sections of the older population, and particularly for their experiences of work and retirement.

Within the context of the global ageing population, rural women are a particular group of interest. Women in rural areas are at the intersection of two trends at the global level – the higher share of women in the ageing population and the declining gender gap in labour force participation in both developed and developing countries (see Burholt and Scharf, Chapter 6 in this volume). Additionally, the rise in rural women's labour force participation reflects a new trend of these women increasingly taking on the role of cultivator or "farmer" (EP, 2019; ILO, 2016). This is true across the globe, though the underlying dynamics are slightly different depending on the regional context (see Amoah and Phillips, Chapter 7). Research within the global South suggests that increasing poverty results in out-migration of men, thus pushing women to manage household farms as cultivators (Lahiri-Dutt and Adhikari, 2016; Maharjan, Bauer and Knerr, 2012). In contrast in industrialised countries, women are asserting their connection to farming and family farm as an inherent right (Silvasti, 2003; Byrne et al., 2014; EC, 2019). Additionally, in both regions, a significant proportion of women remain in rural areas as unpaid family workers or in temporary/informal paid employment. Most women in the rural workforce continue to be constrained by gender norms and responsibilities that prioritise family and care work combined with limited employment opportunities in the rural sector (EP 2019; FAO, 2010).

In this chapter we focus on the implications of work and retirement for rural women in the context of ageing (see Seppanen et al., Chapter 16 on rural older men). The central question we explore in this chapter is how rural women either as cultivators or as workers in off-farm rural employment navigate their future in terms of work and retirement in the context of changes in welfare systems, particularly in the area of pension policies. We begin with an overview of ageing trends in rural areas, with a focus on women. In the next section, we outline the employment trends for older rural women and outcomes of such employment. The consequences of employment for rural women in terms of their future security are shaped both by the gender norms that fundamentally influence their work experience and by the structure of pension policies. In section 3 we turn to a case study of Ireland to gain deeper insights into women's perception of economic security in old age. Drawing on qualitative data from two studies, we highlight the central concerns perceived by women with respect to retirement and economic stability. In the conclusion we draw lessons for the ongoing reforms of pension systems globally.

Women, ageing and rural employment rural women and ageing

As highlighted in Chapter 2, a growing policy concern is that the proportion of older people in the population is projected to increase steadily over the next 30 years. This holds true when considering the trends in women's population. The share of the globe's population comprising older women is also projected to increase – from 12% in 2011 to 19% in 2050 (Stevens, Mathers and Beard, 2013). Table 8.1 indicates the rising trend in older female population between 2000–18 across the globe, with the highest levels in Europe and North America. In contrast across the global south, the share of older women ranges from half to one-sixth of the European and North American rates. It should be noted despite this lower proportion in the global south, China and India alone are home to nearly one-third (33%) of the world's older female population in 2019 (World Bank, 2019). Moreover regions in the global south, with the exception of Sub-Saharan Africa, had growth rates in older women's share of the population that equal or outpace the growth rate of Europe and North America.

Table 8.1 Proportion of women over 65 in female population

	2000	*2010*	*2018*	*% Change (2000–18)*
EU-28	18.3	20.1	22.2	21.31
LCA	6.3	7.5	9.2	46.03
ME&NA	4.7	5	5.6	19.15
SSA	3.2	3.3	3.4	6.25
China	7.4	8.7	12.1	63.51
India	4.7	5.5	6.6	40.43
North America	14.2	14.6	17.4	22.54
World	7.82	8.57	9.88	26.34

Source: World Bank Data (2019)

Globally many older women will continue to live in rural areas despite a steady increase in urbanisation. According to research by United Nations Economic Commission for Europe UNECE (2017), the gender distribution of the 65+ rural population is on average 54% women and 46% men in the UNECE region.[1] In the age group of 80+, this increases to 64% women and 36% men, with even higher ratios in some East European countries. In the Global South countries, the same trend is observable with older women constituting a sizeable proportion of the rural population. This is particularly evident in China – rural women constituted 60% of the 80+ population in 2010 and are projected to be even higher by 2050 (UNDESA, 2019).

Rural women's labour force participation

Women's participation in the rural economy is surprisingly similar across both the Global North and South. For example in the EU they represent 45% of the economically active rural population and about 40% of them work on family farms (EP, 2019). Similarly, in most of the developing countries across Africa, Asia and Latin America, women's participation rate varies from 20% (in MENA and South Asia regions) to 64% (Sub-Saharan Africa) (UN Women, 2015: 76). More critically, the participation rate with age does not decline in much of the Global South as many need to work in a context of poor social protection policies that ensure a stable income in old age. In parts of Sub-Saharan Africa, the participation rate of those aged 60 to 64 remains at a high level of 60% or more (UNFPA, 2012: 56).

Employment status and informality

Rural employment for women includes farming, self-employment in trade and other enterprises providing goods and services, or wage work either in rural enterprises or agriculture (FAO, 2010; ILO, 2016). Additionally, women's employment is characterised by informality within the rural sector (ILO, 2018a). The particular pattern of distribution of informality varies depending on the structure of economies across regions. Unpaid work in family enterprises (whether agricultural or non-agricultural) is particularly high in the regions of Asia and parts of Latin America, where agriculture continues to be a highly patriarchal sector of production. For example, in Egypt 85% of women's informal employment is accounted by unpaid work on the family farm (FAO, 2010). In the Sub Saharan African region, the vast majority of women are farmers on land where they have leasing rights, with a smaller proportion working as own-account workers in non-agricultural enterprises. Regarding rural wage employment (agricultural or non-agricultural) recent data from a number of countries from Africa, Asia and Latin America indicate that women are far less likely than men to participate in rural wage employment (ILO, 2018b).

While agriculture is a smaller sector in North America and Europe, surprisingly nearly 30% of farms across the EU-28 are managed by women. However, many of these women farmers face the same constraints as in the Global South having less access to, control over and ownership of land and other productive assets compared to their male counterparts. More importantly they are not viewed as innovators, despite having ideas as marketable as men's, by predominantly male stakeholders (EP, 2019:35). Rural women are not confined to agriculture in the global north as women are equally involved in off-farm employment within the rural development sector. Among these, tourism is a growing sector with women constituting 58% of the labour force in tourism in rural areas. However much of this employment is characterised by informality; it is seasonal, part-time and often involves temporary or zero-hour contracts.

Economic security for older rural women?

The rise in employment of older rural women however does not automatically translate into economic security in old age nor individual agency in retirement planning. First, many older women are concentrated in precarious, low-skilled and low-paid employment putting them at a greater risk of poverty. Secondly, gender norms, particularly with regard to caregiving, result in women working longer hours as older women are more likely to undertake unpaid caregiving roles than older men, further limiting their regularity and level of income (UNFPA, 2012). Third rural women in most societies have little or no land ownership, as noted previously, limiting their access to credit, their potential productivity and ability to navigate the risk of poverty. A recent report of the Food and Agriculture Organization (FAO) suggests that women represent less than 20% of all landholders globally (FAO, 2010), ranging from less than 5% in North Africa to 15% in Sub-Saharan Africa. In Europe this rises to about 30%, with women having farms half the size of men landholders (EC, 2019).

The precarious nature of rural employment (farm and off farm), the high level of unpaid work and lack of access to resources together undermine older women's access to pensions (Duvvury et al., 2012). While extended working life is seen by the OECD as a potential strategy to deal with the increasing pressure on countries' pension system (Ní Léime and Ogg, 2019), there is less focus on whether pension systems account for the specific gendered vulnerability of older workers. On the one hand, globally women have less coverage in contributory pension systems; the nature of women's employment restricts their ability to make regular, adequate payments. On the other hand, social pensions or non-contributory social transfers are effective in extending pension coverage and reducing household's risk of poverty (UNFPA, 2012). In South Africa the old age pension has achieved coverage of 80% of the older population and has improved nutritional outcomes in households where women receive the pension (Duflo, 2003).

In sum, older women in rural areas globally are increasingly economically active. However, older women face the cumulative effects of gender discrimination throughout their lives with lower earning capacity and limited access to rights to land ownership, thus contributing to their vulnerability in older age. The constraints facing older rural women in accessing pensions and planning retirement, which are shaped by the lived experience of women in a specific gender regime and pension system, vary.

Case study of Ireland: challenges for work and retirement

Addressing the focus of this chapter, this section considers rural women workers in Ireland and the challenges affecting their work and retirement. It draws on experiences of women farmers and rural women workers. The analysis is from two qualitative studies undertaken in Ireland on women and property ownership and on older women's access to pensions. The study on the experiences of women farm property owners is based on a qualitative dataset using the Biographic Narrative Interpretive Method (BNIM) over the time period of 2010–2011 (Watson, 2018). The study of women and pensions in Ireland was also conducted in 2010–11, and included both focus groups and individual interviews with rural women – a detailed account of the methodology is available in a report (Duvvury et al., 2012).

For the purpose of this chapter, we begin by presenting national trends regarding women's employment in rural Ireland. We will then focus on rural women's pensions in Ireland, before presenting challenges affecting work and retirement and three themes that emerged from the two empirical studies

Women's employment in rural Ireland

In Ireland, women's employment in rural economic production has been traditionally invisible. Women have become more visible on family farms recently as a result of their critical contribution to the family income from off-farm employment. The 2008 National Farm Survey in Ireland reported that on 56% of farms, the farmer and/or spouse held an off-farm job: 40% of these jobs were held by the farmer and 35% of spouses worked off-farm (Teagasc, 2008). Those who remain employed in rural Ireland are primarily in service sectors such as food and accommodation and wholesale and retail sectors (Duvvury et al., 2012). Women's off-farm work increasingly sustains farms, and women are more likely to be the main source of income in farming households (Kelly and Shortall, 2002). More recently, women are becoming more directly involved as farm managers or as entrepreneurs in the non-agricultural rural economy (Macken-Walsh, 2009; Watson, 2018). Partly this is due to a slow increase in women's property ownership with new schemes for joint ownership. Just over 12% of farm property owners in Ireland are women, and most of these women are aged 55 and older (62%), most likely inheriting property from their husbands later in life (CSO, 2012).

Rural women and pensions in Ireland

In the past, Ireland had relatively low levels of labour force participation especially for older women. The labour force participation rates of women aged 60–64 was only 27.2% in 2005 increasing to 38.5% for women aged 60–64 by 2016 (compared to 57.1% for men) (Eurostat, 2017). This was partly due to traditional gender norms where women were expected to perform most of the unpaid caring roles in society. This was even more pronounced in rural areas where there were limited employment opportunities for women.

Women in Ireland are therefore more likely to be dependent on the lower non-contributory state pension with approximately two thirds of those in receipt of this pension being women (DEASP, 2018). Men are more likely to be in receipt of the full earnings-related contributory pension.

Recent changes in pension policy have made it even more difficult for women to build up full contributory pensions. In 2012, the bands were changed and linked more closely to years spent in paid employment (Bassett, 2017). As a result of these changes people need to work for at least ten years (as compared to five years previously) to receive even a minimum pension. Women have fewer years paid pension contributions since they still bear the primary responsibility for caring work. Another change that has been introduced is that state pension age has been increased to 66 and will further increase to 67 in 2021 and 68 in 2028. There is evidence that those who work in physically demanding occupations, such as those working in rural occupations find it more difficult to continue working until this age, particularly if they have developed chronic, work-related health conditions (Ní Léime and Street, 2019).

The fact that there is often limited employment available for women working in rural areas, means that many women may have access only to low-paid, precarious or seasonal employment in running guest houses, working in retail or hotels or as health care assistants (although some are employed in better-paid employment such as teaching and nursing) (Duvvury et al., 2012). This limits their ability either to contribute to private pensions or to build up sufficient contributions to receive the full state contributory pension and many women in these occupations have no occupational pension coverage (Duvvury et al., 2012). Many women who worked on farms or indeed small family businesses were previously not entitled to pensions in their own right and were regarded as "relatives assisting" and their spouse was given an allowance for their pension. In more recent years, women had to be able to prove with documentation that they had actually performed work on the farm (Duvvury et al., 2012). It is against this backdrop that we examine the pension position of older women workers in rural areas.

Challenges affecting work and retirement

In the qualitative data three major themes emerged as central concerns for women regarding access to pensions and active retirement planning. These are caring, precarious employment and recognition of labour.

Caring

There were (and still are but to a decreasing extent) strong gender norms that women are the primary providers of unpaid care for children and dependent family members in the accounts of the women interviewed (Duvvury et al., 2012). Women being primary carers was particularly prevalent in rural Ireland where there was a traditional expectation that women who married a farmer would care for their parents-in law. This, together with the expectation that women provided primary care for children often meant that they were unable to engage in paid employment outside the home, as the following interview excerpt indicates:

> INT: 'So, you were providing care for your mother-in-law?'
> RESP: 'She was there in the house when I got married, so that was an understood thing, you know. So, with three kids, how could you work? No way . . .' (Interview with farm woman, who subsequently worked part-time).
> (Duvvury et al., 2012)

Another farm woman reported a similar situation when asked whether she worked off-farm:

> Now in'72, I had three children and grandaddy in bed. The other man to be fed. . . . Where would you be going? (Interview with farm woman).
> (Duvvury et al., 2012)

However, some rural women with well-paid secure employment were able both to care for children and continue in paid work – for example one woman with a public service managerial occupation was able to afford professional childcare, even though she had five children. Her husband was also working but was as involved as she was in childcare.

Some women farmers in rural Ireland were able to balance the pressure of the non-traditional role for women of full-time farming with traditional care work – for example, one woman operated her own family farm while caring for a child, her parents and her parents-in-law:

> At the end of the day it is the person at home on the farm that has to look after the parents
> (interview with owner of inherited farm).

She had the support of her husband and, as a farm property owner she had the resources to outsource some of the responsibilities of care for an ill parent so she could continue to work full-time.

Precarious employment

Some of the rural women interviewed were involved in precarious or seasonal employment in the hospitality, retail or health sectors and this meant their income was vulnerable and subject to reduction. The interviews were carried out during a period of recession in Ireland (2008–2013), when there were cutbacks in insecure publicly funded jobs. One woman had worked in the community and voluntary sector and had just been made redundant prior to the interview and had ceased making (private) pension contributions as a result. Other participants included health care assistants and childcare workers who had very low earnings and couldn't afford to contribute to pensions:

> There's usually not two euros spare to go to a pension. Can't even save (Interview with childcare worker).
>
> (Duvvury et al., 2012).

Moreover, childcare in Ireland continues to be expensive and contributes to an inability to pay into a private pension:

> Somebody may be paying for a house; paying for childcare. Where will they get room for a pension like? (Interview with retail worker)
>
> (Duvvury et al., 2012).

Another common occupation employment for rural women is health care assistant, an occupation that pays low wages. This is a particular challenge because employees receive even lower wages when a client does not need care due to hospitalisation or death – i.e., payment is cut until the client is replaced. Thus, their income can be precarious from week to week leading to an inability to contribute consistently to a private pension (Ní Léime and Street, 2019).

Similarly, some farm women interviewed operated diversified businesses on their farms and were exposed to fluctuations in the economy that affected their income and the viability of their enterprises. For example, after the economic downturn in 2008, one farm woman streamlined her on-farm tourism business by closing her bed and breakfast, decreasing her workload on the farm and eventually returning to her previous career off farm:

> I can stay in a hotel for so little and you're doing B and B [guest house], you should be able to do it twice is cheap. So I thought I wasn't going to do that, it was such hard work, that I wasn't going to open the B and B [guest house] (interview with farm business owner).
>
> (Watson, 2018)

While rural entrepreneurs can adapt to market fluctuations, the insecurity of a consistent guaranteed income can affect the ability to contribute financially to their personal pension funds. Securing employment off the farm offered a more secure, pensionable income stream but also meant she could cut down on the physical work of her on-farm business and step back from farming. Even for rural entrepreneurs who own on-farm businesses, a clear challenge is the implication of continued physical labour on the farm on ageing and ability to continue farm work. For some women the positive benefit of farm property ownership can be negated with the demands on their health and being forced to find alternatives aside from their cultivator role. This would suggest that flexibility to move between farm and off farm is key to realising economic security in old age.

Farm property ownership recognition of labour

Reflecting international trends on unpaid work in family enterprises, a challenge facing rural women in Ireland is access to pensions through recognition of their labour on the farm. As mentioned previously, farm women who are not visible on official documentation for the family farm are not eligible for contributory pensions. By formalising their labour on the farm, women can plan for their retirement from farming and secure independent income from their pensions. For example, one of the rural women interviewed worked on a family farm owned by her husband, she put her name on assets shared with her husband to formalise her visibility and recognition of her labour on the farm in an official capacity, enabling her to claim a pension and establish her future security.

> When I come to that contributory pension, to have something for myself. To be independent. To dress myself, to feed myself, to keep my car going (interview with manager of a family farm)
>
> (Watson, 2018)

The informality of rural women's labour internationally, highlights the lack of pension rights and agency in future planning for women without access to resources such as farm property. To formalise her labour contributions within the pension system, one woman farm owner kept control of her farm property after marriage by keeping her natal surname.

> My mother, farming over the years with my father, she kind of had no rights, she had no pension, she had nothing. She was a nothing, and I felt myself that this was going to happen to me if I changed my name. (interview with owner of inherited farm)
>
> (Watson, 2018)

This case highlights the benefits of economic security for women who own their own farm property, making their labour contribution visible within the pension system. Keeping her farm separate from her husband's farm, and in her own name, ensured the asset remained in her control with independent decision making about the use of land and the option to improve her financial position in the future through the sale of the farm and associated assets.

Conclusion

The findings from the qualitative studies highlight the constraints older rural women face in finding secure, pensionable employment and building their future pension fund, both of which influence future retirement planning. The central conclusion of the case study that rural women are at a systematic disadvantage is relevant worldwide – most pension systems are characterised by the lack of recognition of women's labour on farms. Additionally, few pension systems have specific measures to address the precarity of rural employment, or the care costs borne by women given patriarchal norms (a note of hope is the growing incorporation of measures providing credits for care work in different countries).

Our exploration of the situation of rural women in relation to farming and agriculture globally in the context of gender norms where women are still primary carers indicates that women are still disadvantaged. Our case study of Ireland suggests that recent reforms of pension systems linking the level of the contributory state pension more closely to participation in paid employment are likely to disadvantage rural women worldwide. The research findings highlight the need to ensure that work such as caring for children and older people is appropriately valued and remunerated. Strengthening women's access to and control of property (as shown in the case study) is critical for rural women, as it provides the space for choice-based retirement planning in a context of inadequate pensions. A social pension, which is universal and based on residency, would address many of the disadvantages rural women face such as precarity of employment. For rural women to have security and agency in their old age, the discourses on ageing, pensions and retirement need to be informed by the challenges faced by rural women to suggest effective policies and programmes relevant to women. This chapter contributes to the field of critical rural gerontology and emphasises the need to interrogate apparently benign policy changes from a critical gendered perspective to tease out the heightened implications of such policies for women in rural areas.

Note

1 UNECE stands for UN Economic Commission for Europe and the region covers Europe, North America, Central Asia and Western Asia (Israel).

References

Bassett, M. (2017). *Towards a Fair State Pension for Women Pensioners*. Dublin: Age Action.

Byrne, A., Duvvury, N., Macken-Walsh, A. and Watson, T. (2014). Finding room to manoeuvre: gender, agency and the family farm. In: B. Pini, B. Brandth and J. Little, eds., *Feminisms and Ruralities*. Lanham, MD: Lexington Books.

CSO (2012). *Census of Agriculture 2010: Main Results*. Dublin: Central Statistics Office.

Department of Employment Affairs and Social Protection (DEASP). (2018). *Statistical Information on Social Welfare Services 2017*. Available at: [Accessed 7 September 2018].

Duflo, E. (2003). Grandmothers and granddaughters: old-agepensions and intrahousehold allocation in South Africa. *The World Bank Economic Review*, 17(1), pp. 1–25.

Duvvury, N., Ní Léime Á., Callan, A., Price, L. and Simpson, M. (2012). *Older women workers' access to pensions: Vulnerabilities, perspectives and strategies*. Galway: Irish Centre for Social Gerontology.

European Commission. (2019). *Females in the Field: More Women Managing Farms across Europe*. Available at: https://ec.europa.eu/info/news/queens-frontage-women-farming-2019-mar-08_en [Accessed 7 September 2019].

European Parliament (EP). (2019). *The Professional Status of Rural Women in the EU Policy*. Brussels: Department for Citizens' Rights and Constitutional Affairs at the request of the FEMM Committee.

Eurostat. (2017). *Farmers in the EU: Statistics*. Available at: https://ec.europa.eu/eurostat/statistics-explained/index.php/Farmers_in_the_EU_-_statistics [Accessed 7 September 2019].

Food and Agriculatural Organization. (2010). *Gender Dimensions of Agricultural and Rural Employment: Differentiated Pathways Out of Poverty – Status, Trends and Gaps*. Available at: www.fao.org/3/i1638e/i1638e.pdf [Accessed 7 September 2019].

International Labour Organisation (ILO). (2016). *Women at Work, Trends 2016*. Geneva: ILO.

International Labour Organisation (ILO). (2018a). *Women and Men in the Informal Economy: A Statistical Picture*, 3rd ed. Geneva: ILO.

International Labour Organisation (ILO) (2018b). *World Employment Social Outlook: Trends for Women, 2018*. Geneva: ILO.

Kelly, R. and Shortall, S. (2002). Farmers wives: women who are off-farm breadwinners and the implications for on-farm gender relations, *Journal of Sociology*, 38(4), pp. 327–43.

Lahiri-Dutt, K. and Adhikari, M. (2016). From sharecropping to crop-rent: women farmers changing agricultural production relations in rural South Asia. *Agriculture and Human Values*, 33(4), pp. 97–1010.

Macken-Walsh, A. (2009). *Barriers to Change: A Sociological Study of Rural Development in Ireland*. Athenry: Teagasc Rural Economy Research Centre.

Maharjan, A., Bauer, S. and Knerr, B. (2012). Do rural women who stay behind benefit from male out-migration? A case study in the hills of Nepal. *Gender, Technology and Development*, 16(1), pp. 95–123.

Ní Léime, Á. and Ogg, J. (2019). Introduction to special issue, gendered impacts of extended working life on the health and economic wellbeing of older workers in Europe. *Ageing & Society*, 39(10), pp. 2163–2169. DOI: 10.1017/SO1686X1800180.

Ní Léime, Á. and Street, D. (2019). Working later in the USA and Ireland: implications for precariously and securely employed women, *Ageing & Society*, 39, pp. 2194–2218.

OECD. (2019). *Pensions at a Glance* (database), 2018 ed. Paris: OECD Pensions Statistics.

Silvasti, T. (2003). Bending borders of gendered labour division on farms: the case of Finland, *Sociologia Ruralis*, 43(2), pp. 154–166.

Stevens, G. A, Mathers, C. D. and Beard, J. R. (2013). Global mortality trends and pattern in older women, *Bulletin of the World Health Organization*, 91, pp. 630–639.

Teagasc. (2008). *National Farm Survey 2008*. Athenry: Rural Economy Research Centre.

UN Women. (2015). *Transforming Economies, Realising Rights: Progress of the World's Women, 2015–16*. New York: UN Women.

UNECE. (2017). *Older Persons in Rural and Remote Area* (Policy Brief on ageing No 18). Available at: www.unece.org/population/ageing/policybriefs.html [Accessed 24 July 2019].

UNFPA and Help Age International. (2012). *Ageing in the 21st Century: A Celebration Uand A Challenge*. New York; London: UNFPA and Help Age International.

United Nations, Department of Economic and Social Affairs, Population Division. (2019). *World Urbanization Prospects: The 2018 Revision (ST/ESA/SER.A/420)*. New York: United Nations.

Watson. (2018). *Altering legacies as 'a farmer in my own right': married women's experiences of farm property ownership in Ireland*. PhD Thesis, National University of Ireland, Galway.

World Bank. (2019). *Population Data, 65 and above Population*. Available at: https://data.worldbank.org/indicator/SP.POP.65UP.To [Accessed 24 July 2019].

9 Rural-urban migration of older people

Mobility, adaptation and accessibility

Yang Cheng, David R. Phillips, Mark W. Rosenberg and Rachel Winterton

Introduction

Like many other countries in East Asia and elsewhere, China has a rapidly growing older population. The number of older people was 52.25 million in urban areas and 66.56 million in rural areas in 2010. The total number of the older population reached 166.58 million in 2018. Rapid demographic ageing is occurring in the context of rapid urbanisation, with the consequence of an increasing older rural-urban migrant population. In 2000, China's capital city, Beijing, had a "floating population" of 2.56 million in Beijing (China's capital city), who included 63,700 older migrants. The "floating population" covers migrants without Beijing household registration, a system introduced in the 1950s as a method to regulate rural-urban migration, and by 2010, it had increased to 7.04 million, and older migrants had increased to 238,000 (3.39% of the total migrant and 9.7% of the total older population) (National Bureau of Statistics of People's Republic of China, 2003, 2012, 2020). These changes in rural demography in both developed and developing countries are discussed in Chapter 2 in this volume. The destinations of older migrants in China are consistent with those of the younger generations (Wu, 2013). In Beijing, nearly 90% of older migrants moved to the urban functional extension area and the new urban developing area, which are newly developed areas in Beijing (Yi et al., 2014).

This chapter discusses the experiences of Chinese older rural-urban migrants in order to understand migration behaviours that may be applicable to many low and middle-income countries. Older rural-urban migrants are a particular vulnerable group who have received comparatively little research attention in developed and developing countries. This chapter sheds light on future service planning for the rural-urban older migrants, especially to improve their health and well-being. In an international context where rural aged care service provision is being continually rationalised and centralised (Winterton et al., 2019), these findings will demonstrate how rural older people may become disenfranchised and marginalised within urban settings should they choose to, or be forced to relocate.

The chapter begins with a short review of the relevant literature. This is followed by a description of a qualitative study within Beijing, upon which the analysis is based. Thematic findings are then presented before these findings are discussed with reference to existing studies and future directions.

Older adult urban-rural migration, vulnerabilities and accessibility

Increased attention has been paid to older migrants within the international literature. For example, rural retirement migration (RRM) and its impact on rural social sustainability have been studied in developed countries (Philip, Macleod and Stockdale, 2013; Winterton et al., 2019). Factors influencing migration among older migrants have also been studied in the sub-Saharan African country of Malawi (Kendall and Anglewicz, 2018). However, most of the research focuses on urban-rural migration in developed countries rather than rural-urban migration in developing countries.

Family factors play an important role in older people's migration decision making in China, compared with the greater importance of work-related factors for younger migrants. Older rural populations often migrate to cities to follow their children and provide childcare, or to access care and support as they age (Wu, 2013). Many older people move to live with families (Dou and Liu, 2017) and/or for residential care facilities (Cheng et al., 2012).

Mobility is a key concept in migration research and transportation geography. It links the movements of many things, including people, ideas and objects, across different spatial scales in relation to forms of places, stopping, stillness and relative immobility (Cresswell, 2011). A person's capacity to be mobile is dependent upon multiple social, political, cultural and economic contextual variables (Cheng et al., 2019). Bustamante et al. (2002) identifies "mobile vulnerability" as a social condition where the human rights of migrants are violated. Multiple vulnerabilities associated with mobility have been identified among migrants, associated with language and cultural barriers, ethnicity, race or low socio-economic status. The mobile vulnerabilities that older migrants experience, and their adaptive capacity, influence their likelihood of staying in their new destinations (Bustamante et al., 2002).

The concept of accessibility has been broadly applied in the study of health care delivery (Joseph and Phillips, 1984). Rosenberg (1983) interpreted access as having two components: economic and physical access. The former incorporates the ability to purchase health care service, whereas the latter involves the ability to overcome the cost of distance in using health care service. In Andersen's (1995) revised Behavioral Model of Health Services Use, access incorporates potential access (enabling resources), realised access (actual use of health service), equitable access (occurring when demographic and need variables account for most of the variance in utilisation) and inequitable access (occurring when social structure, health beliefs and enabling resources determine who gets medical care) (Andersen, 1995).

Data and methods

We collected 45 semi-structured in-depth interviews with older people in four sites in Beijing (Figure 9.1). The four sites are all located in the urban function extension area of Beijing which has the largest number and the highest proportion of older migrants among the city's four functional areas. Participants were aged between 55 and 86 years, and their length of tenure in Beijing ranged from two months to 30 years.

Beichenfudi community (A) is established as a social housing project with low housing prices and private rental costs in 2011. The population aged 60 years and over in this community comprised 1950 people, 13% of its total residents.

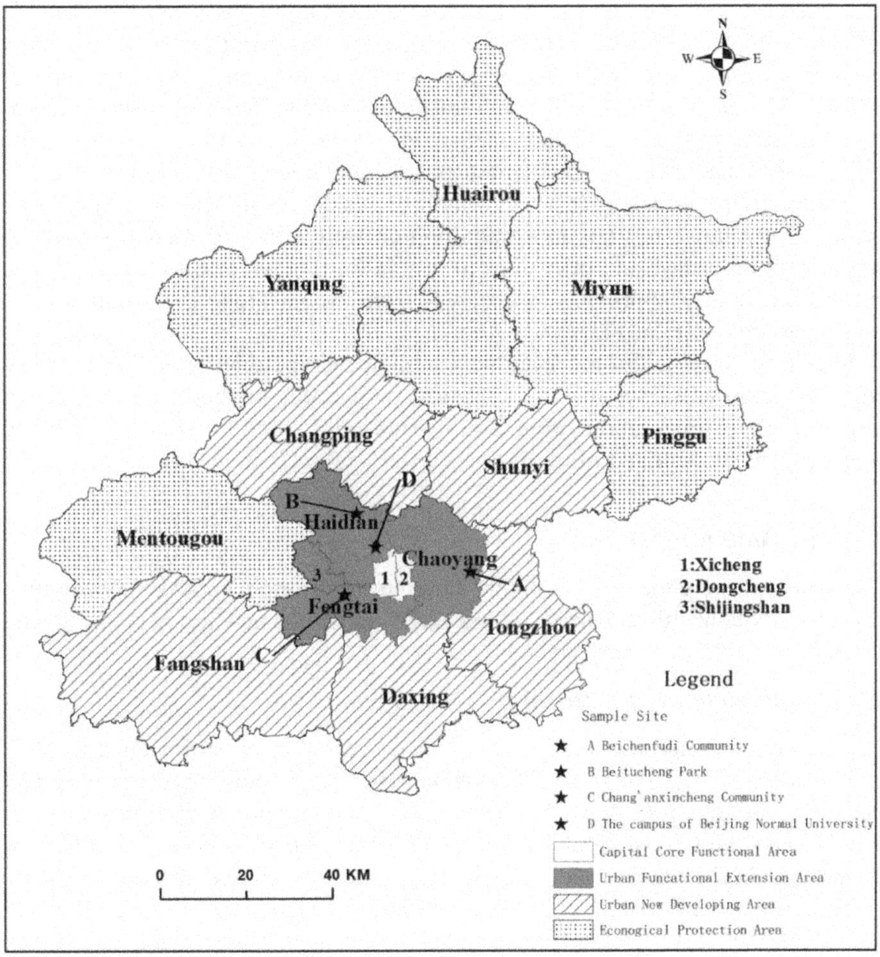

Figure 9.1 The location of case study sites in Beijing, China

Beitucheng Park (B) is a public park built on the historical site of the ancient city wall in Yuan Dynasty (1271–1368AD). It provides open space for recreation for the public, which can be easily accessed by public transportation. Many older people living nearby carry out social activities, such as singing, dancing and Tai Ji in the park. Chang'anxincheng community (C) is one of the first social housing projects built in Beijing in 2002. The population of those aged 60 years and over in the community comprised 2100 people, 17.5% of the total residents. The campus of Beijing Normal University (D) is located in the Haidian District. Some families of university faculty, staff members and retirees live on campus, as do some contract workers responsible for cleaning and gardening in the university. Many contract workers are rural-urban migrants, including some older migrant workers.

Convenience sampling was used to recruit participants in the four study sites with a particular time focus on the mornings and afternoons when older people gather for recreation. Interviews were conducted in April 2017 by six research assistants who approached potential participants to ascertain if they met the age criteria (55+) and had migrated from rural areas to urban Beijing with rural household registration (without Beijing urban household registration). Demographic information (age, income, level of education, occupation before migration, marital status, living arrangements and length of residence in Beijing) and data on reasons for migration, impact of migration on well-being, adaptation strategies, health care utilisation and intensions to stay in Beijing were collected during the interviews. Ethical approval was received from the La Trobe University human ethics committee (HEC17/005).

Interviews were audio-recorded with participants' permission and transcribed into Mandarin and English. The constant comparative method was applied for data analysis (Strauss and Corbin, 1998). Transcripts were first open-coded, and themes were summarised from the concepts emerging from the open-codes.

Rural-urban migration

Three themes were identified from the interview transcripts: mobile vulnerability, accessibility to health care services and adaptation to the new living environment.

Mobile vulnerability of older rural-urban migrants

Mobile vulnerability is one of the key themes reported by participants. It refers to the vulnerability that older migrants experience affecting their mobility and social conditions in the form of physical, socio-economic, institutional and cultural aspects. Mobile vulnerability has a significant impact on people's decision to migrate, and behaviours associated with migration, as well as their ability to adapt (Cheng et al., 2019). Many older rural-urban migrants had moved to Beijing and lived with their adult children to provide care for their grandchildren when they are physically capable, while some of the older migrants moved to live with their adult children for care and support when their physical mobility declined. Due to changing vulnerability, some older migrants reported that their intention to stay in Beijing depended on their physical mobility and their children's decision on whether they and their children would stay in Beijing or move back home.

Accessibility to transport affects older people's physical mobility. However, most participants reported that they always depend on their adult children for transfers and financial support on travel expenses, which reflects their vulnerability in terms of physical mobility and financial affordability.

> I was from Yanqing [one of the rural districts of Beijing]. I have two daughters. One lives in Yanqing and the other lives in the city. I live with their families in turn. Every time I move, they drive me to the other daughter's home. . . . If I didn't have the two daughters, I would have died already.
>
> (C8, 80 years old, living in Beijing for 20 year)

Economic vulnerability affects older rural-urban migrants' mobility in several ways, due to low incomes and high cost of living and health care and small accommodation, in Beijing, which make many participants financially dependent on their adult children. Meanwhile, social welfare benefits are still currently tied to a person's household registration rather than their actual living place. After migration, some older rural-urban migrants experience vulnerability when accessing health care services, which limits their institutional mobility and again increases their dependency on adult children. Consequently, older rural-urban migrants may have to decide to live in Beijing on a temporary basis due to their lack of economic and institutional mobility, which eventually influences their sense of belonging in Beijing and willingness to stay.

Older rural-urban migrants also experience vulnerability in the form of social exclusion, which impacts on their success of living in new environments. Several individual characteristics such as poor health status, language (dialect) barriers, low education, short length of stay in Beijing, and introverted personalities, can affect migrants' social belonging and participation. The vulnerability older migrants experience restricts their social integration and willingness and comfort in and outside of the communities they know best more than their physical capability to be mobile. Some older migrants who migrated to care for grandchildren reported their busy caregiving responsibilities and household chores restricted their time and opportunity to develop social networks in the new living environment. They experienced mobile vulnerability in adapting to the changes in the living environment and social integration in their new places.

Cultural vulnerability related to mobility is tied to a sense of filial piety and family care support in traditional Chinese culture. Findings indicated that some older rural-urban migrants' expect their adult children to help them make arrangements for care and may be provide care and support to them directly. The high dependency on adult children is affected by the strong sense of filial piety and reinforced by the lack of formal support. However, they were also worried about the burden their care placed on their adult children and they expressed a willingness to move back to their hometown if their adult children wanted them to. Given the cultural expectation that adult children care for their parents and their levels of dependency, older migrants' mobility was contingent on their children's preferences and needs.

> The apartment in my hometown is not sold yet. If we are too much burden for my daughter in the future, we will go back home to move into residential care facilities. If we move to a residential care facility in Beijing, given our small pension, we don't think we can get into a good one. We will take my daughter's advice.
>
> (D6, 64 years old, living in Beijing for 13 years)

Accessibility to health care services

Accessibility to health care services was also a theme identified from the transcripts and interpreted from spatial, economic and institutional perspectives by the older migrants. Older rural-urban migrants evaluated the spatial accessibility to health care facilities in Beijing differently, depending on their accessibility to health care services in their hometowns. Those who had accessed health care services easily using the clinics in their villages reported better spatial accessibility to health care services in their hometowns, as services were in walking distance from their homes. However, some of the participants reported better spatial accessibility to health care facilities after moving to Beijing.

> It is much more convenient here than in the rural areas. It takes hours to get to a hospital in the rural areas. Here it only takes for 10 minutes. The community we live in here is close to the Peking University Third Hospital.
>
> (B5, 72 years old, living in Beijing for 25 years)

Older migrants generally reported better quality of health care services in Beijing but also long waiting times and financial issues in accessing health care services. Participants stated that costs of medicine are higher in Beijing compared to their hometowns. Older migrants' access to pensions and health care insurance are determined by where one's household registration is. If they hold a rural *hukou* (by older rural-urban migrants) it means they cannot receive public social welfare in Beijing, so they have to pay out-of-pocket to cover health care expenses (Liu and Feng, 2015; Smart and Smart, 2001). In 2014, the Urban Residents' Pension Plan and the New Rural Social Pension System were merged into a unified basic pension insurance plan for both rural residents and non-employed urban residents. However, the benefits payable under this plan are low. The average national pension monthly payment is only 81 RMB (US$13); by contrast, the average household consumption per month in rural areas was 896 RMB in 2016 (National Bureau of Statistics of People's Republic of China, 2017). By the end of 2013, the central government of China had set three types of health care insurances, the Urban Employee Basic Medical Insurance, Urban Resident Basic Medical Insurance and New Rural Cooperative Medical Insurance, which cover 95% of urban employed workers, non-employed residents and rural residents respectively. However, even with these social welfare improvements, urban and rural residents in various regions still experience great disparities in access to social welfare benefits. Older rural-urban migrants reported difficulties in obtaining reimbursement for hospitalisation costs if using hospitals outside the regions where they are registered (Liu and Sun, 2016). They have to pay privately for out-patient expenses which are not covered by the New Rural Cooperative Medical Insurance they are enrolled in.

> There are too many people waiting to see a specialist. It takes a week to see a doctor for my stomach problem. Does the disease get delayed? It was much easier to see a doctor in my hometown. There was not much waiting time. Here I need to get a waiting number when I first see the doctor, and I need to get another waiting number when I receive the examination results.
>
> (B11, 73 years old, living in Beijing for six years)

Family support clearly plays an important role in mitigating vulnerability in accessing health care services for older rural-urban migrants in Beijing. Their adult children who have settled in Beijing provide physical, financial and emotional support for accessing health care in Beijing, including paying the costs of health care services and transportation, and also accompanying them to see a doctor. Living with their adult children also improved older migrants' capacity for health literacy and health behaviours. Some participants reported that they had taken their families' suggestions on health promotion and started to complete annual physical examinations after they moved to Beijing.

> I went to see a doctor as I had pain in my lower back and legs. I went to a hospital in Dongsi Street. It is far from here and it took me an hour waiting to see a doctor. There were so many people in the hospital and some of them were sitting there on a fluid drip. Every time my son takes me to see a doctor. I totally depend on my son and I am not used to it. It is like I am a blind person and I bring many difficulties for my children.
>
> (B9, 65 years old, living in Beijing for half a year)

Adaptation to the new living environment

In adapting to challenges around mobile vulnerabilities and difficulties in accessing health and social care services, older migrants exercise significant agency and develop a number of effective coping strategies. Participants reported it was more practical for them to improve their social mobility by developing coping strategies. For example, they put efforts into developing social interactions with other residents.

> There are many older people coming here to care for the grandchildren. We get in touch with others. Sometimes, we all come out with the kids and we get together to chat a little bit.
>
> (C1, 67 years old, living in Beijing for one year)

The new social connections with the local residents, or other migrants, helped older migrants feel a greater sense of belonging within their communities. Older migrants also maintained connections with their families in their hometowns via communication technologies, such as cell phones and WeChat (a Chinese app for video calls), which helped them receive emotional support and to remain connected with their hometown social networks. This helped overcome the physical distance between Beijing and their hometowns and supports them in adapting to their new life post-migration, even after a longer period of residence in Beijing, as they could keep in touch.

> I don't think my migration affect my connection with my relatives in my hometown. I still go back to visit. . . . I call them every 10 days or two weeks.
>
> (B6, 64 years old, living in Beijing for four months)

Personal attitudes towards population ageing, such as whether older migrants should continue to participate in social activities, engage with others, lead physically active lives and maintain autonomy and independence, also have significant impacts on decision-making related to adapting to the new environment (Kuh et al., 2014). This fits with the concept of "active ageing", long advocated by WHO (2002). Active ageing among older rural-urban migrants appears beneficial for the maintenance of physical and socio-cultural mobility to mitigate the vulnerability that they experience.

> My children all moved to cities. I moved to Beijing for work and I am a gardener now. I have less workload and make more money. As a farmer, physical work (gardening) is a way to keep fit. The other good thing is that I got to live in the capital and can visit many places of interest and parks.
>
> (D10, 61 years old, living in Beijing for three years)

To cope with the challenges in accessibility to health care services post-migration, older migrants often bring medicine from their hometown and go back to their hometown for health care services if needed. These actions are motivated by the lower out-of-pocket costs for health care and their familiarity with doctors in local hospitals which make them more comfortable using health care services. The older rural migrants also go to pharmacies in Beijing to buy medicines for common illnesses such as colds instead of going to see a doctor first, and as a result save the time and cost of seeing doctors in hospitals.

The physical, economic, institutional and socio-cultural aspects of mobile vulnerability that older migrants can experience and their adaptation to new living environments and access to health care services clearly affect their intention to stay. The adaptation process reflects the role of agency in developing coping strategies for meeting the challenges related to migration, which also affects older migrants' mobile vulnerability.

Discussion and conclusion

With the rapid urbanisation process, the number of older rural-urban migrants is growing in the megacities in China, as well as in other developing countries. Older rural-urban migrants are a cohort that are under-researched and under-considered in policy making. In developed countries, researchers have raised concerns about older people's migration from urban to rural areas after retirement and transnational older migrants. This study responds to this knowledge gap and focuses on older rural-urban migrants in the Chinese context, adding new insights into the study of rural ageing. Older migrants are generally not the cohorts that megacities desire to attract and wealthier, better educated migrants appear to be generally preferred (Liu and Feng, 2015). Consequently, older migrants often experience mobile vulnerability and are often concentrated on the peripheries of cities, with limited ability to access social and health care services. Older rural-urban migrants experience various forms of mobile vulnerability including physical, economic, institutional and socio-cultural aspects related to their vulnerable positions. Vulnerability in physical mobility was reflected by the decline in older people's physical health status and their dependence on adult children. With the rapid socio-cultural changes in China, traditional cultural norms may be changing with a lessening of instrumental support. But this may sometimes be substituted by financial support due to the limited availability of family care resources and distances from adult children (Phillips and Feng, 2015). Older migrants are also often financially very dependent on their adult children for housing, living and health care expenses. Their economic vulnerability also affects their level of authority in the household when co-residing with their adult children after migration, which has also been reported by other studies (Phillips and Feng, 2018). Meanwhile, older migrants can experience institutional vulnerability, being not able to fully enjoy public social welfare benefits after migration because of their lack of urban *hukou*, which also affects their economic mobility. Vulnerability in socio-cultural mobility comes from factors such as lack of spare time, low education levels and language barriers (different dialects). Older migrants therefore often have to reconstruct their identity and develop coping strategies to communicate with local people and adapt socio-culturally to the urban environment.

In terms of accessibility to health care services, older migrants reported they were highly dependent on their adult children for transport and financial support to use health care services in Beijing. The central government has long devoted efforts to introduce social welfare reform to provide a safety net for older people with rural household registrations, but older migrants with rural household registrations are still not yet able to enjoy the equivalent health and social benefits within urban environments (Wen and Wallace, 2019). Old-age pension benefits, health care resources and care policies do not address the particular demands of older rural-urban migrants (Mou et al., 2013; Zhang and Goza, 2006).

To address these challenges, actions can be taken at all levels. At the macro level, national policies on social welfare play a significant role in mitigating older migrants' vulnerabilities (Ciobanu, Fokkema and Nedelcu, 2017). Future policy making should aim for the provision of universal social welfare benefits to cover the older population and address their needs based on their age instead of their *hukou* status. At the meso-level, there is the potential to increase/harness community resources for the provision of care and support to older migrants. For example, evaluation of older migrants' care needs can be conducted at the community level and community support can be delivered based on the evaluation. Improving community services was also identified as an effective solution to help older migrants reconstruct their social support networks and promote their communication with local people (Liu and Feng, 2015). The micro level refers to individuals actions such as active ageing and adaptation to the migration experience.

This study has some limitations. For example, it has a relatively small sample and participants were recruited in public spaces, which suggests that they may be more physically and socially mobile than those who are trapped in their homes. Therefore, although it provides qualitative insights, this study cannot be considered representative of all older rural-urban migrants living in Beijing. While the case study focuses on China, its implication for policy making at micro, meso and macro levels to improve mobility and adaptation of older migrants may also be relevant to other countries with similar economic and political systems. It identifies key issues and questions and provides critical insights into whether and how older rural-urban migrants can maintain their health and quality of life in a new urban environment.

Acknowledgements

This research was partly supported by the China Studies Research Centre, La Trobe University, Australia and Project no. DR182B of the Direct Grant scheme of Lingnan University, Hong Kong, China.

References

Andersen, R. M. (1995). Revising the behavioral model and access to medical care: does it matter? *Journal of Health and Social Behavior*, 36(3), pp. 1–10.

Bustamante, J. A. (2002). Immigrants' vulnerability as subjects of human rights. *International Migration Review*, 36(2), pp. 333–354.

Cheng, Y., Rosenberg, M.W., Wang, W.Y., Yang, L.S. and Li, H. R. (2012). Access to residential care in Beijing, China: making the decision to relocate to a residential care facility. *Ageing & Society*, 32(8), pp. 1277–1299.

Cheng, Y., Rosenberg, M., Winterton, R., Blackberry, I. and Gao, S. (2019). Mobilities of older Chinese rural-urban migrants: a case study in Beijing. *International Journal of Environmental Research and Public Health*. 16, p. 488.

Ciobanu, R. O., Fokkema, T. and Nedelcu, M. (2017). Ageing as a migrant: vulnerabilities, agency and policy. *Journal of Ethnic and Migration Studies*, 43(2), pp. 164–181.

Cresswell, T. (2011). Mobilities I: catching up. *Progress in Human Geography*, 35(4), pp. 550–558.

Dou, X. and Liu, Y. (2017). Elderly migration in China: types, patterns, and determinants. *Journal of Applied Gerontology*, 36(6), pp. 751–771.

Joseph, A. E. and Phillips, D. R., (eds). (1984). *Accessibility and Utilization: Geographical Perspectives on Health Care*. London; New York: Harper and Row.

Kendall, J. and Anglewicz, P. (2018). Migration and health at older age in rural Malawi. *Global Public Health*, 13(10), pp. 1520–1532.

Kuh, D., Karunananthan, S., Bergman, H. and Cooper, R. (2014). A life-course approach to healthy ageing: maintaining physical capability. *Proceedings of the Nutrition Society*, 73(2), pp. 237–248.

Liu, Q. and Feng, L. (2015). The empirical analysis on the social interaction of the older migrants in Shenzhen. *Chinese Journal of Gerontology*, 18, pp. 5347–5349 (in Chinese).

Liu, T. and Sun, L. (2016). Pension reform in China. *Journal of Aging and Social Policy*, 28(1), pp. 15–28 (in Chinese).

Mou, J., Griffiths, S. M., Fong, H. and Dawes, M. G. (2013). Health of China's rural-urban migrants and their families: a review of literature from 2000 to 2012. *British Medical Bulletin*, 106, pp. 19–43.

National Bureau of Statistics of People's Republic of China. (2003). *Tabulation on the 2000 Population Census of the People's Republic of China*. Beijing, China: China Statistics Press (in Chinese).

National Bureau of Statistics of People's Republic of China. (2012). *Tabulation on the 2010 Population Census of the People's Republic of China*. Beijing, China: China Statistics Press (in Chinese).

National Bureau of Statistics of People's Republic of China. (2017). *China Statistical Yearbook 2017*. Beijing, China: China Statistics Press (in Chinese).

National Bureau of Statistics of People's Republic of China. (2020). *National Data*. Accessed January 16, 2020. Retrieved from: http://data.stats.gov.cn/easyquery.htm?cn=C01 (in Chinese).

Philip, L., Macleod, M. and Stockdale, A. (2013). Retirement transition, migration and remote rural communities: evidence from the Isle of Bute. *Scottish Geographical Journal*, 129(2), pp. 122–136.

Phillips, D. R. and Feng, Z. (2015). Challenges for the aging family in the People's Republic of China. *Canadian Journal on Aging*, 34(3), pp. 290–304.

Phillips, D. R. and Feng, Z. (2018). Demographics and aging. In: W. P Wu and M. W. Frazier *Handbook of Contemporary China*. London: Brill Academic Publishing, pp. 149–1071.

Rosenberg, M. W. (1983). Accessibility to health care: a North American perspective. *Progress in Human Geography*, 7(1), pp. 78–87.

Smart, A. and Smart, J. (2001). Local citizenship: welfare reform urban/rural status, and exclusion in China. *Environment and Planning A*, 33(10), pp. 1853 – 1869.

Strauss, A. and Corbin, J., (eds). (1998). *Basics of Qualitative Research: Procedures and Techniques for Developing Grounded Theory*. Thousand Oaks, CA: SAGE Publications.

Wen, C. and Wallace, J. (2019). Toward human-centered urbanization? Housing ownership and access to social insurance among migrant households in China. *Sustainability*, 11, p. 3567.

Winterton, R., Butt, A., Jorgensen, B. and Martin, J. (2019). Local government perspectives on rural retirement migration and social sustainability. *Australian Geographer*, 50(1), pp. 111–128.

World Health Organization (WHO). (2002). *Active Ageing: A Policy Framework*. Geneva: WHO.

Wu, Y. (2013). One-child policy and elderly migration. *Sociological Studies*, 4, pp. 49–73, 243 (in Chinese).

Yi, C., Zhang, C., Wu, S., Gao, M. and Liang, H. (2014). Spatial restructuring of senior population in Beijing from 2000–2010. *Urban Develop Studies*. 21, pp. 66–71 (in Chinese).

Zhang, Y. and Goza, F. W. (2006). Who will care for the elderly in China? A review of the problems caused by China's one-child policy and their potential solutions. *Journal of Aging Studies*, 20, pp. 151–164.

10 Policy and program challenges in delivering health and social care services to rural older people

Nina Glasgow and Stefanie Doebler

Introduction

This chapter synthesises recent literature on health and social care service delivery to older persons in rural areas of the United States (US) and United Kingdom (UK) and indicates how the situations described present challenges (and opportunities) for maintaining health and well-being in rural communities. The analytical approach we take is to place the US and UK in parallel to identify important similarities and differences in how the respective nations provide health and social care to their elderly rural populations. By placing the two nations in parallel, we can identify fundamental similarities and differences in how they provide health and social care services in rural areas during a time of fiscal austerity (Shucksmith and Brown, 2016; Shucksmith, 2019).

We compare health and social care services in the US and the UK because their structures of welfare capitalism differ, and governmental devolution is much greater in the US. We employ a multi-scalar approach (Brown et al., 2019) that provides a window into how national policies filter down to affect local-level health and social care services in both nations. Examining how policies and programs affect health and social care service delivery in ageing rural regions is merited because the smaller populations and sparseness of settlement in rural compared to urban places increase the difficulty and per capita cost of delivering services (Goins and Krout, 2006). Rural communities are those most likely to be medically underserved (Glasgow et al., 2004), and they often lack medical care specialists and advanced equipment (Hanlon and Kearns, 2016). Studies report recent increases in rural hospital closures in both the US and UK (Kaiser Family Foundation, 2016; Walsh, O'Shea and Scharf, 2019; Yeginsu, 2019). Policies developed at national levels often fail to address the diversity within and across rural communities (Hanlon and Kearns, 2016). The US rural population is diverse but is on average older, poorer and more socially isolated than the urban population (Glasgow and Berry, 2013; Glasgow and Brown, 2012). By contrast, poverty in the UK has typically been lower in rural compared to urban places. Recent research using the British Household Panel, however, showed that 50% of rural households experienced at least one year of poverty between 1991 and 2008 compared with 55% of urban households. Hence, rural poverty in the UK, while lower than urban, is not rare (Vera-Toscano, Shucksmith and Brown, 2020).

The older age structures of rural populations in the US, UK and many other countries (Glasgow and Berry, 2013; Keating, 2008) have implications for designing health and social care services for older rural residents (Brown et al., 2019; Thiede et al., 2017). Thiede et al. (2017) found that the number of services in rural communities, including health care, increase with each year of median age up to about age 40, but then decline after rural populations reach a median age of about 46 years. The ways in which such places deliver services are complex and diverse. Diversity exists even within this subgroup of rural places. Hence, one-size-fits-all approaches to services delivery tend to disadvantage particular types of places. In particular, rural places lacking linkages with other, often larger places, and places at higher scales in the intergovernmental system, find service provision for older residents particularly challenging. The multi-scalar organisation of ageing-related services approach provides an effective window into this complexity. This perspective is shaped by relational social science (Jones, 2009), which focuses on collaborations among institutions and organisations within rural communities, and their relationships with organisations and institutions at different spatial scales (Brown et al., 2019). The relational framework includes the nation state and lower levels of government, non-governmental organisations and private businesses. The UK's more centralised intergovernmental system than that of the US provides an interesting comparison of rural services organisation and delivery. Examining service access through this perspective provides insights into rural localities relationships with extra-local entities, including the national-level government, that improve access to health and social care services for older rural people.

Hanlon and Poulin (Chapter 4 in this volume) examine the health status of rural elderly populations. Keating et al. (Chapter 5), Burholt and Scharf (Chapter 6) and Phillips and Amoah (Chapter 7) examine health services and other issues related to rural ageing in differing global contexts. Here, we focus on *rural health and social care services* in the US and UK. The term *health care services* refers generically to the whole of the health care provision infrastructure including hospitals, nursing homes, health clinics, physicians and other providers. Community-based *social care services* include social work, personal care, adult protection and other support services. The purpose of social care is to safeguard and promote the welfare of vulnerable persons and to help older individuals adjust to and cope with problems they may experience as they advance in age. Our use of the term *rural* recognises that different countries conceptualize rural differently and define it differently in official government statistics (Shucksmith and Brown, 2016). We do not attempt to conduct a methodologically harmonised comparative analysis, but, as noted earlier, we place the two nations in parallel.

Governance in the US is highly devolved from the national level to states and from states to localities. Except for providing partial funding, the national government is not deeply involved in providing social care services in the US. Rather, the US devolves these responsibilities to states and localities. In contrast, the UK national government is responsible for the conduct of national affairs. While more devolved than in the past, the UK federal government retains more power and authority over health, education and other government programs than does the US (Jones, 1974). The UK reorganised relationships between the national government and local authorities in 1974, ceding some functional responsibilities to local authorities. Compared to the US, however, local policy and program administration remains more centralised in the UK.

Theoretical framework

The community level multi-scalar organisation perspective provides a framework for examining how community structures operate internally, and how they link with institutions and organisations at other spatial scales. While this hybrid perspective focuses on inter-place linkages across spatial scales, it also recognises the continuing importance of individual places in service delivery (Jones, 2009). The framework is useful in investigating the "agency" communities employ in securing, in this case, health and social care services for older residents. Similarly, Walsh, O'Shea and Scharf (2019) refer to community "agency" in the milieu of older rural residents' advantages and disadvantages in access to services in Ireland and Northern Ireland.

This perspective aligns with the "neo-endogenous development" perspective on community vitality and "agency" (Shucksmith, 2009). The multi-scalar organisation of services perspective has roots in earlier community sociology exemplified by *The Community in America* (Warren, 1963), that identified horizontal (within) and vertical (outside) relationships between local leaders and external actors. Flora and Flora's (2003) "community social capital" perspective is also consistent with our multi-scalar approach. The multi-scalar perspective expands community sociology's historical roots by enhancing the focus on relational aspects of internal and external ties across various institutional spheres at different spatial scales. Many observers have noted that many rural communities are too small and/or isolated to support a full complement of services. The multi-scalar approach acknowledges these limitations, while also indicating that the limitations of small size, peripheral geographic location diminish, if a place is efficiently organised and has effective relationships with external places and organisations.

Health and social care services for older rural residents

As indicated earlier, because of its federal system of governance, localities have more control over service delivery in the US than the UK. This is *not* to say that local authorities have no power in the UK, or that the national government is powerless in the social and health care arenas in the US. Moreover, we are not arguing that having greater local authority necessarily results in older rural persons receiving superior access to care. Local control may be desirable in many ways, but it can also result in substantial inequality between communities.

Health and social care in the US

In 1965, the US passed the Older Americans Act to deliver community-based social care services to older people across the country. In 1973, the US revised the law to create Area Agencies on Aging as the planning arm to deliver social care services to older Americans. The US federal system funnels funds allocated to the program through state governments. State governments, in turn, establish Area Agencies on Aging (AAAs) as multi-county regional planning districts that design multiple services to help older people maintain independent living while they age-in-place. Senior centres, key institutional actors in this multi-level system, are widespread across the US. Senior centres typically receive funding from the AAA, local government and nominal fees from service recipients. Senior centres often serve as delivery points for AAA services. Figure 10.1 describes the interaction between place-based institutions and multi-scalar relational structures that deliver health and social care services for older people in rural communities of the US (Brown et al. 2019).

Research using this framework found that older residents obtain services directly from local government agencies (e.g., the health department), the local hospital, local businesses, the faith community (e.g., food pantries) and other non-profit organisations, and *indirectly* through the AAA (Brown et al., 2019).[1] The majority of older rural (and urban) residents do not interact directly with the AAA that serves their multi-county area, but local representatives serve on advisory boards to recommend ageing-oriented services needed in particular communities. As shown in Figure 10.1, access to AAA programs and resources often depends on a place having an effective senior centre. Across the US, however, senior centres run the gamut from offering little more than a weekday site for congregate meals to offering health management and other social care services (Goins and

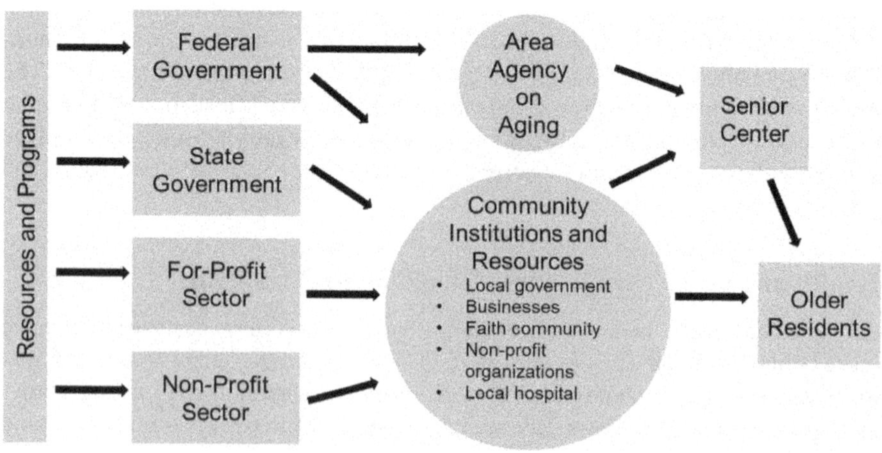

Figure 10.1 Multi-scalar organisation of ageing-related social care services delivery in US rural areas

Krout, 2006). Rural communities with an effective senior centre linked to the AAA and other external institutions typically provide a greater variety of health and social care services locally.

Access to a local hospital is also critically important to older rural residents (Glasgow et al., 2004; Warner et al., 2016). An accelerated number of rural hospital closures in the US in recent years (Kaiser Family Foundation, 2016) has left many older rural residents with few options other than to travel great distances to obtain hospital care and physicians' specialty care. As the COVID-19 global pandemic unfolds, this deficiency has become an even greater problem, imperilling the lives of both elderly and non-elderly rural residents severely affected by the disease (Wilson, 2020).

The presence of a hospital itself, however, does not guarantee local access to quality medical care. Hospitals must cooperate with local government, other local institutions and the private sector to develop strategic plans for providing health care services locally. For example, community leaders in a geographically isolated but prosperous farming community studied by Brown et al. (2019) developed a regional hospital that served the community's own residents and residents of surrounding small rural communities. They used a multi-pronged approach, including developing a transit system to shuttle residents of nearby rural communities to and from the regional hospital for health care. Community leaders showed "agency" in creating a hospital that was performing above expectations for a small rural community.

Mergers between big city hospitals and small rural hospitals are increasingly common in the US, and they represent one approach to achieving economies of scale in small communities. While such arrangements may diminish local control, they also show how multi-scalar relationships, such as joining a regional medical system, can enhance health care access in rural communities.

The use of telemedicine services is another way that linking across geographic scales and organisational boundaries can enhance health care services available to older rural residents (Grigsby and Goetz, 2004). However, relatively few rural communities in the US have developed effective telemedicine capacities (Ward, Ullrich and Mueller, 2014). With the emergence of the COVID-19 pandemic, the federal government very recently passed the CARES Act that included increased funding for telehealth programs for Medicare recipients, rural health clinics and other vulnerable groups (Wicklund, 2020). This could have a long-term unanticipated positive consequence for older rural residents' access to health care.

Health and social care in the UK

Since its establishment in 1945, the National Health Service (NHS) has been the main provider of state funded primary health care in the UK. The NHS is a founding pillar of the British welfare state (Taylor-Gooby and Wallace, 2009). Every British citizen is entitled to free health care provided by the NHS. However, *availability of care is not the same as access to care*. If people lack the resources to travel to health providers, or live in an area lacking convenient and reliable public transport, services become inaccessible (Doebler, 2016; Doheny and Milbourne, 2012). Providing access to sufficient, reliable public transport for the UK's rural older population so they can reach general practitioners, other health care services and pharmacies is vital to ensuring health and well-being. Lack of transport, however, is not the only limiting condition affecting access to health care in the rural UK. As Walsh, O'Shea and Scharf (2019) showed in Northern Ireland, communities differed in the effectiveness and capacity of local health and social care infrastructure. Some rural community residents indicated satisfaction with services, while residents of other rural communities reported not having adequate access to care. The availability of NHS services also differs across UK communities. Walsh, O'Shea and Scharf (2019) reported that recent hospital closures in some rural communities made rural residents vulnerable to health care exclusion. Similarly, a recent article in the *New York Times* on Cumbria in England characterised residents as having "shrinking lives in the rural countryside" due to poor transport and hospital closures (Yeginsu, 2019). Although none of the earlier studies used a multi-scalar relationships perspective, it may have been a useful perspective for examining differences across rural communities.

According to the Care Act of 2014, the UK's social care objective is to keep people independent and out of residential care. This has created new demands and expectations, but funding has not kept pace. Local governments manage and oversee social care provision in the UK, with accountability to the Department for Communities and Local Government (Exworthy et al., 2017).

Within communities, organisations deliver a range of social care services. While health care through the NHS is free of charge, social care provision is fee-for-service and the costs vary between local councils (Exworthy et al., 2017). Humphries et al. (2016) found that access to care depends increasingly on what people can afford rather than on what they need. Moreover, Humphries et al. (2016) found that social care providers are struggling to retain staff, maintain quality and stay in business. With social care managed locally, the study found that local authorities must make difficult choices about where to make reductions. Inter-local cooperation, consistent with the multi-scalar perspective, because it reduces the per capita cost of services, would seem to be a viable option for keeping social care services available in rural communities. A "go it alone attitude" will probably result in a loss of services in all but the wealthiest communities.

Under the Conservative-Liberal Democrat coalition government (2010–15), health and social care governance underwent reforms attempting to integrate and coordinate services (Exworthy, Powell and Glasby, 2017; Humphries, 2015). One aim of the concept of Integrated Care (IC) was to facilitate different actors working together (National Collaboration for Integrated Care and Support, 2013). The multi-scalar perspective suggests integrated health and social care would particularly benefit older rural populations, where access to services is more restricted than in urban and suburban areas. However, the concept of Integrated Care (IC) in the UK remains elusive. Exworthy, Powell and Glasby (2017) and Humphries (2015) criticised the IC reforms of the UK's previous and current national governments for lacking efficiency. They argued that some reforms implemented with the

Health and Social Care Act of 2014 led to more (rather than less) fragmentation of responsibilities for health and social care and to confusion among practitioners.

Recent changes in national policies affecting rural older populations: US and UK comparisons

The US is the only highly industrialised nation of the world that does not have a national health service. In 1965, however, President Lyndon Johnson signed Medicare and Medicaid legislation into law. Medicare provides health care insurance to all US citizens 65 years of age or older. Medicare is not means-tested, and it is available to any citizen who applies for Medicare upon reaching age 65. Medicaid covers health care services for indigent non-elderly individuals and disabled individuals of all ages. While neither Medicare nor Medicaid covers all health care expenses of those enrolled, the two programs together provide a degree of social safety net protection for elderly, poor and disabled individuals. Health care provided through Medicare represents an expansionist period in US history in policies favourable to the elderly population (Hudson, 2015). More recently, conservative politicians have contested these programs and have frequently tried to cut Medicare funding. To date, their efforts have been largely unsuccessful. Rural and urban hospitals, nursing homes, physicians and other health care professionals receive reimbursement for services provided to Medicare and Medicaid recipients. Reimbursements for services under the two programs have been especially important for the *survivability* of rural hospitals, clinics and nursing homes (Kaiser Family Foundation, 2016). Consistent with the multi-scalar perspective, these national policies have benefitted vulnerable rural residents and vulnerable rural communities.

The Affordable Care Act (ACA), passed in 2010 and implemented in 2014, is the only major piece of health care legislation passed at the national government level in the US since Medicare and Medicaid in 1965. The ACA's purpose is to provide subsidies to previously uninsured low-income non-elderly adults to purchase insurance through government-run exchanges. In 2013, the year before the ACA's implementation, 44.4 million non-elderly adults in the US had no health insurance, representing 16.8% of that population (Kaiser Family Foundation, 2019). With implementation of the ACA, by 2016 the number and proportion of non-elderly adults without health insurance dropped to a low of 26.7 million, or 10.0%, of that group.

In 2017, Republicans in Congress, with support from President Trump, came within one vote of repealing the Affordable Care Act. Attempts by Republicans to repeal the ACA led to confusion among citizens. Thus in 2018, 27.9 million, or 10.4%, of non-elderly adults were uninsured, which was an uptick from 2016 (Kaiser Family Foundation, 2019). Before the ACA's implementation, rural residents faced 26% higher odds of having no health insurance than urban residents (Hummer et al., 2004). Rural residents stood to gain more than urban from passage of the Affordable Care Act. Recent years' rapid increase in the number of rural hospital and nursing home closures, however, likely relates to conservative rural states being those that were more likely to refuse Medicaid expansion offered under the ACA (Kaiser Family Foundation, 2019). This is another example of how decisions made at higher governmental levels affect rural communities.

During the 1990s and 2000s, UK Labour governments pursued a liberal approach to welfare provision. Research has shown that New Labour's policies dramatically reduced poverty in both rural and urban areas, with the impact being especially great among rural households (Vera-Toscano, Shucksmith and Brown, 2020; Commission for Rural Communities, 2007). This approach changed radically when the Conservative–Liberal Democrat coalition government took over in 2010. The coalition government and later the Conservative government pursued a neoliberal approach (Grimshaw and Rubery, 2012) and implemented wide-ranging austerity measures, including cuts to welfare state provision, the NHS, disability and unemployment benefits (Reeves et al., 2013). Policies of marketisation and privatisation of health care led to a reduction of the NHS workforce and an increase in private health care providers (Pownall, 2013). Reeves et al. (2013) emphasize that austerity cuts to public services in the UK since 2008 disproportionately affected the disabled and poor and adversely affected the UK's public health via cutting prevention and treatment programs. Since 2008, evidence indicates that austerity cuts also adversely affected the UK's public transport provision, especially in rural areas (Veeneman et al., 2015). Cuts to health care provision, disability benefits and public transport are likely to affect remote rural more than urban populations due to the larger geographical distances they need to travel to access health and social care services (Milbourne, 2015).

The UK national government proposed that community and voluntary groups playing a greater role in providing services to disadvantaged and vulnerable groups within society would buffer its austerity policies. Public policies cut the third sector as well, however, and austerity undermined their actual and potential contributions to the welfare system. Austerity calls for community and volunteer groups to fill gaps in services, but their own resources also suffered under austerity (Milbourne, 2015).

The British exit from the European Union is a crucial challenge for health and social care services in the UK, regardless of rural/urban residence (Burdett and Fenge, 2018; Fahy et al., 2017; Shucksmith, 2019; Simpkin and Mossialos, 2017). Deleterious effects may include a labour deficit due to the loss of non-UK service workers. European Structural Investment (ESI) funding of transport infrastructure likely will disappear (Brien, 2018). Rising costs of medicines, shortages and delays in delivery due to problems with transport and border controls and staffing shortages in rural hospitals and clinics are likely scenarios (Aziz, 2019; Chu and Spring, 2019). Some scholars believe that the UK's central government has not sufficiently addressed predictable challenges, especially as they play out across the nation's multi-scalar geography.

Conclusions

National level changes in social policy can have differential effects across the spatial landscape, depending on variability in community capacity and the nature of multi-scalar relationships linking communities. Hence, "go it alone" systems of health and social care services are likely to produce spatial inequality wherein the rich get richer, and poor communities spiral downward. Communities with effective local government and other institutions will be better able to maintain services by mobilising resources internally and from elsewhere in the spatial system. For example, in the US, the ACA offered expansion of Medicaid funding to state governments. Many of the most rural, most conservative states refused the funding, which contributed to increased closures of rural hospitals. In the UK, local authorities manage Social Care Act programs, but national level withdrawal of funds will disproportionately disadvantage rural communities that have lower fiscal capacity to retain services.

We have examined health and social care services through a multi-scalar lens. As diagrammed in Figure 10.1, rural places should be able to provide a wider range of health and social care services when embedded in virtuous, win-win relationships with other communities at the same, or higher, level of a nation's settlement system. Multi-scalar relationships, however, are not always "virtuous" and can result in increased inequality and exclusion from health and social care services. Remote rural communities, in particular, lag in providing "age friendly" services for their elderly populations. (Warner, Xu and Morken, 2016). Even if they obtain their fair share of resources from the regional AAA, smaller places with limited tax bases, relatively low fiscal capacity and weaker governance institutions are often unable to produce adequate services for older residents in the US. Moreover, given the extreme degree of governance devolution in the US where the local (and private) share of the cost of health and social services have grown over time, poorer places are increasingly excluded from essential services.

While the UK and the US are similar in ways, they also differ. The UK has a national health service; the US lacks health care for all. The UK gives higher priority to the well-being of rural people and communities than does the US. Rural older persons in the UK have been less vulnerable compared to older rural residents in the US, although recent evidence suggests this may be changing. Regardless of differences, rural communities in both countries are too small to provide all health and social care options locally. Accordingly, access depends on effective collaboration among communities at different, frequently higher, spatial scales. Inter-local collaborations level differences between places, thereby reducing spatial inequality.

Note

1 The AAA is a gateway organization that helps seniors on the local level. AAA provides 19 different social care services including, but not limited to, adult day care, family caregiving support, supplemental nutrition, legal assistance, home energy assistance and residential repair (Revore, 2017).

References

Aziz, Z. (2019). GP practices are struggling: And Brexit will only make things worse. *The Guardian*, 26 March. Available at: https://www.theguardian.com/society/2019/mar/26/my-gp-practice-struggling-brexit-worse-staff-shortages

Brien, P. (2018). *UK Funding from the EU* (Briefing Paper. 7847). London: House of Commons Library.

Brown, D. L., Glasgow, N., Laszlo, K., Sanders, S. and Thiede, B. C. (2019). The multi-scalar organization of aging-related services in US rural places. *Journal of Rural Studies*, 68, pp. 219–229.

Burdett, T. and Fenge, L-A. (2018). Brexit: the impact on health and social care and the role of community nurses. *Journal of Community Nursing*, 32, pp. 62–65.

Chu, B. and Spring, M. (2019). Unprecedented drug shortage linked to Brexit. *BBC News*, 20 March. Available at: https://www.bbc.com/news/health-47646193

Commission for Rural Communities. (2007). *Working Age Benefit Claimants in Rural England, February 2000 to February 2006: State of the Countryside Update*, CRC WEB03. Cheltenham: Commission for Rural Communities.

Doebler, S. (2016). Access to a car and the self-reported health and mental health of people aged 65 and older in Northern Ireland. *Research on Aging*, 38(4), pp. 453–76.

Doheny, S. and Milbourne, P. (2012). Modernization and devolution: delivery services for older people in rural areas of England and Wales. *Social Policy and Administration*, 47(5), pp. 501–519.

Exworthy, M., Powell, M. and Glasby, J. (2017). The governance of integrated health and social care in England since 2010: great expectations not met once again? *Health Policy*, 121(11), pp. 1124–1130.

Fahy, N., Hervey, T., Greer, S., Jarman, H., Stuckler, D., Galsworthy, M. and McKee, M. (2017). How will Brexit affect health and health services in the UK? Evaluating three possible scenarios. *The Lancet*, 390(10107), pp. 2110–2118.

Flora, C. and Flora, J. (2003). Social capital. In: D. L Brown and L. Swanson, eds., *Challenges for Rural Communities in the Twenty First Century*. University Park: Pennsylvania State University, pp. 214–247.

Glasgow, N. and Berry, E. H., (eds.). (2013). *Rural Aging in 21st Century America*. Dordrecht: Springer.

Glasgow, N. and Brown, D. L. (2012). Rural aging in the United States: trends and context. *Journal of Rural Studies*, 28, pp. 422–431.

Glasgow, N., Wright Morten, L. and Johnson, N. E., (eds). (2004). *Critical Issues in Rural Health*. Iowa: Blackwell Publishing.

Goins, R. T. and Krout, J. A., (eds). (2006). *Service Delivery to Rural Older Adults: Research, Policy and Practice*. New York: Springer.

Grimshaw, D. and Rubery, J. (2012). The end of the UK's liberal collectivist social model? The Implications of the coalition government's policy during the austerity crisis. *Cambridge Journal of Economics*, 36(1), pp. 105–126.

Grigsby, W. and Goetz, S. J. (2004). Telehealth: What promise does it hold for rural areas? In: N. Glasgow, L. W. Morten and N. E. Johnson, eds., *Critical Issues in Rural Health*. Iowa: Blackwell Publishing, pp. 237–350.

Hanlon, N. and Kearns, R. (2016). Health and rural places. In: M. Shucksmith and D. L. Brown, eds., *Routledge International Handbook of Rural Studies*. London: Routledge, pp. 62–70.

Hudson, R. B. (2015). Politics and policies of aging in the United States. In: K. Ferraro and L. George, eds., *Handbook of Aging and the Social Sciences*, 8th ed. Cambridge, MA: Academic Press, pp. 441–459.

Hummer, R. A., Pacewicz, J., Wang, S. and Collins, C. (2004). Health insurance coverage in nonmetropolitan America. In: N. Glasgow, L. W. Morton and N. E. Johnson, eds., *Critical Issues in Rural Health*. Iowa: Blackwell Publishing, pp. 197–209.

Humphries, R. (2015). Integrated health and social care in England – progress and prospects. *Health Policy*, 119(7), pp. 856–859.

Humphries, R., Throlby, R., Holder, H., Hall, P. and Charles, A. (2016). *Social Care for Older People: Home truths*. London: The Kings Fund and the Nuffield Trust.

Jones, G. (1974). Intergovernmental relations in Britain. *The ANNALS of the American Academy of Political and Social Science*, 416, pp. 181–193.

Jones, M. (2009). Phase space: geography, relational thinking and beyond. *Progress in Human Geography*, 33, pp. 487–506.

Kaiser Family Foundation. (2016). *A Look at Rural Hospital Closures and Implications for Access to Care: Three Case Studies*. Available at: http://Kff.org/report-section/a-look-at-implications-for-access-to-care-three-case-studies-issue-brief/

Kaiser Family Foundation. (2019). *The Uninsured and the ACA: A Primer – Key Facts about Health Insurance and the Uninsured Amidst Changes to the Affordable Care Act: How Many Uninsured?* Available at: www.kff.org/uninsured/fact-sheet/key-facts-about-the-uninsured-population/

Keating, N., (ed.). (2008). *Rural Ageing: A Good Place to Grow Old*? Bristol: Policy Press.

Milbourne, P. (2015). Austerity, welfare reform and older people in rural places: competing discourses of voluntarism and community? In: M. Skinner and N. Hanlon, eds., *Ageing Resource Communities: New Frontiers of Rural Population Change, Community Development and Voluntarism*. London: Routledge, pp. 74–88.

National Collaboration for Integrated Care and Support. (2013). *Integrated Care and Support: Our Shared Commitment*. London: National Collaboration for Integrated Care and Support.

Pownall, H. (2013). Neoliberalism, austerity and the Health and Social Care Act 2014: the coalition government's programme for the NHS and its implications for the public sector workforce. *Industrial Law Journal*, 42(4), pp. 422–433.

Reeves, A., Basu, S., McKee, M., Marmot, M. and Stuckler, D. (2013). Austere or not? UK coalition government budgets and health inequalities. *Journal of the Royal Society of Medicine*, 106(11), pp. 432–436.

Revore, T. (2017). *19 Free Services for Seniors or Their Caregivers: Social Media Scoop for Seniors*. Available at: https://seniornet.org/blog/19-free-services-for-seniors-or-their-caregivers/

Shucksmith, M. (2009). Disintegrated rural development? Neo-endogenous rural development, planning and place-shaping in diffused power contexts. *Sociologia Ruralis*, 50(1), pp. 1–14.

Shucksmith, M. (2019). Rural policy after Brexit. *Contemporary Social Science*, 14(2), pp. 312–326.

Shucksmith, M. and Brown, D. L., (eds). (2016). *Routledge International Handbook of Rural Studies*. London: Routledge.

Simpkin, V. L. and Mossialos, E. (2017). Brexit and the NHS: challenges, uncertainties and opportunities. *Health Policy*, 121(5), pp. 447–480.

Taylor-Gooby, P. and Wallace, A. (2009). Public values and public trust: responses to welfare state reform in the UK. *Journal of Social Policy*, 38(3), pp. 401–419.

Thiede, B. C., Brown, D. L., Sanders, S. R., Glasgow, N. and Kulcsar, L. J. (2017). A demographic deficit? Local population aging and access to services in rural America, 1990–2010. *Rural Sociology*, 82(1), pp. 44–74.

Veeneman, W., Augustin, K., Enoch, M., Faivre d'Arcier, B., Malpezzi, S. and Wijmenga, N. (2015). Austerity in public transport in Europe: the influence of governance. *Research in Transportation Economics*, 51, pp. 31–39.

Vera-Toscano, E., Shucksmith, M. and Brown, D. L. (2020). Poverty dynamics in rural Britain 1991–2008: did labour's social policy reforms make a difference? *Journal of Rural Studies*, 75, pp. 216–228.

Walsh, K., O'Shea, E. and Scharf, T. (2019). Rural old age social exclusion: a conceptual framework on mediators of exclusion across the life course. *Ageing & Society*. Epub ahead of print 22 July. DOI: 10.1017/S0144686X19000606.

Ward, M., Ullrich, F. and Mueller, K. (2014). *Extent of Telehealth Use in Rural and Urban Hospitals*. RUPRI Center for Rural Health Policy Analysis. Ames: Iowa State University. Available at: https://europepmc.org/article/med25399469

Warner, M. E., Xu, T. and Morken, L. J. (2016). What explains differences in availability of community health-related services for seniors in the United States? *Journal of Aging and Health*, 29(7), pp. 1–22.

Warren, R., (ed.). (1963). *The Community in America*. Chicago, IL: Rand McNally.

Wicklund, E. (2020). *CARES Act Expands Telehealth Coverage for Medicare, FQHCs and the VA*. Available at: https://mhealthintelligence.comnews/cares-act-expamds-telehealth-coverage-for-medicare-fqhcs-and-the-va.

Wilson, R. (2020). Rural America braces for the Coronavirus. *The Hill*, 4 March. Available at: https://thehill.com/homenews/state-watch/491032-rural-america-braces-for-coronavirus

Yeginsu, C. (2019). 'This is all we can afford': shrinking lives in the English countryside. *New York Times*, 8 September. Available at www.nytimes.com/2019/05/13/world/europe/cumbria-uk-austerity-cuts.html

11 Rural ageing, housing and homelessness

Maree Petersen

Introduction

There is a lack of understanding of older people's precarious housing circumstances in rural communities around the world, in countries like Australia, which is the focus of this chapter. Older people living in precarious housing – housing that is unaffordable, insecure and unadaptable for their future needs are at risk of homelessness (Grenier et al., 2016; Morris, 2013; Petersen and Parsell, 2015). Alongside older people living homeless, this vulnerable group face extreme disadvantage, exacerbated by rurality.

Precarity as experienced by older people was historically overlooked in Australia's housing and homelessness policy and practice. Research alongside advocacy by welfare organisations has provided a growing understanding of the pathways to homelessness. Homelessness is arguably more visible in urban areas, and most research and service delivery focuses on the experiences of people sleeping rough, women and children escaping violence and the concerns of young people in Australia's largest cities (Coleman and Fopp, 2014; Darab, Hartman and Holdsworth, 2018). Despite this, increasing knowledge of older people's housing vulnerability highlights both the commonalties in rural and urban locales, as well as the distinctive issues in rural and remote parts of Australia (Petersen et al., 2014). However, the urban centric focus in Australia's housing policy results in a lack of knowledge of the experience of older people in rural communities. Australia is a vast country, with distinctive structural, distal, climatic and cultural impacts that result in place based differences associated with older people being at risk of homelessness. And yet, national issues such as housing unaffordability and tenancy laws resulting in no fault evictions equally impact on older people living in in rural areas (Petersen et al., 2014).

Housing in Australia is strongly orientated towards home ownership. Older people traditionally had high rates of home ownership, with higher proportions in rural Australia (Australian Bureau of Statistics, 2019). A similar pattern exists for rural seniors in the UK (Wilson, 2016). However, there is a lack of attention to patterns of home ownerships among older rural residents in many countries where owner occupation is high. The high rates of home ownership in Australia cloud the living circumstances and needs of increasing numbers of older people living in precarious housing circumstances and at risk of homelessness. There is a major transformation occurring in Australia's housing system with growing numbers of indebted middle aged and older Australians falling out of home ownership. Private rental and social housing is unable to supply secure, affordable housing to meet the needs of people on low incomes (Morris, 2016; Wood and Ong, 2017). The seriousness of this transformation is compounded by the structural disadvantage many rural older Australians experience. The social determinants of disadvantage evident in income, education, employment and access to services is more profound for rural Australians (Australian Institute of Health and Welfare, 2018a). The population in Australian rural communities is ageing. Further, the higher proportion of older people in rural communities is accentuated with older residents being less likely to move, the in-migration of retirees and out-migration of young people seeking education and employment (Feist, 2016; Winterton and Warburton, 2011). The disadvantage many older Australians have experienced over their life course in rural communities can accumulate into precarious housing (Petersen and Parsell, 2015).

The following discussion focuses on the nature of older people's homelessness and critically examines the focus on homeless counts and how this obscures the wider issues of housing precarity. International research is reviewed showing the distinctive pathways into later life homelessness with particular attention to the experiences of older people in rural locales. The chapter concludes by examining the nature and adequacy of policies and services for rural older people and concludes that a lack of policy integration hampers prevention.

Defining homelessness and marginal housing in Australia

A range of homelessness definitions exist across the Western world (Minnery and Greenhalgh, 2007), and are linked to differing purposes including census enumeration, legislation and policy guidelines that shape service provision. Often understood as a person sleeping rough or staying in a shelter, homelessness is wider than this literal interpretation in many western countries. The definition known as the European Typology of Homelessness and Housing Exclusion asks "what is it to have a home?" and considers physical, social and legal domains (Edgar and Meert, 2006). In Australia, the definition of homelessness has changed over time and remains contested given it is used to assess eligibility for a service and in census counts. This results in different estimates of the numbers of people living homeless.

Homelessness is a relative concept that is meaningful in relation to the housing conventions of a particular culture (Chamberlain and MacKenzie, 2014). With high rates of home ownership, Australia's definition is linked to what is considered a minimum community standard of housing namely a small, self-contained unit with a kitchenette and bathroom. Service providers utilise the guideline of "inadequate access to safe and secure housing" which includes people living in low quality housing, temporary accommodation or a severely overcrowded house as homeless (Chamberlain and MacKenzie, 2014). Criticisms centre on the classification of homelessness being considered separately from a person's view of their situation. In addition, the creation of a dichotomy between homeless and therefore not homeless arguably results in a lack of focus on the large number of older people living insecurely in private rented housing (Jones and Petersen, 2014).

A dichotomy is less useful than viewing homelessness as a continuum involving degrees of adequacy, security and control over housing (Jones and Petersen, 2014). As such being at risk of homelessness and living in marginal housing are important considerations. Many older people living in the private rental market, with very limited housing security, or unaffordable rents are at risk of homelessness. Significant numbers of older low income private renters fear eviction and unaffordable rent increases (Morris, 2016; Petersen and Parsell, 2015) and live in areas where there is an undersupply of affordable rental housing (Stone, Reynolds and Hulse, 2013). In Australia, tenants can be evicted once the lease expires despite no breach. This lack of security in Australian rental tenure is in contrast to many international countries and results in increasing numbers of older people living with the risk of homelessness (Petersen and Parsell, 2015). Marginally housed older people including those living in rundown boarding houses with limited facilities and poorly maintained caravans are living circumstances to bring to considerations of homelessness. Alongside the inclusion of being at risk of homelessness due to unaffordable rent or being marginally housed it is important the viewpoints of older people are considered. Older people in rural and remote Australia commonly live in houses similar to a shed with a camp like kitchen, in a place they are strongly attached to, and do not consider themselves homeless. Further, Aboriginal families living in multigenerational homes on country, where an older person is the tenant holder, differ in their perception of overcrowding (Memmott and Nash, 2014).

Causes of homelessness and housing precarity

Discussions of the causes of homelessness dominate media stories, scholarship and policy responses; yet cause is a difficult and contested matter. A substantive focus in research is the identification of individual factors that place older people at risk of homelessness. Up until the late 1970s, many researchers and policy makers relied on individual factors to explain the causes of homelessness (Johnson and Jacobs, 2014). However, it is important to acknowledge that a housing loss sits in a complex interconnection between individual influences such as trigger events, risk factors, cumulative life events and changing structural conditions over time. An example is a major health event may not be a trigger for an older adult if affordable and accessible social housing is available. Poverty and inequality are inextricably linked to homelessness (Grenier et al., 2016). There has been increasing attention to the interaction between structural factors and the individual processes which result in people vulnerable to precarity and homelessness. Therefore, it is important to be clear of what presents as individual factors leading to homelessness and underlying structural explanations.

Individual explanations

The individual pathway into homelessness for older people include a loss of a spouse through divorce or death, a relation who shared expenses and provided support; domestic violence; family breakdown; and loss of employment (Crane et al., 2005; Gonyea et al., 2010; McFerran, 2010). International evidence highlights the first time homeless in later life includes more women, people with high levels of education, a work history and better physical and mental health (McDonald et al., 2007, Shinn et al., 2007). Poor mental and physical health, precarious work history, involvement with corrections and traumatic life changes are associated with people who have lived precariously for a long period to time (Crane et al., 2005; Petersen and Parsell, 2015). Empirical research demonstrates that older people who are homeless or at risk of homelessness have estranged or conflicted relationships with family (Grenier et al., 2016).

Structural explanations

There is broad consensus that structural factors, in particular a shortage of affordable housing, are the fundamental drivers of homelessness, with personal issues increasing a person's vulnerability to structural factors and therefore homelessness (Fitzpatrick, 2005). In countries with relatively robust social housing sectors there is little evidence of older people's homelessness (Petersen, 2015).

There are a number of structural factors commonly linked to older people being at risk of homelessness or living homeless. First, home ownership in Australia is declining. Home ownership, once seen as the primary asset that kept low income people out of poverty, is now seen as a "crumbling pillar" of social insurance in Australia (Yates and Bradbury, 2010). The proportion of older Australians who own their own home outright is decreasing and increasing numbers of low income older homeowners experience housing stress as a consequence of mortgage payments (Ong et al., 2019). Couples who separate in later life and relinquish home ownership are often unable to re-enter the housing market as a single person (Morris, 2013).

Second, Australia's aged pension assumes most Australians will retire as outright home owners with no mortgage payments or rental payments (Yates, 2015). With unaffordable rents and a fixed low income of the aged pension, older renters experience extreme disadvantage with little or no disposable income. Further, more than 70% of older renter householders are single adults, and of these most are women (Yates, 2015). For most older people in Australia renting is not a choice but a necessity (Beer et al., 2011), and those reliant on the aged pension almost never have the financial means to change their housing circumstances from private rental to home ownership. In the context of demographic change in Australia the social and policy implications of the weakening of home ownership as social insurance are evident in the short and long term (Yates, 2015).

Table 11.1 Rent for single aged pension households in rural and regional Australia

	Rent as a share of income	*Relative unaffordability*
Rest of NSW	38%	Severely unaffordable
Rest of Victoria	36%	Unaffordable
Rest of Queensland	53%	Severely unaffordable
Rest of South Australia	29%	Moderately unaffordable
Rest of West Australia *	63%	Extremely unaffordable
Rest of Tasmania	31%	Unaffordable

Source: SGS Economics and Planning (2018)

Note: RAI scores for Quarter 2 2018
* Due to unavailability of data, median rents for all dwellings rather than one bedroom is used

Third, there is a shortage of affordable, accessible housing. Across all regional areas in Australia, rents for single people on the aged pension are measured as unaffordable, that is households are paying 30% or more of their income on rent (see Table 11.1). While rural and regional private rents are marginally more affordable than Australian cities, market rates vary greatly due to differing economic conditions and housing markets.

The vast geography of Australia compounds the structural forces that impact on homelessness, namely housing affordability, employment and housing policy (Wood et al., 2015). Australia's decade long mining boom, 2002 to 2012, profoundly affected housing affordability due to large population increases in rural and remote towns (Haslam McKenzie and Rowley, 2013; Warren, McDonald and McAufille, 2017). Private rental housing during this period was unaffordable for residents on a low income, including the aged pension. In contrast, there are rural areas with less expensive rents, and older people migrate to rural towns and acreage in former farming areas for less expensive rental housing.

Last, the residual nature of the social housing sector further limits an individual's housing options. Access to social housing varies in rural areas, and in some cases the only properties available are for families, with no single accommodation available (Atkinson, 2015). Social housing will only be a small part of the solution for vulnerable older people as it is considered a residual form of housing provision in Australia and the stock has hardly increased over the last decade (Wood and Ong, 2017).

Thus, most older non-homeowners have to rely on the private rental sector for accommodation (Morris, 2016). Despite a history of working, the burden of high private rental payments, results in accumulated disadvantage and precarity in later life. How renting can be sustained for older people reliant on the aged pension is uncertain, given private rental tenancy is detrimental to health and well-being (Morris, 2016; Petersen et al., 2014; Wood et al., 2015). Faced with eviction or unaffordable rents on fixed low income of the pension, older people commonly experience extreme stress and health consequences (Morris, 2016; Petersen and Parsell, 2015).

Pathways into housing precarity and homelessness

The use of a housing pathways framework, with attention to the transitions and trajectories over the life course, provides an understanding of the processes leading into and out of homelessness (Chamberlain and Johnson, 2011). Further, the pathway approach enables consideration of locale and thereby accounts for similarities and differences between the experience of homelessness in urban, rural and remote areas.

Understanding the pathways into homelessness for older people in rural Australia is enhanced by national studies encompassing both urban, rural and remote locales (Petersen et al., 2014; Darab, Hartman and Holdsworth, 2018; Warren, McDonald and McAufille, 2017). International research has consistently highlighted two pathways to later life homelessness: people experiencing homelessness for the first time at an older age and people ageing with a long history of homelessness (Crane et al., 2005; McDonald et al., 2007; Shinn et al., 2007). Australian studies reinforce these findings with first time homelessness in later years being prominent in rural areas. There are also other pathways linked to the nature of work in rural areas (Petersen et al., 2014) and the history of colonisation continuing to have direct and indirect impacts on the lives of Aboriginal and Torres Strait Islander people (Memmott and Nash, 2014: 155).

A national Australian study collaborating with key service organisations included 25% of participants (144 of a sample of 561) from regional and rural areas in the first stage of the study and nine of the 20 in-depth interviews with rural and remote service providers in the second stage (Petersen et al., 2014). The three pathways identified in this study are set in the following section, highlighting the distinctive rural issues. The inclusion of urban, rural and remote areas of Australia not only provides an understanding of different geographies and their respective structural contexts but also demonstrates the interplay of gender and culture.

The first pathway, just under 70% of participants, comprised people who throughout their life course had "conventional" links to housing and were experiencing housing disruption for the first time in their lives. Most people in this group had rented privately, some had secured home ownership or a mortgage in the past; all found themselves in a housing crisis for a variety of reasons late in life. They faced unaffordable rents, eviction notices, inaccessible housing that did not permit functional changes, or were unable to continue to live with family. The unsustainability of remaining to live with family included situations of carer stress, a lack of resources and elder abuse. Elder abuse is an important consideration resulting in older people being at risk of homelessness as a house (including a mortgaged house) is a major asset and adult children seek to attain ownership fraudulently.

Interviews with rural service providers highlighted the critical housing incidents where frail older people could not remain in their rented housing including caravans due to outside amenities. There are very limited housing options available for older people who have to move given unaffordable private rents and the residual nature of social housing. Unable to secure alternative housing, rural workers account that emergency accommodation in a motel is their only option, albeit a short term one. There were further examples, of "grey nomads", people who sold their home and travel in retirement. However, with the onset of a health crisis, unable to continue living in a van, few resources, limited affordable housing and in a rural area with no support networks, a risk of homelessness ensues.

Of the group with a conventional housing history just over a half were older women at risk of homelessness. The women had raised a family and worked, and some were previous home owners. In recent years, there is an increase in older Australian women living in cars and parking in national parks and cemeteries, housesitting and staying in temporary insecure accommodation in rural areas (Darab, Hartman and Holdsworth, 2018; Petersen, 2015).

The conventional housing pathway also includes older Aboriginal people who had lived long term in rented housing on country and were forced to seek alternate accommodation due to poor health, frailty, family violence or the stresses of living in multi-generational crowded housing. There were examples of families unable to continue to care for older adult due to overcrowding and carer stress. The issue of overcrowding within Aboriginal families requires specialised consideration and acknowledgement of the continuing effects of colonisation on Aboriginal and Torres Strait people. It is put forward that crowding should not continue to be conceptualised as simply high-density, nor assume that stress and annoyance will automatically arise in high-density situations (Memmott and Nash, 2014). The need to reconceptualise overcrowding was confirmed by interviews with service providers, accounting the connection to family and land is paramount for older Aboriginal people.

The second pathway involved people who had throughout their working lives been transient. Distinctive life patterns were evident for this group of older people in rural and remote areas of Australia. Many in this group worked in different places as farm workers and employees on communities often with temporary housing supplied and as such had a tenuous link to a "home" and community. Both men and women had worked as farm hands, shearers or in clerical roles. Again, health needs are commonly a trigger for needing stable accommodation, as they were unable to continue working. The lack of affordable housing results in this transient group unable to secure accommodation. Aboriginal people in this group travelled to meet kinship responsibilities and cultural obligations as well as following seasonal work or relocating for access to health resources.

The third broad pathway comprised people who lived with ongoing housing disruption and included individuals residing in marginal housing and people sleeping rough along riverbanks and in bush settings. Marginally housed older people commonly live in inadequate or inappropriate dwellings, including sheds, old farm houses and makeshift caravan parks near rural sporting ovals (Atkinson,

2015; Petersen et al., 2014). This group included Aboriginal people experiencing high levels of social exclusion and dislocation from kin and home communities associated with colonisation, poverty, poor access to housing and mainstream services, mental illness or chronic health problems requiring regular hospitalisation.

Policy responses to older people's housing needs

The recognition of older people in housing and homelessness policy and service provision is arguably less than other cohorts. The assertion that services are underdeveloped, with little attention to prevention is stronger for rural people living with homelessness. Older people are underrepresented as clients of homelessness services, and with far fewer services in rural Australia there are serious issues with accessing assistance. Mainstream homelessness services are less likely to attend to the specialist needs of older people (Jones and Petersen, 2014). Further, older people experiencing homelessness for the first time in later life have limited experience with welfare agencies. They view their circumstances as a housing difficulty, do not identify as homeless and are unlikely to access homelessness services (Petersen and Parsell, 2015). In Australia, a specialist outreach service for older homeless people, Assistance with Care and Housing (ACH), is part of the policy portfolio responsible for community aged care. This small program is considered very effective with flexible one to one support and thereby prevents premature entry to residential aged care (Petersen et al., 2014). However, there are important gaps in ACH services in regional and rural areas (Fielder and Faulkner, 2019). West Australia, Australia's largest state, has only four ACH services, one covering a vast remote area. Further, the capacity of all homelessness agencies to assist older people at risk of homelessness is dependent of the availability of affordable housing, service-integrated housing and residential aged care (Jones and Petersen, 2014).

With the number and proportion of older Australians in rural areas increasing and many requiring formal support (Australian Institute of Health and Welfare, 2018b), it is critical to understand the role of housing in the implementation of community aged care. Housing is an essential platform for healthy ageing, and yet the achievement of housing security alongside appropriate community aged care for older people is poorly understood (Petersen and Parsell, 2020). The disjuncture that exists between aged care, housing and homelessness sectors has serious consequences for older people at risk. Housing is paramount for ageing at home, yet housing circumstances are not assessed as part of community aged care. Living in private rental housing, at risk of homelessness has serious consequences for the effective implementation of aged care policy. In the past, older Australians benefited from well-designed social housing, but now as a residual form of housing with less stock in rural areas, there is limited scope for providing ageing in place. For rural older residents, the impact is worse given the challenges in providing community care in sparsely populated areas (see Glasgow and Doebler, Chapter 10 and Urbaniak, Chapter 18 in this volume).

There are strengths in the service sector, with examples of best practice models, including service-integrated housing in rural locales for people who have lived homeless and culturally appropriate share housing for older Aboriginal women living in the country (Petersen, 2015). Further, the flexibility of the ACH model enabling workers to tailor services to local need is considered effective (Petersen et al., 2014). However, service provision in rural Australia is variable and scant (Jones and Petersen, 2014). Further, the sectors of housing, community aged care, health, social security (responsible for the aged pension), law (responsible for tenancy regulations) and Aboriginal and Torres Strait Islander Affairs are not integrated. All are vital for the housing security of older people. Further, there is also little emphasis on prevention. Extension of the ACH outreach model, alongside specialist gateway services that link older people to services, combined with increases in affordable housing and service-integrated housing would likely reduce the prevalence of precarious housing and homelessness among older people in rural Australia.

Conclusion

Housing, not just shelter, but as home, is not available to increasing numbers of older people living in rural Australia. Without a home they are unable to have the routine and security to live their lives or manage their health and well-being. The concept of home comes into sharpest relief in the context of precarious housing and homelessness for older people living in rural communities. These issues remain largely hidden nature in Australia and internationally and bought into sharper focus given the economic and social disadvantage associated with rural life across many countries (see Milbourne, Chapter 25). There is much to be done to bring older people living in poverty, in precarious housing, into policy architecture and service provision.

The security a home brings, alongside the ability to manage health and age in place, is in contrast to the healthy ageing policy (see Hanlon and Poulin, Chapter 4). Aged care and health policy are important areas of policy given the ageing of the population. Yet, for older people living precariously, ageing at home is characterised by insecurity, anxiety and for some fear. The aim of policy to have older people manage their well-being and independence at home is increasingly out of reach for a considerable number of older people living in precarious housing in rural communities.

References

Atkinson, T. (2015). Finding the elderly homeless in regional Victoria. *Parity*, 28(2), pp. 27–28.

Australian Institute of Health and Welfare. (2018a). *Housing Assistance in Australia 2018* (Cat. no.HOU 296). Available at: www.aihw.gov.au/reports/housing-assistance/housing-assistance-in-australia-2018/contents/housing-in-australia

Australian Institute of Health and Welfare. (2018b). *Older Australians at a Glance* (Cat. No. AGE 87). Available at: www.aihw.gov.au/reports/older-people/older-australia-at-a-glance/contents/summary [Accessed 15 August 2019].

Australian Bureau of Statistics (ABS). (2019). *Housing Occupancy and Costs, 2017–2018* (No. 4130). Canberra, ACT: Australian Bureau of Statistics.

Beer, A., Tually, S., Rowland, S., Haslam McKenzie, F., Schlapp, J., Birdsall-Jones, C. and Corunna, V. (2011). *The Drivers of Housing Supply and Demand in Australia's Rural and Regional Centres* (Report no. 165). Adelaide, SA: Australian Housing and Urban Research Institute.

Chamberlain, C. and Johnson, G. (2011). Pathways into adult homelessness. *Journal of Sociology*, 49(1), pp. 60–77.

Chamberlain, C. and MacKenzie, D. (2014). Definition and counting: Where to now? In: C. Chamberlain, G. Johnson and C. Robinson, eds., *Homelessness in Australia*. Sydney, NSW: UNSW Press, pp. 71–99.

Coleman, A. and Fopp, R. (2014). Homelessness policy: benign neglect or regulation and control? In: C. Chamberlain, G. Johnson and C. Robinson, eds., *Homelessness in Australia*. Sydney, NSW: UNSW Press, pp. 11–29.

Crane, M., Byrne, K., Fu, R., Lipmann, B., Mirabelli, F., Rota-Bartelink, A., Ryan, M., Shea, R., Watt, H. and Warnes, A. (2005). The causes of homelessness in later life: findings from a 3-nations study. *Journal of Gerontology: Social Sciences*, 60B(3), pp. 152–159.

Darab, S., Hartman, Y. and Holdsworth, L. (2018). What women want: single older women and their housing preferences. *Housing Studies*, 33(4), pp. 525–543.

Edgar, B. and Meert, H. (2006). *Fifth Review of Statistics on Homelessness in Europe*. Brussels: European Federation of National Organisations Working with the Homeless (FEANTSA).

Feist, H. (2016). *Ageing in Rural Australia* (Australian Ageing Agenda no. Jul/Aug, 10).

Fitzpatrick, S. (2005). Explaining homelessness: a critical realist perspective. *Housing, Theory & Society*, 22(1), pp. 1–17.

Fielder, J. and Faulkner, D. (2019). *'One Rent Increase from Disaster': Older Renters Living on the Edge in Western Australia*. Available at: www.oldertenants.org.au/sites/default/files/ageing_on_the_edge_wa_report_a40819.pdf [Accessed 20 August 2019].

Gonyea, J. G., Mills-Dick, K. and Bachman, S. (2010). The complexities of elder homelessness, a shifting political landscape and emerging community responses. *Journal of Gerontological Social Work*, 53(7), pp. 575–590.

Grenier, A., Barken, R., Sussman, T., Rothwell, D., Bourgeois-Guerin, V. and Lavoie, J-P. (2016). A literature review of homelessness and aging: suggestions for a policy and practice-relevant research agenda. *Canadian Journal on Aging*, 35(1), pp. 28–41.

Haslam McKenzie, F. M. and Rowley, S. (2013). Housing market failure in a booming economy. *Housing Studies*, 28 (3), pp. 373–388.

Johnson, G. and Jacobs, K. (2014). Theorising cause. In: C. Chamberlain, G. Johnson and C. Robinson, eds., *Homelessness in Australia*. Sydney, NSW: UNSW Press, pp. 30–47.

Jones, A. and Petersen, M. (2014). Older people. In: C. Chamberlain, G. Johnson and C. Robinson, eds., *Homelessness in Australia*. Sydney, MSW: UNSW Press, pp. 135–154.

McDonald, L., Dergal, J., Cleghorn, L. (2007). Living on the margins: older homeless adults in Toronto. *Journal of Gerontological Social Work*, 49(1–2), pp. 19–46.

McFerran, L. (2010). *It Could be You: Female, Single, Older and Homeless*. Woolloomooloo, NSW: Homelessness NSW.

Memmott, P. and Nash, D. (2014). Indigenous homelessness. In: C. Chamberlain, G. Johnson and C. Robinson, eds., *Homelessness in Australia*. Sydney, NSW: UNSW Press, pp. 155–178.

Minnery, J. and Greenhalgh, E. (2007). Approaches to homelessness policy in Europe, the United States, and Australia. *Journal of Social Issues*, 63(3), pp. 641–656.

Morris, A. (2013). The trajectory towards marginality: how do older Australians find themselves dependent on the private rental market? *Social Policy and Society*, 12(1), pp. 47–59.

Morris, A. (2016). *The Australian Dream: Housing Experiences of Older Australians*. Melbourne, VIC: CSIRO Publishing.

Ong, R., Wood, G., Cigdem-Bayram, M. and Salazar, S. (2019). *Mortgage Stress and Precarious Home Ownership: Implications for Older Australians* (Report no. 319). Melbourne, VIC: Australian Housing and Urban Research Institute.

Petersen, M. (2015). Addressing older women's homelessness: service and housing models. *Australian Journal of Social Issues*, 50(4), pp. 419–438.

Petersen, M. and Parsell, C. (2015). Homeless for the first time in later life: an Australian study. *Housing Studies*, 30(3), pp. 368–391.

Petersen, M. and Parsell, C. (2020). The family relationships of older Australians at risk of homelessness, *British Journal of Social Work*, bcaa007.

Petersen M., Parsell, C., Phillips, R. and White, G. (2014). *Preventing first time homelessness amongst older Australians*, Report no.222. Australian Housing and Urban Research Institute, Melbourne, VIC.

SGS Economic and Planning. (2018). *Rental Affordability Index*. Available at: www.sgsep.com.au/application/files/8015/4336/9561/RAI_Nov_2018_-_high_quality.pdf [Accessed 10 August 2019].

Stone, W., Reynolds, M. and Hulse, K. (2013). *Housing and Social Inclusion: A Household and Local Area Analysis*. Melbourne, VIC: Australian Housing and Urban Research Institute.

Shinn, M., Gottlieb, J., Wett, J. L., Bahl, A., Cohen, A. and Ellis, D. (2007). Predictors of homelessness among older adults in New York City, *Journal of Health Psychology*, 12(5), pp. 696–708.

Warren, S., McDonald, D. and McAufille, D. (2017). Homelessness in rural and regional Queensland mining communities, *Parity*, 30(9), pp. 41–42.

Wilson, B. (2016). *Older People in Rural Areas: Income and Poverty Paper*. Rural England. Available at: https://ruralengland.org/wp-content/uploads/2016/04/Final-report-Income-and-Poverty.pdf

Winterton, R. and Warburton, J. (2011). Does place matter? Reviewing the experience of disadvantage for older people in rural Australia. *Rural Society*, 20(2), pp. 187–197.

Wood, G., Batterham, D., Cigdem, M. and Mallett, S. (2015). *The structural drivers of homelessness in Australia 2001–11,* Report no.238. Australian Housing and Urban Research Institute, Melbourne, VIC.

Wood, G. and Ong, R. (2017). The Australian housing system: a quiet revolution, *The Australian Economic Review*, 50(2), pp. 197–204.

Yates, J. (2015). Living income and asset poor in retirement. In: CEDA, ed., *The Super Challenge of Retirement Income Policy*. Melbourne, VIC: Committee for Economic Development of Australia, pp. 65–81.

Yates, J. and Bradbury, B. (2010). Home ownership as a (crumbling) fourth pillar of social insurance in Australia, *Journal of Housing and the Built Environment*, 25(2), pp. 193–211.

12 Rural ageing and transportation

How a lack of transportation options can leave older rural populations stranded

Stine Hansen, K. Bruce Newbold, Darren M. Scott, Brenda Vrkljan, Amanda Grenier and Kai Huang

Introduction

Declining fertility levels, increasing life expectancy and rapid urbanisation have been some of the dominant demographic processes shaping the global population in recent decades. Population ageing, and particularly ageing in the rural context, has emerged as challenges for policy makers as shifts in the age profile of residents require different needs in terms of services, including transportation. For the ageing rural population, reliance on the personal automobile, as opposed to other mobility options such as public transportation, is greater (Collia, Sharp and Giesbrecht, 2003; Hjorthol, Levin and Sirén, 2010; Newbold et al., 2005; Rosenbloom, 2001). Research has also demonstrated older adults are less likely to use public transit than younger generations, even when it is available (Newbold and Scott, 2018; Scott et al., 2009). Indeed, driving remains the most common and preferred mode of transportation among older adults (Kim and Ulfarsson, 2004; Páez et al., 2007; Scott et al., 2009), enabling shopping, commuting, completing daily tasks and socialisation (Dahan-Oliel et al., 2010; see also Hennessy and Innes, Chapter 17 in this volume). In rural areas and small towns, however, public transit options are limited (or non-existent), irregular and/or too far to easily reach, particularly in poor weather conditions (Clarke et al., 2015; Pucher and Renee, 2005). Additionally, active transportation options may not be feasible in rural areas given greater distances between locations and limited infrastructure (i.e., sidewalks or bike lanes), reinforcing the reliance on the personal automobile.

In recent years, a number of different models have been developed to look at areas that lack transportation options, including the notion of "transit deserts". Similar to the well-established concept of "food deserts", transit deserts are defined as areas that contain a large transit-dependent population who have limited automobile access but live in locations where transit does not meet the demand (Jiao and Dillivan, 2013; Jiao, 2017). Extending the concept of transit deserts to alternate locations, including rural areas and small towns with limited transportation options, can be problematic. In the case of rural areas, the critical difference is that the population is comparably smaller and spread across a larger geographic area as compared to the urban-oriented cases of transit deserts described by Jiao and colleagues. Nonetheless, the need for transportation options beyond the personal vehicle in rural areas is critical, even as the reliance on this form of transportation for daily trips and commuting to work continues to grow due, in part, to the lack of viable alternatives.

It may be more appropriate to think about how transportation decisions and choices in rural and small urban locations are shaped in terms of access to and availability of mobility options, or what Scott and Axhausen (2006) refer to as "mobility tools". Such options (tools) can include access to personal vehicles, public transit, active transit and ride-sharing options. The lack of mobility options is especially limiting among those populations who are dependent on "public transit" (i.e., "captive riders") including the young, old, low income and/or or those who are unable or choose not to drive (Grengs, 2001). The negative impacts of limited transit options may be further magnified among older adults with limited financial resources, family/friends, or health issues.

With limited or no public transit and few active transport options, residents of small towns and rural areas often lack a diversity of mobility options, constraining transportation choices, particularly for older adults. Given these constraints, this chapter examines the concept and use of mobility options as it applies to older people living in rural and small towns.

Methods

With approximately 17% of Canada's population residing in rural areas or small towns (Statistics Canada, 2016), Canada's rural population is older and ageing faster than their urban counterparts (Dandy and Bollman, 2008). The results shared in this chapter summarize qualitative interview data that examined the driving and transportation experiences of older rural adults, with the current analysis set among the small towns and rural areas surrounding the city of Hamilton, Ontario. Located approximately 60 kilometres (km) west of Toronto, Hamilton had a 2016 population in excess of 530,000. The city has a well-established public transit system, along with GO Transit, which is a regional public transportation service that is comprised of both bus and rail options.

Participants were recruited from small towns or rural areas proximate to Hamilton. Two of the communities are defined by Statistics Canada (2017) as "small population centers" with 2016 populations of 1,869 (14.7% aged 65+) and 5,489 (16.6% aged 65+), respectively. The other two communities from which participants were recruited had populations less than 1,000 and are defined by Statistics Canada as "rural". All communities were at least 20km from Hamilton, with the furthest community approximately 30km from Hamilton's city centre. Most of these communities had no access to any form of public transportation despite being part of the larger city/region that provides public transit. Some areas previously had services, for example, buses and trains, but lost these services over time as the personal car dominated travel choices and as cutbacks to public transit were made by service providers and funding agencies. In many of these places, populations are growing rapidly due to more affordable housing prices in rural areas; however, the influx of people has not translated into renewed investments in publicly available transportation services.

Recruitment was conducted during the first 6 months of 2018. Purposeful sampling was utilised to recruit drivers and non-drivers, aged 65+, living in small towns or rural areas in proximity to Hamilton. Several strategies were utilised to ensure people with different driving statuses were recruited, including drivers and non-drivers. Participants were recruited through email distribution lists for older adults living in the study area as well as through the use of posters that advertised the need for volunteers for the study. Recruitment material was posted at grocery stores and at rural Hamilton Public Library branches, and snowball sampling was utilised to recruit additional participants. Snowball sampling proved to be the most successful recruitment method, as many of the participants either had large social circles or were active in their communities. All respondents were screened for age (i.e., aged 65+) and confirmation of their rural residency.

In total, 17 semi-structured interviews were completed, with some people interviewed alone and others as couples, bringing the total number of participants to 23 (see Table 12.1). The average age was 76 years. Most were females (n = 17) and most participants were current drivers (n = 19), with four non-drivers. Several

Table 12.1 Characteristics of driver and non-driver research participants

Participant	*Gender*	*Age*	*Driving status*	*Health*
1	Female	86	Non-driver	Good
2	Female	83	Driver	Very good
3	Female	74	Driver	Excellent
4A	Female	72	Driver	Excellent
4B	Male	74	Driver	Very good
5	Female	70	Driver	Excellent
6A	Female	80	Driver	Excellent
6B	Male	83	Non-driver	Fair
7	Female	75	Driver	Good
8	Female	77	Driver	Excellent
9A	Female	71	Driver	Very good
9B	Male	74	Driver	Good
10	Female	77	Driver	Excellent
11A	Female	70	Driver	Very good
11B	Male	68	Driver	Good
12A	Female	75	Driver	Very good
12B	Male	78	Driver	Good
13	Female	76	Driver	Excellent
14	Female	88	Non-driver	Very good
15	Female	81	Non-driver	Very good
16	Female	68	Driver	Very good
17A	Female	72	Driver	Good
17B	Male	73	Driver	Good

participants self-regulated their driving behaviour by limiting their driving at night, in severe weather or on limited access highways.

The semi-structured interview guide included questions that explored participants' modes of transportation, transportation habits, "typical" daily transportation needs, as well as questions about their health, including diagnosis with chronic diseases and their mental health. If participants indicated they were non-drivers (former drivers), they were asked about their reasons for driving cessation, their current transportation options and whether or not they had planned ahead for driving cessation. Current drivers were asked about their current driving behaviours and whether they were planning for the day that they could no longer drive. The duration of the interviews ranged from 30 to 90 minutes and were audio-recorded with the consent of each participant. Hand-written notes were also taken to supplement the audio recordings, including notes on emotional states of participants. The study received ethics clearance from the authors' home institution.

Interviews were analysed and coded using thematic analysis (Braun and Clarke, 2006). Coding was done by the lead author and reviewed by other team members. The notes were then arranged into themes that emerged from the analysis. Given our interest in exploring mobility options in rural communities, the discussion for this chapter is as follows: (1) the reliance on the personal automobile as the primary form of transportation, and (2) the lack of mobility alternatives, including public and active transit options.

Results

The personal automobile as the primary mobility tool

Mobility options (or tools) include the personal automobile, public transit, ride-sharing and active transit options such as walking and biking. However, the personal car was the primary means of transportation for older adults living in rural areas and small towns, with respondents identifying the lack of public transport and their consequent reliance on the personal automobile for out-of-home activities:

> There is no bus here. Nothing. If you live here, you have to have a car, you have to be able to drive.
>
> (Participant 10, female driver aged 77)

> Everything here revolves around driving, really . . . there is no other option. Out here we have nothing. Driving is crucial if you live in a small community that is outside the city.
>
> (Participant 9B, male driver aged 74)

> You can't live in [community 1], you can't live in [community 2 or community 3] without a car. It's not like you can walk down to town.
>
> (Participant 4A, female driver aged 72)

In most cases, respondents relied heavily on their cars for daily transportation needs, including shopping, doctor's appointments, meeting friends, entertainment, volunteering and church. For most, there were few other transportation options.

Absence of Alternative Mobility Options

Living in rural areas or small towns typically means that the personal automobile was the primary mobility tool available to respondents. Beyond the personal automobile, other mobility options were usually not available, not an option, or not used. Although a few respondents had limited access to bus service, public transit was not an option for most participants. Even in the community of Waterdown (which is a part Hamilton with a 2016 population of approximately 20,000, and thus a "small population center") which has some bus service, the lack of mobility options was still problematic. While offering the opportunity for travel to destinations within the community, bus stops were noted as being too far from one's place of residence. Other participants found it equally difficult to access bus services.

> The other thing is too, it's quite a tiny stretch, it would take me probably 20 minutes to walk up to the highway to get the bus and they do stop if you put your hand out. There is no specific bus stop up there or anything. They just come along the highway and if they see somebody waiting and they wave them down, then they stop for them.
>
> (Participant 8, female driver aged 77)

> And if I was walking out to the highway to catch a bus it would take me probably more than half an hour to walk out.
>
> (Participant 10, female driver aged 77)

Furthermore, there were poor connections by transit to other communities, reinforcing the need to rely on their car, with another participant describing the situation where they lived:

> The closest we get [to public transportation] is to the bottom of the steep hill to [community]. . . . It is hard if you are a kid in [respondent's community] because your parents are probably not going to drive you down to get the bus . . . there should be a bus because lots of people, it's getting bigger all the time just like every other little outlying area but what can you do?
>
> (Participant 8, female driver aged 77)

Participant 8's response also recognised that it was not just the old that were overlooked in terms of other transportation options, but that younger populations also faced a lack of transit options in these rural locations, a theme that was picked up by other respondents:

> For a young person living out here you can't get to a job, if you have an after school activity you have to be picked up or drive yourself, like from high school.
>
> (Participant 4A, female driver aged 72)

Many participants felt that they were so close, yet so far from any needed services, such as banking or the post office, with the personal car offering the only option for shopping and other errands. For example, respondents noted the added challenge of scheduling trips out of the home given the timing and availability of buses, along with challenges with obtaining or understanding bus schedules, with a participant noting:

> But again it is getting information out and trying to explain to seniors how to read the schedules because they are not all . . . seniors aren't comfortable all the time going online.
>
> (Participant 5, female driver aged 70)

Respondents also noted problems with being able to physically access public transit, if it was available. Problems included distance to bus stop or physically boarding a bus, waiting for buses in the winter (or on hot, humid days) and the difficulty of returning home from shopping.

> Of course when you go grocery shopping, then you got to cart all that stuff back if you were on a bus.
>
> (Participant 2, female driver aged 83)

Ultimately, the bus was not a viable transit option even among the few that had limited access, and most respondents had no transit service whatsoever.

Beyond public transportation, taxis and ride-sharing options such as Uber or Lyft were limited and expensive, with respondents commenting on the lack of available public transit and cost of using a taxi:

> So yeah, so no public transit, taxi obviously super super expensive so not really available.
>
> (Participant 17B, male driver aged 73)

Some respondents indicated they would rely on family and friends to meet their trip needs, particularly among those that either did not drive or self-regulated their driving rather than paying for rides. However, reliance on family and friends was problematic, with respondents describing how asking for a ride made them feel that they were imposing or being a burden to others:

> I guess that we could ask somebody in the family but everybody is busy. Our son in law works, our daughter works, she's out of town half the time,

> she flies to . . . she's flying to . . . [continues to talk about daughter's work]. I don't know that we would have a lot of choices other than a taxi, would we?
>
> (Participant 6A, female driver aged 83)

Active transportation options, including walking and biking, both of which could potentially address transportation needs, were not commonly used. When asked about these options, respondents were concerned about their safety when out on the road due to lack of infrastructure:

> There is no sidewalk and there are people that bike up there and there are enormous trucks that come up there so I don't think it is a very good place for walking.
>
> (Participant 10, female driver aged 77)

Although some of the respondents owned bicycles, they were typically viewed as something for recreation or exercise and not for commuting or transportation purposes.

Declining health and reduced personal mobility in later life further precluded use of transport options beyond the personal automobile, including active transport options.

> But then I've had major back surgery and knee surgery and that's really limited us. That's why I don't walk very far; I can't.
>
> (Participant 17B, male driver aged 73)

> The trouble with me is that I don't walk very well when I'm outside so if I had to walk, that's not . . . if I had to do it I would but it's not a positive for me.
>
> (Participant 7, female driver aged 75)

While it is concerning that the people living in small towns and rural areas do not have access to public transportation or affordable alternatives to the personal car, it is perhaps more concerning that the majority of participants had not planned for driving cessation. Some acknowledge that not being able to drive would result in them having to move; however, moving would be expensive and difficult for some to do.

Conclusions

Concurrent with rapid urbanisation and low fertility levels, rural populations are ageing rapidly around the globe. For many older adults living in rural areas, the preference is to age in place, enabling a continuation of their preferences and lifestyle choices (see Menec and Novek, Chapter 14 and Wiles et al., Chapter 15). At the same time, an increasing number are experiencing changes in their driving or may no longer have licenses altogether due to changes in their health or other concerns (Kim, 2011; Waldorf, 2001), hence implying an increasing need for transportation options beyond the personal automobile. Moreover, the demand for transportation services in rural areas will increase as populations continue to age and the type and nature of required services will also change over time. The challenge for policy makers, therefore, is to ensure that those living in rural areas have access to services, including transportation options that can support them when choosing to age in place.

Results are consistent with literature identifying the personal car as the preferred mobility tool or option among older adults in rural areas (Kim and Ulfarsson, 2004; Páez et al., 2007; Scott et al., 2009; Vrkljan et al. 2018). Even if other mobility options are available, including active transit, ride-sharing, family and friends, or public transportation options, results from this study suggest older adults are less likely, unable or unwilling to use them. Instead, most prefer the convenience of using their own car. For older adults in rural areas and small towns, the lack of transportation options are problematic given a lack of infrastructure and public investment, a lack of knowledge or experience in using other modes, personal health and mobility constraints, cost or the inconvenience of options. The lack of resources is magnified among those with more limited financial resources and/or health issues.

Consequently, access to mobility options is not only defined by physical access but also monetary access as well, with older adults concerned with the cost of alternative modes of transportation. While participants were aware of ride-sharing services, including private services such as Uber or Lyft or public services, such as Hamilton's Disabled and Aged Regional Transportation System (DARTS, operated by Hamilton's transit system), they were not seen as daily transport options. In fact, ride-sharing services were not raised by some respondents, which may reflect their unfamiliarity with such services, or that the service did not reach these areas. In addition to the cost of such services (even if somewhat less than fares charged by taxis), the non-use of ride-sharing services reflected unease with the service, a lack of knowledge and/or its availability (or lack thereof). Those at greatest risk are those who are frail and in poor health, particularly individuals whose health already limits their ability to drive, people with limited financial resources and people without large social networks that can aid with transportation.

Although the issues faced by those in rural areas are not directly aligned with those raised in the context of transit deserts, as defined in Jiao's (2017) work, the needs of individuals living in areas that lack a variety or range of mobility options can place even more pressure on the need to use the personal automobile. While the demand for transportation and alternatives to the personal automobile are growing, the spatial characteristics of rural areas, including the lack of population density, limited infrastructure and large distances, place greater emphasis on the need for viable transportation alternatives beyond the personal automobile. Ride-sharing services may be an emerging alternative but the challenges of creating a viable service remain due to geography, limited populations and distances between locations, which can increase the cost of offering such a service. For public services, such as DARTS, planning and scheduling tasks by the service provider are complicated. For private services, such as Uber, providing adequate subsidies for drivers can be a challenge. In fact, there is one example of a largely rural municipality in Southern Ontario (Innisfil) that contracted Uber to provide ride-sharing services to its population.[1] However, knowledge of how to access or use such ride-share options remains low among older adults (Leistner and Steiner, 2017; Payyanadan and Lee, 2018; Vivoda et al., 2018). Internationally, road and transportation infrastructure will also significantly impact availability of choices beyond the personal car.

Transportation issues facing the ageing population are magnified among those living in rural areas, where the loss of a license magnifies the risk of reduced out-of-home activities and poor health outcomes given the lack of other transportation options. Loss of licensure in this age group has been linked to reduced out-of-home activity levels (Brown, McLaughlin and Vivoda, 2017; Bryanton, Weeks and Lees, 2010; Buys and Carpenter, 2002; Curl et al., 2014; Davey, 2007; Dickerson et al., 2017; Marottoli et al., 2000; Sacker et al., 2017), increased social isolation and loneliness, loss of independence and declining quality of life (Hwang and Hong, 2018; Marottoli et al., 2000), poorer health status (Edwards et al., 2009), higher rates of depression (Fonda, Wallace and Herzog, 2001) and decreased ability to care for oneself and access to health care (i.e., Arcury et al., 2006; Mattson, 2011). Without adequate policies and programs to ensure transportation connections that enable physical access to services and the broader community, the health and well-being of older adults in rural areas is threatened. However, it should be noted that there is no one-size-fits-all solution.

Finally, limitations should be noted. First, we did not capture individuals who live in more remote rural locations and are faced with even greater distance and travel challenges associated with accessing services given the proximity of respondents to Hamilton. The results are also less applicable in international locations where transportation options and infrastructure are more constrained. Second, although the study was open to drivers and non-drivers, we were only able to recruit four individuals who identified as non-drivers, all of whom had driven previously, and we were unable to recruit never-drivers. Third, participants had reasonably high levels of health. The profile of our respondents speaks to the potential "independence" of this population, which may offer a perspective different from those who would be less mobile, more frail or in poorer health. Finally, we did not consider how use changed over time with respect to shifts in public investment in travel options. This paper highlights the dependency on the personal automobile shared by many older adults and the need to develop viable alternatives, particularly for those living in rural areas.

Acknowledgements

The authors gratefully acknowledge funding support from the McMaster Institute on Aging (MIRA) and the Labarge Centre for Mobility in Aging.

Note

1 See, for example, www.uber.com/cities/innisfil/partnership/, and www.citylab.com/transportation/2019/04/innisfil-transit-ride-hailing-bus-public-transportation-uber/588154/ (Accessed 23 May 2019).

References

Arcury, T. A., Preisser, J. S., Gesler, W. M. and Powers, J. M. (2006). Access to transportation and health care utilization in a rural region. *The Journal of Rural Health*, 21(1), pp. 31–38.

Braun, V. and Clarke, V. (2006) Using thematic analysis in psychology, *Qualitative Research in Psychology*, 3(2), pp. 77–101.

Brown, K., McLaughlin, S. and Vivoda, J. (2017). Driving cessation and social engagement: participation in activities away from the home setting. *The Gerontologist*, 56(3), pp. 364–379.

Bryanton, O., Weeks, L. E. and Lees, J. M. (2010). Supporting older women in the transition to driving cessation. *Activities, Adaptation and Aging*, 34, pp. 181–195.

Buys, L. R. and Carpenter, L. (2002). Cessation of driving in later life may not result in dependence. *Australasian Journal on Ageing*, 21, pp. 152–155.

Clarke, P. J., Yan, T., Keusch, F., and Gallagher, N. A. (2015). The impact of weather on mobility and participation in older US adults. *American Journal of Public Health*, 105(7), pp. 1489–1494.

Collia, D. V., Sharp, J. and Giesbrecht, L. (2003). The 2001 national household travel survey: a look into the travel patterns of older Americans. *Journal of Safety Research*, 34, pp. 461–470.

Curl, A. L., Stowe, J. D., Cooney, T. M. and Proulx, C. M. (2014). Giving up the keys: how driving cessation affects engagement in later life. *The Gerontologist*, 54(3), pp. 423–433.

Dahan-Oliel, N., Mazer, B., Gélinas, I., Dobbs, B. and Lefebvre, H. (2010). Transportation use in community-dwelling older adults: association with participation and leisure activities. *Canadian Journal on Aging*, 29(4), pp. 491–502.

Dandy, K. and Bollman, R. D. (2008). Seniors in rural Canada (Catalogue no. 21–006-X). *Rural and Small Town Canada Bulletin*, 7(8), Ottawa, ON: Statistics Canada.

Davey, J. A. (2007). Older people and transport: Coping without a car, *Ageing & Society*, 27, 49–65.

Dickerson, A. E., Molnar, L. J., Bédard, M., Eby, D. W., Berg-Weger, M., Choi, M., Grigg, J., Horowitz, A., Meuser, T. and Myers, A. (2017). Transportation and aging: an updated research agenda to advance safe mobility among older adults transitioning from driving to non-driving. *The Gerontologist*, 59(2), pp. 215–221.

Edwards, J. D., Lunsman, M., Perkins, M., Rebok, G. W. and Roth, D. L. (2009) Driving cessation and health trajectories in older adults. *Journal of Gerontology: Medical Sciences*, 64, pp. M1290–M1295.

Fonda, S. J., Wallace, R. B. and Herzog, A. R. (2001). Changes in driving patterns and worsening depressive symptoms among older adults. *Journal of Gerontology: Social Science*, 56, pp. S343–S351.

Grengs, J. (2001). Does public transit counteract the segregation of carless households? *Transportation Research Record*, 1753, pp. 3–10.

Hjorthol, R. J., Levin, L. and Sirén, A. (2010). Mobility in different generations of older persons: the development of daily travel in different cohorts in Denmark, Norway and Sweden. *Journal of Transport Geography*, 18(5), pp. 624–633.

Hwang, Y. and Hong, G. R. (2018). Predictors of driving cessation in community-dwelling older adults: a 3-year longitudinal study. *Transportation Research Part F*, 52, pp. 202–209.

Jiao, J. (2017). Identifying transit deserts in major Texas cities where the supplies missed the demands. *The Journal of Transport and Land Use*, 10(1), pp. 529–540.

Jiao, J. and Dillivan, M. (2013). Transit deserts: the gap between demand and supply. *Journal of Public Transportation*, 16(3), pp. 23–29.

Kim, S. (2011). Transportation alternatives of the elderly after driving cessation. *Transportation Research Record*, 2265, pp. 170–176.

Kim, S. and Ulfarsson, G.F. (2004). Travel mode choice of the elderly: effects of personal, household, neighborhood, and trip characteristics. *Transportation Research Record*, 1894, pp. 117–126.

Leistner, D. L., Steiner, R. L. (2017). Uber for seniors? Exploring transportation options for the future. *Transportation Research Record*, 2660, pp. 22–29.

Marottoli, R. A., de Leon, C. F., Glass, T. A., Williams, C. S., Cooney, L. M. and Berkman, L. F. (2000). Consequences of driving cessation decreased out-of-home activity levels. *The Journals of Gerontology Series B: Psychological Sciences and Social Sciences*, 55(6), pp. S334–S340.

Mattson, J. (2011). Transportation, distance, and health care utilization for older adults in rural and small urban areas. *Transportation Research Record*, 2265, pp. 192–199.

Newbold, K. B. and Scott, D. M. (2018). Insights into public transit use by Millennials: the Canadian experience. *Travel Behaviour and Society*, 11(1), pp. 62–68.

Newbold, K. B., Scott, D. M., Spinney, J. E. L., Kanaroglou, P. and Páez, A. (2005). Travel behavior within Canada's older population: a cohort analysis. *Journal of Transport Geography*, 13, pp. 340–351.

Páez, A., Scott, D., Potoglou, D., Kanaroglou, P. and Newbold, K. B. (2007). Elderly mobility: demographic and spatial analysis of trip making in the Hamilton CMA, Canada. *Urban Studies*, 44(1), pp. 123–146.

Payyanadan, R. P. and Lee, J. D. (2018). Understanding the ridesharing needs of older adults. *Travel Behavior and Society*, 13, pp. 155–164.

Pucher, J. and Renee, J. L. (2005). Rural mobility and mode choice: evidence from the 2001 National Household Travel Survey. *Transportation*, 32(2), pp. 165–186.

Rosenbloom, S. (2001). Sustainability and automobility among the elderly: an international assessment. *Transportation*, 28, pp. 375–408.

Sacker, A., Ross, A., MacLeod, C. A., Netuveli, G. and Windle, G. (2017). Health and social exclusion in older age: evidence from Understanding Society, the UK household longitudinal study. *Journal of Epidemiology and Community Health*, 71, pp. 681–690.

Scott, D. M. and Axhausen, K. W. (2006). Household mobility tool ownership: modeling interactions between cars and season tickets. *Transportation*, 33(4), pp. 311–328.

Scott, D. M., Newbold, K. B., Spinney, J. E. L., Mercado, R. G., Páez, A. and Kanaroglou, P. S. (2009). New insights into senior travel behavior: the Canadian experience. *Growth and Change*, 40(1), pp. 140–168.

Statistics Canada. (2016). *Focus on Geography Series, 2016 Census*. Available at: https://www12.statcan.gc.ca/census-recensement/2016/as-sa/fogs-spg/Facts-can-eng.cfm?Lang=Eng&GK=CAN&GC=01&TOPIC=1 [Accessed 27 June 2019].

Statistics Canada. (2017). *Population Centre and Rural Area Classification 2016.* Available at: www.statcan.gc.ca/eng/subjects/standard/pcrac/2016/introduction [Accessed 17 December 2019].

Vivoda, J. M., Harmon, A. C., Babulal, G. M. and Zikmund-Fisher, B.J. (2018). E-hail (rideshare) knowledge, use, reliance, and future expectations among older adults. *Transportation Research Part F: Traffic Psychology and Behaviour*, 55, pp. 426–434.

Vrkljan, B., Cammarata, M. Marshall, S., Naglie, G., Rapoport, M., Sangrar, R., Stinchcombe, A., Tuokko, H. (2018). Transportation mobility. *Canadian Longitudinal Study on Aging (CLSA) report on Health and Aging in Canada. Findings from Baseline Data Collection 2010–2015*. Ottawa, ON: Public Health Agency of Canada & Employment and Social Development Canada – Government of Canada.

Waldorf, B. (2001). Anticipated mode choice following driving cessation. *European Research in Regional Science*, 11, pp. 22–40.

13 Rural community development in an era of population ageing

Laura Ryser, Greg Halseth, Sean Markey, Neil Hanlon and Mark Skinner

Introduction

Economic and demographic change is pervasive across small communities within developed economies (Halseth and Ryser, 2018). Increasingly, rural places are left to deal with these changes on their own, and with their own resources. This includes coping with the implications of population ageing (Skinner and Hanlon, 2016). As small communities attempt community development approaches to address population ageing within an era of significant economic and political restructuring (see also Heley and Woods, Chapter 3 in this volume), local responses may be truncated by critical challenges. This chapter connects resource frontier ageing and economic restructuring processes with opportunities for place-based community development to develop communities for people of all ages and stages of life. Drawing upon the case of Tumbler Ridge, British Columbia (BC), Canada, it highlights two critical areas – contested identities and policy incoherence from senior governments – that are working against place-based development in ageing resource-based communities.

Rural restructuring

Economic restructuring in Canada's smaller communities is transforming local populations and presenting challenges for services and infrastructure. In this section, we explore economic and demographic transformations in resource-dependent communities before exploring how place-based development can be mobilised to capitalize on opportunities associated with ageing.

Economic change

In many OECD countries, the context of rural communities is different today compared to when resource development expanded in the post-World War II period. Many small communities are shaped by staples-dependent economies (Argent, 2017; Horsley, 2013). Staples theory describes the reliance of these communities on extracting and exporting raw natural resource commodities to advanced manufacturing economies (Innis, 1933). Historically, coordinated public policies supported investments in better quality housing, amenities and neighbourhoods (Markey et al., 2016) in order to improve the quality-of-life needed to attract and retain young workers and their families.

As price takers, exposure to global commodity markets intensifies the risk of economic crisis for resource-based communities, with more frequent and deeper boom and bust periods (Tonts, Plummer and Lawrie, 2012). For example, once the recession in the early 1980s took hold, national governments reduced expenditures and their role in rural and remote places. Communities were left with limited infrastructure and program investments to renew their community and economic development strategies. At the same time, industries consolidated their operations, invested in automation while adopting flexible and labour shedding arrangements and pursued fluid flows of capital in order to remain competitive (Halseth et al., 2017). The increasing use of mobile workforces further entrenches the vulnerability of these communities and their economies (Markey et al., 2016).

Demographic change

As a result of the restructuring of resource-based economies, these communities have begun to experience resource frontier ageing (Hanlon and Halseth, 2005). After industrial expansion and growth populated these communities with young families, the recession of the early 1980s initiated a long period of job losses and out-migration (Ryser and Halseth, 2013). Younger workers with less seniority left communities with their families in search of other employment opportunities. The remaining workforce has been "ageing-in-place". Skinner et al. (2014) describe ageing-in-place as a process where there are "older individuals staying on in the communities in which they resided before retirement" and ageing places as "communities that have increasing numbers of older residents" (419) (see also Spina, Smith and DeVerteuil, 2013; see also Wiles et al., Chapter 15).

Many resource-based communities are looking towards the "retirement industry" as one option to enhance their economic resilience (Moore and Pacey, 2003; Wassel, 2011). The retirement industry now includes diverse market opportunities related to mobility, IT, home products and renovations, health care, wellness, recreation and tourism, financial services, education and business consulting – all reflecting different needs, expectations and demands among older residents (Doka, 1992; Eitner et al., 2011). This comes with needs that resource towns typically are not equipped or designed to deal with (Davis and Bartlett, 2008). The physical and social infrastructure may not support the mobility of older residents (Lui et al., 2009). The range of housing, health care, home support and community services in place may not support all ages and stages of life (Ryser and Halseth, 2013). Further, isolated communities may not have the communications infrastructure or capacity to connect seniors with information and regionalised supports (Ryser and Halseth, 2011, 2013).

An ageing workforce, and a movement of baby boomers into retirement, means that resource-dependent communities will need to assess how they can meet the needs of current and future older residents. Drawing upon the "political economy of geographical gerontology", researchers are exploring the spatial implications of policies and distribution of resources for health and social services through the public and private sectors (Andrews, Cutchin and Skinner, 2018; see also Hanlon and Poulin, Chapter 4; Rowles and Cutchin, Chapter 21). Redesigning communities to become age-friendly requires a strengthening of synergies across local stakeholders and different levels of government (Phillips, 2018). Restructuring processes, however, can create conflict and contested identities through competing policies and practices that affects how strategic community development shapes healthy ageing (Skinner and Winterton, 2018b).

During periods of transition, place-based development can be held hostage by contested local identities. Many resource-industry towns, for example, have deeply entrenched identities associated with "masculine" and "high paying" resource sector jobs. The (potential) emergence of a retirement industry, with perceived lower-wages and "care giving" employment opportunities, sets the stage for contesting both economies and identities (Skinner and Winterton, 2018a). Too often, "ageing has been identified as a perceived threat to rural community sustainability" instead of being recognised as a potentially important part of a diversified and comprehensive community economic strategy (Skinner and Winterton 2018a: 139).

Reflecting on economic, political and demographic restructuring in these settings, Hanlon et al. (2014) suggest that

> substantial gaps in services needed to support healthy ageing are therefore commonplace, and these remain in spite of, or perhaps because of, the level of commitment shown by different levels of government and the private sector to accommodate older residents in resource-dependent communities.
> (133) (see also Hanlon and Poulin, Chapter 4; Glasgow and Doebler, Chapter 10)

These structural changes combine with the low economies of scale found in smaller communities to hinder older residents' access to resources; however, in these contexts, older residents may also become valuable voluntary assets to sustain programs and services that support ageing-in-place (Skinner and Winterton, 2018b). By inserting themselves into community development initiatives, older volunteers are asserting their rights to age-in-place and challenging the discourses that negatively affect the age-friendliness of rural settings (Andrews, Cutchin and Skinner, 2018).

Pathways of place-based community development

Community development involves processes that increase the capacity of individuals, groups and the community to collectively assess change, support collaboration and coordinate responses as they re-imagine and re-bundle assets to pursue new aspirations on their own terms. To develop and mobilize this capacity, stakeholders require four key community development foundations, including

(Markey, Halseth and Manson, 2012; see also Winterton and Warburton, Chapter 23; Colibaba et al., Chapter 24):

1. Physical infrastructure (transportation, communication, energy, buildings and facilities);
2. Economic infrastructure (financial capital, market information, networks);
3. Human infrastructure (leadership, workforce education and training and health); and
4. Community infrastructure (services, voluntary sector, participation and culture).

In an era of population ageing, investments in these community development foundations must be strategic. As Markey and colleagues (2016) argue, "in the absence of . . . coordinated senior government and industry investment, communities may pursue place-based development as a way to ground local planning decisions and interact with government and industry from a position of organized strength" (131). Place-based development uses inputs from local actors and community capital assets to support flexible responses and investments to realize new aspirations (Sørensen, 2018). By using local assets, "place-making" initiatives are more likely to foster broader community support for initiatives and retain benefits.

At times, the fragmented nature of place-based development may be indicative of underdeveloped local capacities and synergies but also the lack of supportive top-down policy and program supports (Halseth and Ryser, 2018). Small communities across many OECD countries are confronting resource frontier ageing and economic restructuring that present challenges to local service delivery and infrastructure (see also Glasgow and Doebler, Chapter 10). The challenge for local governments is to depart from "staying the course" of depending on resource-based development and senior governments towards support for new investments that can strengthen community development infrastructure and services for the entire community.

Tumbler Ridge, BC, Canada

Drawing upon 32 interviews, Tumbler Ridge provides an example of how economic restructuring has impacted both population ageing and the challenges of supporting age-friendly community development. Tumbler Ridge was an "instant town" constructed in 1981 to support the development of two open pit coal mines. Located in northeastern BC, the municipality had a population of 4,390 shortly after being established (Statistics Canada, 1986). The population consisted of young families, with a significant portion of residents between the ages of 20 and 39, as well as those under 15 years of age. Few seniors aged 65 years and older lived in town (0.3%) (Statistics Canada, 1986). Strategic investments in place-making were made, with attention to social planning principles and a range of centrally located community and recreational infrastructure in order to foster routine interaction and a strong sense of community (Halseth et al., 2017). By 1991, the population peaked at 4,660 (Statistics Canada, 1991). Pricing conflicts with Japanese steel mills, however, led to several rounds of layoffs in the mining sector. In 2000, Teck announced the closure of the largest mine and that the second mine would close within two years (Sullivan, Ryser and Halseth, 2014). By 2001, the population declined to 1,850 (Statistics Canada, 2001). The viability of the community was threatened by "the loss of more than 70% of local jobs and 65% of the municipal tax base" (Mochrie et al., 2014: 2). In the following sections we present findings related to the economic transition of the community and the emerging retirement sector. We then discuss our two main themes in response to the changes, incoherent top-down policies and contested local identities.

An emergent retirement sector

In response to the economic downturn in Tumbler Ridge, a transition task force, formed within 6 days of the closure announcement, identified the need to secure more stable conditions for the community. A key part of this strategy was a housing sale in order to generate local property tax revenue from local homeowners. The strategy was simply to move the housing stock from corporate rental to private homeowners.[1] An unintended outcome was that a large number of newly retired households from urban centres acquired "retirement homes" at very low cost. At the same time, the retained workforce was ageing-in-place, with an older workforce now concentrated between 35 and 54 years of age. The community suddenly included more residents who were approaching retirement (between 55 and 64 years of age) (13.8%) and more retirees (5.1%) (Statistics Canada, 2011).

A retirement industry emerged in Tumbler Ridge and illustrated one way to diversify the local economy. The initiative was assisted by the availability of low cost housing, a beautiful natural setting and related outdoor amenities, high quality civic infrastructure and several senior-led community development initiatives that drew upon place-based assets. As seniors became engaged in the community, the local government organised a seniors' task force that included older residents in order to conduct extensive community consultation about service and infrastructure needs and to devise a strategy to address those needs. As one local government stakeholder explained:

> When the community shut down in the early 2000s, seniors started moving into the community and we had to rethink much of what Tumbler Ridge was about and how to make the community accessible for seniors. The housing here was built very oddly so it's very difficult to get a stretcher or a wheelchair into the house. Most houses have a lot of stairs. You want your demographic to have a wide variety of ages, cultures. It builds a sense of community. Now that we have seniors and we have various culture groups represented here, we have a lot more depth.
>
> (TR#13)

The local government worked with businesses to develop curb cuts, wheelchair ramps, signage, automatic doors, wider aisles and accessible washrooms and parking.

Building upon the work of the task force, older residents engaged in voluntary organisations such as the Tumbler Ridge Hospice and Respite Care Society, TR Cares, the Lions Club, the Tumbler Ridge Museum Society and the Seniors' Corner. With the support of older volunteers, programs were mobilised to improve the quality-of-life of new seniors in the community, including the Snow Angels snow removal program, a seniors' drop-in centre, meals on wheels and a new shuttle van service to connect seniors with regionalised medical services (Skinner et al., 2014). For a period of time, the local government ensured a seniors' need coordinator was in place to support these endeavours. Hartford Court, a new 12-unit accessible housing complex, was constructed along with outdoor exercise equipment on land donated by the municipality in the downtown core in close proximity to civic amenities and services. Older volunteers have also worked on broader community development investments that support tourism and an improved local quality-of-life, including a campground, a palaeontology museum, trail maintenance and events such as the Emperor's Challenge – a half-marathon along the surrounding mountainous region. As one older resident explained:

> When we first came here, seniors operated the tourism facility, they kept the tourism open. Seniors went into Monkman Park. We kept that park open for a whole year because it was going to be closed, and we just moved in for one and two weeks at a time and maintained that park. We did that for BC Parks. We spearheaded the Cascade trails work. We went in and lived in the ranger cabin and we collected the fees and cleaned the park and the fires and took care of Kinuseo Falls and the outhouses for the whole year.
>
> (TR#10)

These initiatives, however, have been tempered by an incoherence within senior government policies and programs that have not fully realised the supports needed for an ageing population in resource-based communities.

Incoherent top-down supportive policies

The challenges of managing transitions in ageing resource-based communities stems from incoherent public policies and processes that work against efforts to proactively engage older residents and address local planning and service deficiencies. As noted, many of the necessary investments to support population ageing are beyond the mandate and jurisdictions of local government and require top-down support policies and programs. As Halseth and Ryser (2018) state, however, senior governments have been trapped in a period of "reactionary incoherence", whereby there is an incoherent and uncoordinated deployment of state support, owing to decades of withdrawal and cuts inspired by neoliberal policy (Markey, Halseth and Ryser, 2016). The period of reactionary incoherence creates challenges and opportunities for local governments seeking to influence their development trajectory and respond to economic and demographic transition. In the case of Tumbler Ridge, while the local community was working on economic diversification plans in the face of pending coal mine closures, the provincial government did not have a vision for resource-based communities beyond their function as "resource banks" from which wealth could be extracted (Vancouver Sun, 2 March 2000).

Population losses following the mine closures prompted a reduction or closure of many provincial services. An influx of older residents, however, did not result in efficient responses by provincial agencies to recognize the growth and demographic transition that was taking place. Instead, service levels remained entrenched in neoliberal public policy decisions to reduce government expenditures. The result was a lack of health and community services to meet the needs of older residents and support an emerging retirement sector (Halseth et al., 2011). As one stakeholder noted:

> The seniors who had moved in had certainly helped to develop it, and essentially help save it, but that really had not been recognized by the health authorities. That changing demographic had not been recognized in terms of home care, nursing homes, or things that generally would be put in place when you have a totally different social fabric and a different age spectrum living in town. What happened was when the population went down before the seniors moved in, the health authority cut back on services. Then when the population bounces back and demographics change, you have to make the case for things to be reinstated or new services to be brought in, and that's the battle because that's not what happened.
>
> (TR#9)

The restructuring of services, however, was pitted against provincial investments in accessible housing. After opening Hartford Court, two key challenges emerged. First, once the facility was completed, BC Housing awarded the contract to manage and maintain the facility to a business in the regional centre of Fort St. John. Local groups felt excluded from the management and operations of the facility. At times, residents experienced difficulties with timely repairs and maintenance. Second, the effectiveness of investments in Hartford Court were undermined by the absence of broader health care and community services to support older residents. The Province did not have a comprehensive plan across various ministries to create synergies between their investments and services. In addition to these challenges, contested identities across the community have diverted attention away from the retirement sector in favour of investments to attract young workers and their families to renewed mining activities.

Contested identities

Resistance to diversify this staples-dependent economy by supporting a retirement industry was driven by residents and decision makers who continued to look exclusively towards large-scale industrial resource projects (Sullivan, Ryser and Halseth, 2014). Efforts to mobilize the lessons of using a retirement industry to strengthen community development opportunities were not sustained in subsequent boom and bust periods. In 2004, Western Canadian Coal opened a new mine, with three more mines opened in the area in the following five years. The population grew to 2,455 in 2006 and reached 2,720 in 2011. The renewal of mining activities diverted attention from community diversification initiatives that could not match the mining sector in terms of jobs or salaries. There were concerns, for example, about sustaining the local government's $200,000 annual contribution to the museum foundation and UNESCO Geopark that features palaeontology and geological assets to attract tourism. Public concerns have also been expressed that the protection of fossils will have a negative impact on industry and employment. Furthermore, the seniors' task force that was established with a mandate to strategically address the needs of older residents was transforming to tackle broader community development issues. As one older volunteer explained:

> The seniors' task force I see may be in effect for another two years and then I think it will diminish. The more young people come in town, the more we're going to lose that or they're going to switch it to another name and it will get lost. It's become almost just a voicing pad to advise council on what's wrong. We spend a lot of time on roads or dirty yards rather than on the shape of society or infrastructure. It definitely stresses the needs of the town more so than the seniors' needs now.
>
> (TR#3)

Diverging community visions can impede the allocation and mobilisation of assets that would support more diversified community development pathways. Over time, there has been uncertainty over the allocation of space for the Seniors' Corner drop-in centre. After initially being co-located with a youth centre in the community centre, the drop-in centre would eventually be moved to a former church. Following the resignation of the seniors' coordinator, the local government did not hire a replacement. For older residents, seniors' issues were disappearing from local government agenda. Their failed expectations of moving to a retirement community prompted an out-migration of residents. The number of older residents (aged 55 and older) declined by 13.6% from 660 in 2006 to 570 in 2016 (Statistics Canada, 2016). Some felt, however, that the mines could provide more funding, exert more influence and work with local stakeholders to obtain more services from senior levels of government in order to support both community development and an ageing population. As one participant argued:

> Some of the disappointment is that when the mines reopened, the only reason they had an infrastructure there to work with is because the seniors came to town. They did this great housing market sale which resulted in a huge influx of seniors. That saved the town from extinction. When the mines came back, I've been trying to make the point they owe the seniors in a big way; and,

> ironically, while we get some assistance from them, it has been disappointing because these are the people who have benefited the most by having the seniors there and are not that inclined to be as generous as they could be to help these initiatives.
>
> (TR#2)

The exodus of older residents had an important impact on broader place-based community development initiatives. One stakeholder explained how the loss of older residents deteriorated the availability of human capital assets for community groups:

> I see that seniors . . . what they did for this community was to make it a more well-rounded community and they tend to be very active volunteers and contribute a huge amount to the community. It was almost an artificial community when we had the mine workers and no seniors, and the community benefited tremendously from them. But our volunteer base has gone down because so many seniors are gone, so it puts more strain on the rest of us.
>
> (TR#1)

Unfortunately, when coal prices dropped again in 2014, there were many impacts, including closure of three mining operations and a loss of 890 jobs (Halseth, Helm and Price, 2011). As the mining sector collapsed, the population declined to 1,987 in 2016. As the community enters another period of transition, its inability to sustain attention to diversification has placed it in a precarious position, as both unemployed workers and disillusioned older residents have left the community.

Conclusion

Drawing upon the experiences of Tumbler Ridge, BC, this chapter has provided insights into the role of older residents in responding to the needs of an ageing resource-based community and supporting broader place-based development. With an ageing workforce and an ageing population, community needs for services, supports and care are evolving. Attention to reshaping investments in age-friendly infrastructure and services will have important implications to support healthy ageing, active engagement and an improved quality of life. Pursuing place-based community development in an era of population ageing, however, will require a shift within the contested identities of resource-based communities and recognition of the sustained synergies needed to create places that support all stages and ages of life. These contested identities impede how stakeholders are able to re-imagine assets and create more diversified and resilient pathways for the future of these communities.

Incoherent top-down policies further challenge these transitions as local governance processes work to proactively engage older residents and address planning deficiencies. This is especially problematic where policy incoherence fosters fragmentation not only across top-down policies and investments but also within communities when there are limited synergies with local assets and community development initiatives.

Acknowledgements

Funding for this research was provided by the Social Sciences and Humanities Research Council's Insight Development Grant, The transformative role of voluntarism in ageing resource communities: Integrating people, place and community (Skinner), along with support provided by the Canada Research Chairs Secretariat (grants 950-200244, 950-203491 and 950-222604).

Note

1 The housing stock was largely owned by one of the two mining companies and CMHC.

References

Andrews, G, Cutchin, M and Skinner, M. (2018). Space and place in geographical gerontology: theoretical traditions, formations of hope. In: M. Skinner, G. Andrews and M. Cutchin, eds., *Geographical Gerontology: Perspectives, Concepts, Approaches*. London: Routledge, pp. 11–28.

Argent, N. (2017). Trap or opportunity? Natural resource dependence, scale, and the evolution of new economies in the space/time of New South Wales' Northern Tablelands. In: G. Halseth, ed., *Transformation of Resource Towns and Peripheries: Political Economy Perspectives*. Abingdon: Routledge, pp. 18–50.

Davis, S. and Bartlett, H. (2008) Healthy ageing in rural Australia: issues and challenges. *Australasian Journal of Ageing*, 27(2), pp. 56–60.

Doka, K. (1992). When gray is golden: business in an aging America. *The Futurist*, 26(4), pp. 16–20.

Eitner, C., Enste, P., Naegele, G. and Leve, V. (2011). The discovery and development of the silver market in Germany. In: F. Kohlbacher and C. Herstatt, eds., *The Silver Market Phenomenon: Marketing and Innovation in the Aging Society*. London: Springer, pp. 309–324.

Halseth, G., Helm, C. and Price, D. (2011). The rural care needs index: a potential tool for "have-not" communities. *Canadian Journal of Rural Medicine*, 16(4), pp. 134–136.

Halseth, G., Markey, S., Ryser, L., Hanlon, N. and Skinner, M. (2017). Exploring new development pathways in a remote mining town: The case of Tumbler Ridge, BC, Canada. *The Journal of Rural and Community Development*, 12(2/3), pp. 1–22.

Halseth, G. and Ryser, L., (eds). (2018). *Towards a Political Economy of Resource-Dependent Regions*. London; New York: Routledge.

Hanlon, N. and Halseth, G. (2005). The greying of resource communities in northern British Columbia: implications for health care delivery in already-underserviced communities. *Canadian Geographer*, 49(1), pp. 1–24.

Hanlon, N., Skinner, M. W., Joseph, A. E., Ryser, L. and Halseth, G. (2014). Place integration through efforts to support healthy aging in resource frontier communities: the role of voluntary sector leadership. *Health and Place*, 29, pp. 132–139.

Horsley, J. (2013). Conceptualising the state, governance and development in a semi-peripheral resource economy: the evolution of state agreements in Western Australia. *Australian Geographer*, 44(3), pp. 283–303.

Innis, H., (ed.). (1933). *Problems of Staple Production in Canada*. Toronto, ON: Ryerson Press.

Lui, C. W., Everingham, J. A., Warburton, J., Cuthill, M. and Bartlett, H. (2009). What makes a community age-friendly: a review of international literature. *Australasian Journal on Ageing*, 28(3), pp. 116–121.

Markey, S., Halseth, G. and Manson, D. (2012). *Investing in Place: Economic Renewal in Northern British Columbia*. Vancouver: UBC Press.

Markey, S., Halseth, G. and Ryser, L. (2016). Planning for all ages and stages of life in resource hinterlands: place-based development in northern British Columbia. In: M. Skinner and N. Hanlon, eds., *Ageing Resource Communities: New Frontiers of Rural Population Change, Community Development and Voluntarism*. London; New York: Routledge, pp. 131–145.

Mochrie, C., Morris, M., Halseth, G. and Mullins, P. (2014). *Tumbler Ridge Sustainability Plan: Strategies for Resilience: A Framework For Action*. Prince George, BC: Community Development Institute, University of Northern British Columbia.

Moore, E. and Pacey, M. (2003). *Geographic dimensions of aging in Canada 1991–2002*. SEDAP Research Paper No. 97, McMaster University, Hamilton.

Phillips, J. (2018). Planning and design of ageing communities. In: M. Skinner, G. Andrews and M. Cutchin, eds., *Geographical Gerontology: Perspectives, Concepts, Approaches*. London; New York: Routledge, pp. 68–79.

Ryser, L. and Halseth, G. (2011). Mechanisms for delivering information to seniors in a changing small town context. *Journal of Rural and Community Development* 6(1), pp. 49–69.

Ryser, L. and Halseth, G. (2013). So you're thinking about a retirement industry? Economic and community development lessons from resource towns in northern British Columbia. *Community Development*, 44(1), pp. 83–96.

Skinner, M. and Hanlon, N., (eds). (2016). *Ageing Resource Communities: New Frontiers of Rural Population Change, Community Development and Voluntarism*. London; New York: Routledge.

Skinner, M., Joseph, A., Hanlon, N., Halseth, G. and Ryser, L. (2014). Growing old in resource communities: exploring the links among voluntarism, aging, and community development. *The Canadian Geographer*, 58(4), pp. 418–428.

Skinner, M. W. and Winterton, R. (2018a). Rural ageing: contested spaces, dynamic places. In: M. Skinner, G. Andrews and M. Cutchin, eds., *Geographical Gerontology: Perspectives, Concepts, Approaches*. London; New York: Routledge, pp. 136–148.

Skinner, M. W. and Winterton, R. (2018b). Interrogating the contested spaces of rural aging: implications for research, policy, and practice. *The Gerontologist*, 45(1), pp. 15–25.

Sørensen, J. (2018). The importance of place-based, internal resources for the population development in small rural communities. *Journal of Rural Studies*, 59, pp. 78–87.

Spina, J., Smith, G. C. and DeVerteuil, G. P. (2013). The relationship between place ties and moves to small regional retirement communities on the Canadian Prairies. *Geoforum*, 45, pp. 230–239.

Statistics Canada. (1986). *Community Profiles*. Ottawa, ON: Statistics Canada.

Statistics Canada. (1991). *Community Profiles*. Ottawa, ON: Statistics Canada.

Statistics Canada. (2001). *Community Profiles*. Ottawa, ON: Statistics Canada.

Statistics Canada. (2011). *Community Profiles*. Ottawa, ON: Statistics Canada.

Statistics Canada. (2016). *Community Profiles*. Ottawa, ON: Statistics Canada.

Sullivan, L., Ryser, L. and Halseth, G. (2014). Recognizing change, recognizing rural: the new rural economy and towards a new model of rural service. *The Journal of Rural and Community Development*, 9(4), pp. 219–245.

Tonts, M., Plummer, P. and Lawrie, M. (2012). Socio-economic wellbeing in Australian mining towns: a comparative analysis. *Journal of Rural Studies*, 28(3), pp. 288–301.

Vancouver Sun. (2000). Miller powerless to stop death of Tumbler Ridge. *Vancouver Sun*, 2 March, pp. D1, D2.

Wassel, J. I. (2011). Business and aging: the boomer effect on consumers and marketing. In: R. Settersten and J. Angel, eds., *Handbook of Sociology of Aging*. New York: Springer, pp. 351–359.

14 Making rural communities age-friendly

Issues and challenges

Verena Menec and Sheila Novek

Introduction

The global age-friendly cities and communities movement has grown rapidly since the World Health Organization (WHO) initiated the Global Age Friendly Cities Project in 2006 (WHO, 2007). The initiative encouraged cities around the world to develop policies, services and structures that support the health, well-being and participation of older adults. The focus of the WHO project was on cities, with age-friendliness being presented as a solution to two major demographic trends: population ageing and urbanisation (WHO, 2007). The age-friendly movement has since then expanded to include rural communities (WHO, 2018). This shift reflects the recognition that rural environments offer unique challenges and opportunities that shape the experience of ageing (Skinner and Winterton, 2018; Keating, Eales and Phillips, 2013). The objective of this chapter is to critically examine issues that rural communities face in trying to become more age-friendly. We use the term "community" here to refer to rural settlements and "city" to larger urban centres (see Berry, Chapter 2 in this volume for a discussion of the diversity of rural communities).

The global age-friendly cities and communities movement

The age-friendly concept is consistent with the well-established view in gerontology that the fit between the environment and the person plays a key role in older adults' health and quality of life (Lawton and Nahemow, 1973). In the policy domain, the need to create supportive environments for older adults was one of three main objectives of the influential United Nations Madrid International Plan of Action on Ageing (2002). Moreover, various initiatives have focused on making cities or communities better places for people to live (e.g., AdvantAge Initiative) (Lui et al., 2009). Building on this foundation, the WHO (2007) conceptualises an age-friendly city as one in which policies, services, settings and structures enable people to be healthy, safe and participate in society. Age-friendly features span the physical and social environment and include, for example, having access to appropriate and affordable transportation (e.g., transportation for those with mobility impairment), housing (e.g., availability of affordable housing), aspects related to public spaces (e.g., accessible buildings) and community and health services (e.g., availability of home supports) (e.g., WHO, 2007; Plouffe and Kalache, 2010).

The age-friendly movement has grown rapidly in the last decade. At the time this chapter was written, 1000 cities and communities in 41 countries had expressed a formal commitment to becoming age-friendly by joining the WHO's Global Network for Age-Friendly Cities and Communities (WHO, n.d.). In Canada, over 1200 Canadian cities and communities are currently part of provincial age-friendly initiatives (PHAC, n.d.).

That the concept of age-friendliness has been gaining increasing attention on the part of policy makers is perhaps not surprising, as it fits well with several issues that are prominent in the public discourse: (1) population ageing and associated concerns about the sustainability of social and health care systems; (2) ageing in place as a policy solution to not only enhance older adults' quality of life but also to minimize the need for long-term care institutions; and (3) a policy focus on community-based solutions to social and welfare issues due to the rise of neoliberalism and associated retrenchment of the state. As DeFilippis, Fisher and Shragge (2010) put it, "community is, quite frankly, 'hot'" (26).

Age-friendly communities in rural contexts

Despite increasing urbanisation, a substantial number of people live in rural communities – globally, it is estimated that 44.7% of the world's population lives in rural areas (World Bank, 2018). Many rural communities and regions are ageing more rapidly than urban centres due to out-migration of young individuals, coupled with older adults' preference to age in place (Dandy and Bollman, 2008). Within the field of rural gerontology, rural environments are increasingly understood as diverse and dynamic contexts that shape how people experience growing older (Keating and Phillips, 2008; Scharf, Walsh and O'Shea, 2016; Skinner and Winterton, 2018). The changing demographic, social and economic contexts of rural communities present challenges and opportunities for older adults that differ from urban settings.

Rural communities are unique because of their small populations, large distances and often limited economic resources, which can create human and infrastructure deficits, such as limited transportation options and health care services (e.g., Hanlon and Halseth, 2005; Ryser and Halseth, 2012; Winterton and Warburton, 2011). Research shows that although many age-friendly features and barriers are similar in cities and rural communities (e.g., having affordable housing options is important in any context), there are unique features and barriers associated with living in rural communities (WHO, 2007; F/T/P Ministers Responsible for Seniors, 2007). For instance, having volunteer drivers and/or informal networks that provide transportation services was identified as an age-friendly feature in Canadian rural communities. Conversely, over-reliance on family, friends and neighbours to provide transportation services, limited options (no buses or taxis) and the high costs to travel outside of the community were identified as barriers (F/T/P Ministers Responsible for Seniors, 2007).

The constraints and opportunities within rural communities also affect their ability to implement and sustain age-friendly initiatives. Moreover, the broader political, policy and economic context not only impacts existing communities' assets and gaps, that is the extent to which they are age-friendly (or not) to begin with but also shape their ability to address them (Menec and Brown, 2018; Keating et al., 2013; Menec et al., 2011; Menec, 2017) (see also Keating et al., Chapter 5).

Issues and challenges in making rural communities age-friendly

Research shows that a number of factors either facilitate or hinder the implementation of age-friendly initiatives (Menec and Brown, 2018). We highlight here three issues that are particularly relevant to rural communities: human capacity, funding and power dynamics. In discussing these issues, we draw on both the published literature, as well as our experience with the age-friendly movement in the province of Manitoba, Canada, a midwestern prairie province. Manitoba has a population of about 1.3 million, about 60% of which is concentrated into one city, with the remainder living in rural and remote communities that are widely dispersed over about 550,000 square kilometres (Statistics Canada, 2017). Our discussion is also, in part, shaped by the policy and program delivery context in Manitoba, characterised by a federal, provincial, municipal government system, in which municipalities are responsible for local infrastructure (e.g., local roads) and social and recreational services, often in partnership with governmental and non-governmental organisations (e.g., libraries, active living centres for older adults), but not health care services, home care and long-term care institutions.

Human capacity

The WHO (2018) identifies four steps that cities and communities should, ideally, take in becoming age-friendly: (1) participatory age-friendly assessment (e.g., forming a steering committee and creating a baseline profile); (2) developing an action plan; (3) implementing the action plan; and (4) monitoring and evaluating progress. In Canada, the Public Health Agency of Canada has identified similar steps, the Pan-Canadian Age-Friendly Milestones (PHAC, n.d., 2012).

Having the human capacity to work through these steps is a challenge for cities (Menec and Brown, 2018), but even more so for rural communities. Given their small populations, the staff complement in the public sector (e.g., municipal staff, social service sector) and in non-governmental organisations is also small, with staff often stretched thin because of multiple responsibilities or part-time employment. As a result, many communities rely on volunteers to fill service gaps, including those in the health, social and community development sectors (Hanlon, Rosenberg and Clasby, 2007; Skinner, 2008; Skinner et al., 2014; Walsh et al., 2014). Getting people, whether they are paid staff or volunteers, to take on an age-friendly initiative can be difficult. As one participant in our research noted when asked about challenges in forming an age-friendly committee: "The fact that there is just not enough people to go around for all the committees and it's just one more committee" (Menec et al., 2014: 43). The lack of volunteers and/or volunteer burnout has been identified as a major barrier to age-friendly initiatives in rural communities in other studies (Russell, Skinner and Fowler, 2019; Winterton, 2016).

Having researchers involved in age-friendly initiatives can help with some activities, such as community assessments (Garon et al., 2014; Menec et al., 2014). In Manitoba, for example, grant funding from a federal research funding agency allowed our research team to conduct community consultations in communities that were part of the provincial government-led age-friendly initiative. These communities would otherwise not have had the capacity to conduct assessments. Although the assessments were considered helpful by communities in identifying priorities (Menec et al., 2014), relying on research grant funding is not sustainable in the long-run.

Evaluating progress is likely also beyond the capacity of many rural communities. For example, administering surveys, conducting focus group or accessing existing administrative data (PHAC, 2015; WHO, 2015) can be time consuming and costly. A pilot study on the implementation of the WHO's (2015) core indicators of age-friendliness that involved 13 cities and two rural communities around the world, identified data collection as the greatest challenge (Kano, Rosenberg and Dalton, 2018). The 15 sites volunteered and received US $5000 for their participation in the pilot project. One wonders how many rural communities would be able and willing to spend time and money on evaluation in the absence of funding.

Funding

Age-friendly frameworks and implementation guides are remarkably silent about the realities of what it takes for communities to become more age-friendly, including where the funding comes from (F/T/P Minister Responsible for Seniors; PHAC, 2012; WHO, 2007, 2018). This is perhaps not surprising; these documents are necessarily generic and cannot address the nuances of local contexts, such as organisational structures and funding mechanisms. However, it is these realities that communities face and determine whether they are able to effect any changes at the local level and, importantly, whether the age-friendly movement is sustainable.

Research shows that having funding is essential to making communities age-friendly; conversely, lack of funding creates major barriers not only for rural communities but also large urban centres (Brasher and Winterton, 2016; Gallagher and Mallhi, 2010; Russell, Skinner and Fowler, 2019; Sun et al., 2017). As one key stakeholder in our research succinctly summarised it – it all comes down to "the almighty dollar". Lack of funds to implement projects was, aside from human capacity, identified as one of the top two challenges by communities

(Menec et al., 2014). These findings are echoed by members of the Global Network for Age-Friendly Cities and Communities, with three quarters of those who responded indicating that they experienced difficulties in mobilising human and non-human resources (WHO, 2018).

At an orientation for communities that joined the Age-Friendly Manitoba Initiative, the provincial Minister responsible for the initiative stated that becoming age-friendly does not require new money. But making a community more age-friendly is not simply about reallocating existing funds – budgets are already stretched to a maximum for many rural communities – it does require new funding for certain projects. Consider the priorities and examples of proposed actions shown in Table 14.1 that were identified in a consultation, conducted by our research team, in one community. The community had a population of less than 1000 in 2016, 29% of whom were 65 years or older. It is located about 40 kilometres from a larger service centre (with a population of about 2500) and about 150 kilometres from an urban centre.

Some of the recommended actions shown in Table 14.1 may already form part of the municipal budget (e.g., sidewalk maintenance and street signage), and some might be taken on by a community group and volunteers (e.g., intergenerational activities). However, others may require applying for infrastructure grants (e.g., installing automatic doors), with others requiring substantial funding (e.g., developing housing). Partnerships are critical to making many of the projects a reality – and are identified as a key aspect of age-friendly initiatives (Brasher and Winterton, 2016; Garon et al., 2014; Menec et al., 2014; Plouffe and Kalache, 2011). But within a context where human capacity is already stretched, creating and maintaining partnerships in rural communities can create additional pressure. One key stakeholder in our research noted: "Partnerships are great but you have to have the time to put into them". Given these complexities, it is not surprising

Table 14.1 Example of priorities and potential actions in a rural community in Manitoba

Priorities	*Examples of proposed actions*
Increase building accessibility and safety	Replace existing heavy doors of public buildings with automatic doors
Improve safety and accessibility of sidewalks and trails	Ensure that sidewalks are safe through ongoing maintenance
Enforce traffic safety and increase awareness of handicapped parking spaces	Post more speed limit signs throughout the community
Provide community information through various formats	Provide regularly updated service and activity information to seniors in various ways
Increase active participation opportunities and sharing of information	Expand intergenerational activities
Improve transportation options for seniors	Expand current handi-van transportation services
Increase housing options for seniors	Develop condo style housing for independent seniors

that communities have been able to implement small projects, but have more difficulty addressing bigger ones, like increasing transportation options and affordable housing (Menec et al., 2014; Russell, Skinner and Fowler, 2019).

The need for communities to obtain funding is addressed by some organisations and governments that offer grants for age-friendly projects, such as the AARP (2019) and several Canadian provinces (e.g., British Columbia, Ontario, Quebec, Newfoundland and Labrador). As Russell, Skinner and Fowler (2019) in a study of rural communities in Newfoundland and Labrador in Canada describes, grants for age-friendly projects, which are typically small, led to the implementation of small projects, such as social programs and benches, but were not conducive to the sustainability of age-friendly initiatives. Age-friendly committees were sometimes disbanded once grant money was spent. Alternatively, grant funding was depleted by hiring a consultant to conduct a needs assessment, which meant that no funding was available to implement projects. Having to apply for grants also places additional pressure on age-friendly committees, even more so when this task falls to volunteers who lack the necessary skills (Winterton, 2016).

Power dynamics

The concept of power – and empowerment – is vast, and doing it justice (and the controversies within it) is well beyond the scope of this chapter. Here we draw only on the distinction between three dimensions of power (Lukes, 2005). Building on previous conceptualisations (e.g., Dahl, 1957; Bachrach and Baratz, 1962), Lukes (2005) describes: (1) overt power in the context of decisions, whereby one actor prevails over another; (2) covert control by decision makers that involves suppressing or thwarting challenges to their values or interests; and (3) the control of preferences accomplished by shaping perceptions, cognitions and preferences, which results in people not having grievances to begin with.

Power dynamics – within communities and in their relationships with organisations – affect decisions and activities in age-friendly initiatives. For example, who serves on an age-friendly committee (and who doesn't)? Who is invited to participate in community assessments (and who is not?). Who attends them (and who doesn't)? Who makes decisions over what actions are proposed and what is funded (what is left out; what is not funded?).

Rural communities are not exempt from power dynamics (Douglas, 2018; Menec, 2017; Shucksmith, 2018; Skinner and Winterton, 2018). At the local level, given the limited number of staff and often (necessarily) small pool of "super volunteers", one individual can have an enormous impact on an age-friendly initiative. On the positive side, one well-connected individual who champions the initiative and is able to shape decisions is a tremendous asset (Gallagher and Mallhi, 2010; Menec et al., 2014; Russell, Skinner and Fowler, 2019). Given that establishing contact with local decision makers can be easier in a small community than in a large urban centre (Menec et al., 2015a), such a champion can help to rapidly move proposed actions into reality. Overall, our research indicates that rural communities were able to get age-friendly initiatives off the ground and implement projects more quickly than urban centres (Menec et al., 2014) likely, in part, because age-friendly committees were able to influence decisions more easily. In part, this was also due to the nature of the projects – for example, decisions around adding one or two benches are more easily made than decisions over placing benches throughout a city.

Conversely, one person (or a few individuals) can also effectively stall an age-friendly initiative. Such individuals can become gatekeepers and block activities, either actively (e.g., by denying requests) or passively through inertia (e.g., by failing to make decisions) (Menec et al., 2014, 2015a). This can, of course, also happen in an urban centre; nevertheless, a larger population provides more opportunities to engage other people or organisations and, in effect, bypass a gatekeeper than is possible in a small community.

A dominant person or group can also more covertly exert influence over age-friendly initiatives by influencing whose perspectives are reflected in decisions. The community consultations we conducted in Manitoba will serve as an example. We relied on age-friendly committees to invite individuals to the consultations, stressing that attendees should reflect the diversity of older adults in the community, as well a range of key stakeholders (Menec et al., 2014; 2015b). As we did not collect personal information about attendees (the consultations were not designed as research projects), we do not know to what extent this goal was achieved. It is likely, however, that consultations did not reach more marginal groups in the community, such as cultural minorities.

On the one hand, this may be because there is an unconscious bias at play – age-friendly committee members may have invited people who are most easily reached and whom they considered most likely to attend a consultation, targeting "the usual suspects". On the other hand, who is invited (and who is not) may reflect overt or covert assumptions about the community. For example, a community that wants to brand itself as a retirement community for healthy, active older adults is likely to favour the input of younger, healthier residents and ignore the perspectives of frail older adults. In this way, marginalised groups (e.g., those disabled, low income, cultural or ethnic minorities) can be further marginalised not despite of but *because* of age-friendly initiatives (Keating, Eales and Phillips, 2013; Buffel and Phillipson, 2018).

Marginalised groups also tend not to participate in consultations (Fudge, Wolfe and McKevitt, 2007) for a variety of reasons, such as language barriers, the feeling that their input will not make a difference or will be ignored, or not wanting to criticize existing services, even if they don't meet their needs (Fudge, Wolfe and McKevitt, 2007; O'Keefe and Hogg, 1999). Ways to include them have been proposed (Ianniello et al., 2019), but presume that there is a will to do so on the part of the dominant group, which may not be the case. As Fillipis et al. note, "community-based" does not necessarily mean progressive or inclusive. Lukes' (2005) third dimension of power may also be at play, with the dominant group's view of what the community should look like remaining uncontested by minority groups.

Power dynamics are also reflected in rural communities' relationships with regional (provincial in the case of Canada) government. Political and economic decisions favour large urban centres, and rural communities have little recourse or influence over resource allocation (e.g., funding) and policy decisions. One might argue that power differentials would be minimised when governments lead age-friendly initiatives (e.g., as was the case in Manitoba); that is, when there is a top-down approach. However, this is not necessarily the case. For example, a key stakeholder we interviewed expressed the view that while government had effectively kick- started activities at the community level, there was a lack of needed, continuing financial government support, which left them feel abandoned and struggling to continue with the initiative (Spina and Menec, 2015). The process of being involved in the initiative was experienced as disempowering rather than empowering. To be equitable, there needs to be a willingness from all members in a partnership to share responsibilities and work towards a common goal (e.g., Shucksmith, 2018).

Conclusions

In this chapter we have identified some of the complexities of becoming age-friendly in a rural context. We offer three conclusions. First, the discourse on age-friendliness needs to address the challenges communities face in becoming more age-friendly, particularly the resource gaps outlined in this chapter. In doing so, the discourse needs to go beyond the community and must acknowledge the impact of, and the need to influence, broader policy and fiscal decisions. Second, a top-down approach, whereby federal or regional government (e.g., provincial, state) provides leadership for age-friendly initiatives is essential to support rural communities becoming more age-friendly. This must involve a commitment to not only support communities in initiating age-friendly activities (e.g., by providing start-up or small project grants) but also make necessary policy changes and larger investments in rural communities over the long term. Without such a commitment, the age-friendly movement runs the danger of becoming another responsibility offloaded onto communities. Third, age-friendly initiatives are not *ipso facto* inclusive. It behooves decision makers and researchers to be vigilant of power dynamics that may reinforce existing or create new inequities a result of age-friendly initiatives, thus further marginalising certain groups of older adults.

References

AARP. (2019). *About the AARP Community Challenge*. Available at: www.aarp.org/livable-communities/about/info-2017/aarp-community-challenge.html [Accessed 20 August 2019].

Bachrach, P. and Baratz, M. S. (1962). The two faces of power. *American Political Science Review*, 56(4), pp. 947–952.

Brasher, K. and Winterton, R. (2016). Whose responsibility? Challenges to creating an age-friendly Victoria in the wider Australian policy context. In: T. Moulaert and

S. Garon, eds., *Age-Friendly Cities and Communities in International Comparison.* Cham: Springer, pp. 229–245.

Buffel, T. and Phillipson, C. (2018). A manifesto for the age-friendly movement: developing a new urban agenda. *Journal of Aging and Social Policy*, 30(2), pp. 173–192.

Dahl, R. A. (1957). The concept of power. *Behavioral Science*, 2(3), pp. 201–205.

Dandy, K. and Bollman, R. D. (2008). Seniors in rural Canada (Catalogue no. 21–006-X). *Rural and Small Town Canada Analysis Bulletin.* Ottawa, ON: Statistics Canada.

DeFilippis, J., Fisher, R. and Shragge, E., (eds). (2010). *Contesting Community: The Limits and Potential of Local Organizing.* New Brunswick, NJ: Rutgers University Press.

Douglas, J. D. (2018). Governance in rural contexts: toward the formulation of a conceptual framework. *EchoGéo*, 43, pp. 1–20.

Federal/Provincial/Territorial Ministers Responsible for Seniors. (2007). *Age-Friendly Rural and Remote Communities: A Guide.* Ottawa, ON: Public Health Agency of Canada, Division of Aging and Seniors.

Fudge, N., Wolfe, C. D. A. and McKevitt, C. (2007). Involving older people in health research. *Age and Ageing*, 36(5), pp. 492–500.

Gallagher, E., and Mallhi, A. (2010). *Age-Friendly British Columbia: Lessons Learned from October 1, 2007-September 30, 2010.* Beaconsfield, QC: Canadian Electronic Library.

Garon, S., Paris, M., Beaulieu, M., Veil, A. and Laliberté, A. (2014). Collaborative partnership in age-friendly cities: two case studies from Quebec, Canada. *Journal of Aging and Social Policy*, 26(1–2), pp. 73–87.

Hanlon, N. and Halseth, G. (2005). The greying of resource communities in northern British Columbia: implications for health care delivery in already-underserviced communities. *Canadian Geographer*, 49(1), pp. 1–24.

Hanlon, N., Rosenberg, M. and Clasby, R. (2007). Offloading social care responsibilities: Recent experiences of local voluntary organisations in a remote urban centre in British Columbia, Canada. *Health and Social Care in the Community*, 15(4), pp. 343–351.

Ianniello, M., Iacuzzi, S., Fedele, P. and Brusati, L. (2019). Obstacles and solutions on the ladder of citizen participation: a systematic review. *Public Management Review*, 21(1), 21–46.

Kano, M., Rosenberg, P. E. and Dalton, S. D. (2018). A global pilot study of age-friendly city indicators. *Social Indicators Research*, 138, 1205–1227.

Keating, N., Eales, J. and Phillips, J. E. (2013). Age-friendly rural communities: conceptualizing 'best-fit.' *Canadian Journal on Aging*, 32(4), pp. 319–332.

Keating, N. and Phillips, J. (2008). A critical human ecology perspective on rural aging. In: N. Keating, ed., *Rural Ageing: A Good Place to Grow Old?.* Bristol: The Policy Press, pp. 1–9.

Lawton, M. P. and Nahemow, L. (1973). Ecology and the aging process. In: C. Eisdorfer and M. P. Lawton, eds., *The Psychology of Adult Development and Aging.* Washington, DC: American Psychological Association, pp. 619–674.

Lui, C., Everingham, J., Warburton, J., Cuthill, M., and Bartlett, H. (2009). What makes a community age-friendly: a review of international literature. *Australasian Journal on Ageing*, 28(3), pp. 116–121.

Lukes, S., (ed.). (2005). *Power : A Radical View*, 2nd ed. New York: Palgrave Macmillan.

Menec, V. (2017). Conceptualizing social connectivity in the context of age-friendly communities. *Journal of Housing for the Elderly*, *31*(2), pp. 99–116.

Menec, V. H. and Brown, C. (2018). Facilitators and barriers to becoming age-friendly: a review. *Journal of Aging and Social Policy.* Epub ahead of print 15 October. DOI: 10.1080/08959420.2018.1528116.

Menec, V. H., Bell, S., Novek, S., Minnigaleeva, G., Morales, E., Ouma, T., Parodi, J. F. and Winterton, R. (2015a). Making rural and remote communities more age-friendly: experts' perspectives of issues, challenges and priorities. *Journal of Aging and Social Policy*, 27(2), pp. 173–191.

Menec, V., Hutton, L., Newall, N., Nowicki, S., Spina, J. and Veselyuk, D. (2015b). How 'age-friendly' are rural communities and what community characteristics are related to age-friendliness? The case of rural Manitoba, Canada. *Ageing & Society*, 35(1), pp. 203–223.

Menec, V. H., Means, R., Keating, N., Parkhurst, G. and Eales, J. (2011). Conceptualizing age-friendly communities. *Canadian Journal on Aging*, 30(3), pp. 479–493.

Menec, V. H., Novek, S., Veselyuk, D. and McArthur, J. (2014). Lessons learned from a Canadian, province-wide age-friendly initiative: The Age-Friendly Manitoba Initiative. *Journal of Aging and Social Policy*, 26(1–2), pp. 33–51.

O'Keefe, E. and Hogg, C. (1999). Public participation and marginalized groups: the community development model. *Health Expectations*, 2(4), pp. 245–254.

Plouffe, L. A. and Kalache, A. (2010). Towards global age-friendly cities: determining urban features that promote active aging. *Journal of Urban Health*, 87(5), pp. 733–39.

Plouffe, L. A. and Kalache, A. (2011). Making communities age friendly: State and municipal initiatives in Canada and other countries. *Gaceta Sanitaria*, 25(Suppl. 2), pp. 131–137.

Public Health Agency of Canada. (n.d.). *How Can Canadian Communities Become More Age-Friendly*? Available at: www.canada.ca/en/public-health/services/health-promotion/aging-seniors/friendly-communities.html [Accessed 20 August 2019].

Public Health Agency of Canada. (2012). *Age-Friendly Communities in Canada: Community Implementation Guide*. Ottawa, ON: Her Majesty the Queen in Right of Canada.

Public Health Agency of Canada. (2015). *Age-Friendly Communities Evaluation Guide: Using Indicators to Measure Progress*. Ottawa, ON: Minister of Health.

Russell, E., Skinner, M. W. and Fowler, K. (2019). Emergent challenges and opportunities to sustaining age-friendly initiatives: Qualitative findings from a Canadian age-friendly funding program. *Journal of Aging and Social Policy*. DOI: 10.1080/08959420.2019.1636595.

Ryser, L. and Halseth, G. (2012). Resolving mobility constraints impeding rural seniors' access to regionalized services. *Journal of Aging and Social Policy*, 24(3), pp. 328–344.

Scharf, T., Walsh, K. and O'Shea, E. (2016). Ageing in rural places. In: M. Shucksmith and D. L. Brown, eds., *International Handbook of Rural Studies*. London: Routledge, pp. 80–91.

Shucksmith, M. (2018). Re-imagining the rural: from rural idyll to Good Countryside. *Journal of Rural Studies*, 59, pp. 163–172.

Skinner, M. W. (2008). Voluntarism and long-term care in the countryside: the paradox of a threadbare sector. *The Canadian Geographer*, 52(2), pp. 188–203.

Skinner, M. W., Joseph, A. E., Hanlon, N., Halseth, G., and Ryser, L. (2014). Growing old in resource communities: exploring the links among voluntarism, aging, and community development. *The Canadian Geographer*, 58(4), pp. 418–428.

Skinner, M. W. and Winterton, R. (2018). Interrogating the contested spaces of rural aging: implications for research, policy, and practice. *The Gerontologist*, 58(1), pp. 15–25.

Spina, J. and Menec, V. (2015). What community characteristics help or hinder rural communities in becoming age-friendly? Perspectives from a Canadian prairie province. *Journal of Applied Gerontology*, 34(4), pp. 444–464.

Statistics Canada. (2017). *Census Profile. 2016 Census* (Catalogue no. 98-316-X2016001. Ottawa, ON: Statistics Canada. [Accessed 15 January 2020].

Sun, Y., Chao, T.-Y., Woo, J., and Au, D. W. H. (2017). An institutional perspective of "glo-calization" in two Asian tigers: the "structure-agent-strategy" of building an age-friendly city. *Habitat International*, 59, pp. 101–109.

United Nations. (2002). *Madrid International Plan of Action on Ageing*. New York: United Nations.

Walsh, K., O'Shea, E., Scharf, T. and Shucksmith, M. (2014). Exploring the impact of informal practices on social exclusion and age-friendliness for older people in rural communities. *Journal of Community and Applied Social Psychology*, 24(1), pp. 37–49.

Winterton, R. (2016). Organizational responsibility for age-friendly social participation: views of Australian rural community stakeholders. *Journal of Aging and Social Policy*, 28(4), pp. 261–276.

Winterton, R. and Warburton, J. (2011). Does place matter? Reviewing the experience of disadvantage for older people in rural Australia. *Rural Society*, 20(2), pp. 187–197.

World Bank. (2018). *Rural Population (% of Total Population)*. Available at: https://data.worldbank.org/indicator/SP.RUR.TOTL.ZS [Accessed 15 January 2020].

World Health Organization. (n.d.). *Age-Friendly World*. Available at: https://extranet.who.int/agefriendlyworld/ [Accessed 15 January 2020].

World Health Organization. (2007). *Global Age-Friendly Cities: A Guide*. Geneva: World Health Organization.

World Health Organization. (2015). *Measuring the Age-Friendliness of Cities: A Guide to Using Core Indicators*. Kobe: WHO Centre for Health Development.

World Health Organization. (2018). *The Global Network for Age-Friendly Cities and Communities: Looking Back Over the Last Decade, Looking Forward to the Next*. Geneva: World Health Organization.

15 Rural ageing in place and place attachment

Janine Wiles, Robin Kearns and Laura Bates

Introduction

Place is a tightly woven gestalt between location, people and identity (reflected in the term "place-in-the-world"), and place attachment is central to the achievement of healthy ageing. Most older people in rural places prefer to "age in place", or "stay put" (Gilleard et al., 2007). For many this ideally involves remaining in their own homes, or at least the same community. In rural communities, place attachment and the ability to age in place may be challenging. In a context of age-related losses, relationships with place help to sustain and build identity and a sense of self; but in rural areas the significance of place can also be under threat (Wenger, 2001; Winterton and Warburton, 2012) (see Berry, Chapter 2 in this volume). Whereas urban residents might move into supported living in the same neighbourhood or city and retain links with familiar places and people, many rural places preclude this; there are less opportunities to move between independent living and levels of residential care and stay "in place".

In this chapter, we explore relationships between attachment to place and ageing for older residents of rural communities. We examine existing literature on ageing and attachment in rural places and draw on examples from our research. Place attachment is not solely related to intrinsic material qualities of places themselves; we recognise the importance of a sense of belonging, identity and emotional affect and social factors such as connectedness, reciprocity and community involvement, as interrelated elements connected with positive feelings of place attachment. While recognising rural places are not either simply idyllic nor exclusively difficult places in which to age well (Hanlon and Kearns, 2016), we posit place attachment as means/resource and motivation for, and sometimes an outcome of, ageing in rural places. Thus, we emphasise the wide variety and complexity of ways in which people express or develop positive attachment. Whereas for some, social connections may be important, others may prioritise different modes or aspects of attachment. Moreover, not all older adults feel attachment to place in the same ways, or even at all.

Place attachment

Place attachment is an expression of affective and positive person-to-place bonds that evolve through the connections and understandings that people develop with a specific place over time (Oswald and Wahl, 2004; Rowles, 1978; Rubinstein, 1990). This concept is especially relevant to older people: a long life lived in a familiar home and neighbourhood leads to the accumulation of place-related memories and significances, which include interrelated physical, social, temporal and psychological factors (Burholt, 2006).

Attachment to place is associated with feelings of safety and an increased autonomy and independence (potentially meaning less reliance on external support services such as residential care) (Rubinstein and Parmalee, 1992; Wiles et al., 2017). Good knowledge of place facilitates flexibility and adaptation in the face of declining health and independence, including awareness of local services and amenities as well as the availability of practical and emotional support from others (Gitlin, 2003; Kontos, 1998; Lawton, 1985). Social support and integration developed through connections over time can be practical resources for staying in a location (Burholt, 2006). In terms of meaning and identity, place attachment may also facilitate positive feelings of pride, familiarity and belonging, or what Rowles (1993) calls "autobiographical insideness". We thus argue place attachment acts as both *motivation* and *means* for (and sometimes an *outcome of*) ageing in rural places.

Rural place attachment

Place attachment for older people in rural areas may operate differently than for those in more (sub)urban environments (Burholt, 2006). Rural areas have smaller and more dispersed older populations and commonly feature sparse specialist health and social support and other services, funding constraints, limited infrastructure, isolation and poverty and sometimes harsh climates. They are often subject to policy developed and implemented from afar and aimed at urban contexts (Keating, 2008; Weinhold and Gurtner, 2014). Yet being older and a rural resident can mean one is better known and more visible within the community. Compared to their urban counterparts, older rural residents may be more connected to other residents as well as to community-based service providers and often make significant contributions to the well-being of rural places (Winterton and Warburton, 2014). Further, for service providers there can be greater opportunities for coordination across sectors in rural areas (Hanlon and Kearns, 2016), although this can be stretched too thin when too many demands are made, or there is over-reliance on non-governmental services (Skinner and Rosenberg, 2006; Winterton and Warburton, 2012).

Long-standing relationships with place also offer symbolic connections, including shared history and experience of community (Rowles, 1978; Rubinstein, 1989). This depth of connection to locality is crucial when rural communities are experiencing significant long-term change, such as service-sector closures and rural population decline (Joseph and Cloutier-Fisher, 2005; Kearns et al., 2009; Winterton and Warburton, 2012). Counterurbanisation and over-tourism bring additional challenges to the character and identity of rural communities. Expressions of place attachment such as community newspapers, memorial seats and events at community halls can help keep the past alive for older residents, keeping them loyal to a tightly defined yet changing landscape. Hence, when publicly provided amenities like clinics, schools or postal agencies close they can herald not only the end of a service but also sever memories and symbols that connect people to their pasts (Herron and Skinner, 2013; Kearns et al., 2009). Experiences of place attachment are thus undermined by both actual and threatened disruption to the physical or social landscape. Closure of the only general store or local branch of a bank can be as disruptive and distressing as the spectre of a hillside carved up for new housing. Both examples suggest breaches in the fabric of connection to the wider familiar place of residence beyond the home itself. Yet the collective reservoir of place attachment can be mobilised to promote social commitment, often expressed through community involvement and institutional ties to a locality (Keeling, 2001; Manzo and Perkins, 2006).

If issues of isolation and under-resourcing can be toxic to the well-being of rural populations, the countertrend can be the trust and collective enterprise that can emerge in small communities when expertise is shared and solidarity develops (Kearns, 1998). Mobilising across interest groups in defence of one's local place is a common activity in rural communities with benefits beyond the retention of services or amenities under threat, such as re-invigorating community cohesion (Skinner and Joseph, 2007). In times of uncertainty, re-examining what is valued about the locality and acting to maintain its integrity can serve as an anchor, galvanising collective identity and place attachment (see Kearns et al., 2009 re: rural school closure and Collins and Kearns, 2013 re: resistance to a new housing development). Yet repeated and multiple losses of services and resources in rural communities can stretch informal and voluntary support sectors beyond their ability to cope (Hanlon and Halseth, 2005; Skinner et al., 2008), making older residents more vulnerable to displacement or institutionalisation (Cloutier-Fisher and Joseph, 2000). While older residents can offer a longer-term perspective on the ups and downs over time in such struggles, they can also feel distress most acutely when battles are lost (e.g., Kearns and Collins, 2012).

Services and supports play a significant and multi-dimensional role in developing and sustaining place attachment in rural settings. It is challenging and expensive to provide meaningful formal health and social resources like home care or residential aged care in rural places. This is exacerbated in regions of population and industry decline (Skinner and Hanlon, 2016), where older people can rapidly become vulnerable people in vulnerable places (Joseph and Cloutier-Fisher, 2005).

Yet when such facilities are available, they are often highly valued by their communities and play roles well beyond the tangible and functional provision of services. For example, residential facilities and formal services are often complemented by robust informal support systems and social networks of exchange.

These may be personal networks such as exchanges of surplus fruit and vegetables, carpooling for shopping or appointments or getting help from families or neighbours. Institutional and locally appropriate residential care facilities, clinics or post offices in small rural communities can also become focal points for social connection and sense of identity, or *de facto* community centres (Kearns, 1991, 1998).

Well-being in older age is a2chieved through both personal effort and the support of social relationships, in the context of being in-place and feeling at home (Rowles, 1993; Sixsmith et al., 2014). There are social and connective aspects of attachment to place, and many older people in rural communities (especially those who have lived there for a long time) may themselves be "rich" in supportive resources, with strong networks of social exchange and interdependence (Rubinstein and Parmalee, 1992). Place attachment for such older people is often deeply entwined with a lifetime of reciprocity, underscored by their own historical and ongoing contributions to their rural communities (Burholt, 2006). This is what many people value about rural living but, with the onset of health challenges in later life, also what is potentially difficult about it (Winterton and Warburton, 2011).

Ideas and narratives about place itself are often deployed as supportive resources, especially with reference to enabling physical and social characteristics. Yet these characteristics of rural places can create extra challenges, where there may be greater expectations of rural people to contribute or to be more resilient (Blackstock et al., 2006; Meijering et al., 2017). Perceptions of "closeness" and intimacy in rural settings, for example, may further isolate older people who are more transitory, who do not have social or financial resources or who are perceived as being "different" in ways that are not accepted or embraced by their communities (see McCarthy, 2000; Parr and Philo, 2003; Scharf and Bartlam, 2008).

Some older people may be "stuck" in rural places, living in poorly maintained or otherwise unsuitable housing, or unable to capitalise resources to move to more suitable or supportive accommodation (Golant, 2015). Others may have few alternative options for local accommodation if they wish to age in place or increasing difficulties with mobility associated with age-related losses (Cloutier-Fisher and Joseph, 2000; Walker et al., 2012). Nonetheless, attachment to a place (as in locality, community, neighbourhood) may be sufficiently strong to eclipse these issues (Bates et al., 2020).

Thus, we should not assume rural places to be idyllic for ageing in place, nor that everyone can feel strong attachment to rural places; but at the same time, nor should we assume rural places to be devoid of supportive resources. Rather, we acknowledge that for some, rural living conditions may be challenging or even toxic, whereas for others a sense of attachment to place may be a supportive resource; for still others, a sense of attachment may become an impediment to moving out of less-than-satisfactory housing or community conditions (Golant, 2015; Keating, 2008; Scharf and Bartlam, 2008; Winterton and Warburton, 2011).

In the remainder of this chapter, we explore three case study examples from our own research, illustrating the role of attachment to rural places. Each offers insight into the multifaceted and ongoing relationship between place and ageing "well" for older people in rural communities. The first example illustrates long-standing social connections anchored in place and how a sympathetic health care service is moulded around local aspirations and cultural norms. The second highlights the situation of older people based in a semi-rural island community in the context of a metropolitan and tourism-driven housing market. We show that strong bonds with familiar land/seascapes and a robust sense of community lead to tenacity despite challenging costs of living and distance from all but basic services. The third example shows how place attachment among primary care service providers is a resource that can support rural older people's well-being and access to support.

A place-embracing health system in the Hokianga district

Our first case study exemplifies the role of social connection in place attachment and comes from the Hokianga district of northern New Zealand, which is approximately 300km north of Auckland and has a long history of community-based health care. It is the economically poorest part of the Northland region and in the 1990s, 67% of adults were either unemployed or not actively seeking work (Kearns and Reinken, 1994). Over half of the population identified as indigenous Māori. Services in the area have been free to all users (unlike most other places in New Zealand) since the mid-twentieth century, when a "special medical area" was established to address issues of accessibility and acceptability. Pioneering doctor G. M. Smith recognised that it was an unreasonable expectation that financially poor and residentially remote patients travel to a centralised location to receive services, so doctors travelling to community clinics was established as the norm. Since then the district has opted to run its own health service through a health trust with representation from each local community that has an outpatient clinic. A policy of not charging patients for care (unlike most places in New Zealand) has endured (Kearns, 1998).

Research at community clinics in the early 1990s showed they were serving as medical centres and were important as *de facto* community centres. This was confirmed during research when older residents were observed coming to the clinic waiting rooms to meet and engage with others even when they did not need medical attention. Those who did have an appointment had a relaxed approach to clinic attendance and were often observed arriving well before their appointment time and lingering long afterwards "having a yarn". Lively conversations about a range of topics, including community well-being and concern for the threats to the local health service by restructuring processes at work in the wider region, suggested that, for many (and particularly older) people in these otherwise isolated rural communities, "clinic day" offered an opportunity to reconnect with others and exchange community news (Kearns, 1991). It was evident that here, going to the doctor was more than a medical interaction. Not infrequently, spare produce was brought to the clinic waiting rooms and a kuia (female elder) greeted each person in the reception area with a kiss. Far from being passive spaces "containing" people prior to their consultations, these waiting rooms were lively venues for local expression of the significance of community and place in peoples' lives. For long-time "locals", the confidence with which they expressed attachment to, and being at-home in, the clinic reflected the deeper Māori value of *turangawaewae* or having a place to stand.

Social connections are a necessary but not sufficient aspect of place attachment for older people in this rural area, however. For Māori in particular, the land and topographic features are of deep significance in the affective glue that connects people to place. In a *pepeha* (a way of introducing oneself in *te reo me ngā tikanga Māori* – Māori language and protocol), it is customary to identify the mountain and the river one associates with by birth and/or ancestry. This significance flows into expressions of place attachment and efforts by health care providers to accommodate a connection between people and place. Within the community hospital in the township of Rawene, on the shores of the Hokianga Harbour, for instance, there is further evidence of attachment and links between older people and local places. Here, the ten-bed aged care ward is situated in an historic building such that it allows older patients to look out through wide windows across the harbour towards the prominent hills on the north side that are symbols of identity and belonging with their Māori communities. Hence even if they are no longer able to return to their local community across the harbour, place attachment can still be acknowledged and enacted through viewing from afar. This dynamic reflects Jivén and Larkham's (2003) contention that observing the landscape and being within sight of familiar landmarks often informs and reflects place attachment in acute ways.

Older island residents

Our second example comes from Waiheke Island, which is part of Auckland City and 23km from the city centre. The bounded character of an island and its distance from a mainland can have varied implications for older residents' lives. On the one hand, island living can shrink an older person's support network because when friends or family members move off-island, distance, time and financial barriers can develop between the older person and their social networks (Coleman and Kearns, 2015; Róin, 2015). Alternatively, older people may be supported and enabled by the cohesive communities that may often be found in island settings (Burholt et al., 2013; Róin, 2015).

The heightened place attachment that comes from a small-scale island community and strong bonds with familiar land/seascapes can also bolster older islanders' well-being (Coleman and Kearns, 2015; Róin, 2015). But rural island life can be challenging in light of over-tourism and the arrival of new residents who may alter or even destroy the collective identity and community character (Baldacchino, 2012). Tourism may benefit a rural island's economy, but older residents may also feel outnumbered, if not displaced, by visitors who are perceived to change or contradict the island's character (Baldacchino, 2012; Burholt et al., 2013).

Waiheke Island has approximately 8,000 permanent residents. The island's population is ageing, with proportionately more residents aged 55+ than in mainland Auckland. Many older residents are "long-term islanders" who have formed strong place attachments (Coleman and Kearns, 2015). The island is also a desirable retirement destination for those who may have little or no previous connections. The motivations and interests for tourists and older residents overlap in terms of their enjoyment of scenic landscapes and a general sense of well-being experienced in/through island places. Waiheke is characterised by small neighbourhoods, rural open farmland, wineries and vineyards and a mixture of rocky and sandy beaches. Its holiday "feel", tourist appeal and detachment from mainland Auckland's busyness contribute to an idealised island image and identity. Like the long-stay patients at Rawene Hospital, older residents expressed the strength of their place attachment by speaking of the importance of having views of local beaches and the sea. Vantage points from the comfort of their homes offered consolation during times of unwellness or recovery and enabled residents to recall times when they were able to be more actively engaged with coastal activities. At a deeper level, they served as reminders of the more primal rhythms of life. Polly, for instance, reported "think[ing] of myself as simply travelling from shore to shore, rather than thinking 'this hurts' or 'that hurts' or 'when will the pain stop'. I am just on a journey and journeys begin and end" (Coleman and Kearns, 2015).

Recent work with older renters, a minority compared to homeowners, examined how uncertainties related to ageing, housing and community influence experiences of rural islandness in the context of the influx of tourists and new residents (Bates et al., 2019a, 2019b). As interviews proceeded, it became clear that attachment to the island-at-large often rivalled or surpassed attachment to the dwelling itself.

This place attachment encompassed Waiheke's small neighbourhoods, ocean views, proximity to multiple beaches and a strong sense of community. Some of these island and community aspects were said to be unique to Waiheke; as Caroline noted, "the island just has something special about it". Eliza also explained a range of special island characteristics: "I like the sea views, I can walk to the beach. It's an excellent community . . . we are united". Beaches facilitated social interactions and enjoyment of Waiheke's scenic features, thereby contributing to older participants' enjoyment of island life (Bates et al., 2019a).

Such expressions of place attachment and familiarity with the island community are resources older people may draw resilience from (Wiles et al., 2017). Yet, living on the island was also experienced as inherently precarious. Participants explained challenges associated with Waiheke's distance from hospitals and some health care and age-related services and heightened living costs due to tourism popularity and the increased cost of shipping goods from mainland Auckland.

Some participants became tearful while reflecting on how the island had changed since they had first moved to Waiheke. However, a strong sense of home and attachment to the island endured for many participants. They often attributed this place attachment to the same long-standing emotional connections to island places and people that made them upset to observe changes in their community. Participants commonly expressed a desire to remain on the island "no matter what", and the two participants who moved off-island during our research period explained that they would have preferred to stay had there been suitable housing opportunities available to them. Well-established social networks, place familiarity, local knowledge and a strong sense of home are resources through which older renters can develop independence and resilience (Bates et al., 2019b).

Older people and primary care providers

Finally, we present another New Zealand-based study, of primary health practitioners' views on opportunities and challenges of working with people in advanced age (85+) in rural Bay of Plenty communities. Many practitioners had lived and worked in these communities for long periods of time (often several decades) and/or had specifically chosen to work there because they perceived them as attractive. The study shows how place attachment and the experience of "insiderness" of rural older people's supporters, such as primary professionals, can be a beneficial resource for ageing and speaks to the deep connection between rurality, ageing and place attachment (see also Farmer et al., 2003).

Rural primary practitioners readily identified aspects of their practice unique to, and challenging about, being rural. Most had high proportions of older patients because they were living in attractive retirement destinations because a population had aged in place, or because younger people had moved away for work. Many older patients did not have family living nearby, with implications for supportive resources (e.g., help attending appointments) and quality of care (e.g., family to organise and monitor homecare). Practitioners highlighted reduced formal resources within rural communities; fewer options for community-based care, long distances to access specialist services and limited local public transport or taxi options, posing particular challenges for older people who could no longer drive (see Hansen et al., Chapter 12). Yet practitioners also emphasised the supportive and enabling aspects of attachment and ageing in rural places. Many commented that having good knowledge of their older patients and their families was advantageous in providing more contextualised, holistic care. They often had long-term relationships with their older patients. In most practices, doctors, nursing staff, or allied pharmacists made home visits, which they positioned as helping older people to remain in their dwelling place of choice, as one family doctor explained:

> I do house calls. And I find them very, very helpful, because you learn how someone lives, what their house is like. . . . You can also go through all their medications with them and I find that has been very, very revealing.

Networks and collaboration between different primary professional groups developed and grown over time also enhanced quality of service. A nurse noted how their practice has "a good dialogue" with the local pharmacist, who does home visits for medication reviews and also, "they'll let us know if someone hasn't picked up their pills . . . there's that sort of communication which is quite good that way in the community. Being a smaller community, we can offer these things". Several practitioners commented on the paradox that although rural people are spatially isolated, they are less socially isolated than their urban counterparts, with better social connectedness, reciprocity, mutual respect and a sense of belonging all mitigating spatial isolation. A doctor stated:

> In my experience of working in London and in cities in New Zealand, it is much easier for people here . . . because they're still a valued part of the community. Someone will pick them up and take them to things, they still have people who will knock on their door and drop off [food] . . . the isolation is not as bad as I think it can be in the cities even though we're in the middle of nowhere.

Practitioners also identified the importance of their clinic waiting rooms providing opportunities for connection and a sense of belonging. In the same isolated community mentioned earlier, another doctor observed, "I think [the older people] love coming, often they love coming in and they sit in the waiting room for a couple of hours and chat and catch up [with] each other which they quite enjoy. They entertain everyone in the waiting room."

This observation – that there is a social significance to sites of rural primary health care provision – connects to our first example from the Hokianga district.

Thus, service providers perceive rural communities as generative of social capital and connectedness. Practitioners presented this as a resource increasing the quality of care and support provided for older people, despite structural threats and challenges in primary health care.

Conclusion

Our examples of Waiheke Island and the relatively isolated areas of Hokianga and rural Bay of Plenty offered cases in which one could attribute experiences of place attachment to intrinsic material qualities of the places themselves: their boundedness and particular character. However, we contend that there is a risk in over-attribution of place attachment to the tangible qualities of places themselves: rather it is important to recognise the critical role of belonging, social connectedness, reciprocity and community involvement in attachment to place.

Experience of a rural place is deeply and recursively connected to people's identity or "place-in-the-world" (Eyles, 1985). To feel attached to a place is to feel rooted to a locality and contribute to a sense of identity. It also involves having accumulated memories and having been moulded by the mutuality of relationships that the smaller scale of rural communities facilitates. While we can and should celebrate this, we should also be mindful that such attachment is not a given. Nor is place attachment the same for every older person or group of older people in rural places. For those unable to develop a strong place attachment a double jeopardy develops, as not feeling attached to place reinforces or exacerbates other aspects of difference or lack of resources (including privilege and belonging).

Place attachment is a complex process which acts as both a *motivator* and a *means* for ageing in rural places, as well as often being an *outcome* of ageing in such places. A sense of belonging can motivate older people to remain in place. Place attachment can also enhance autonomy and help with adapting to the evolution of circumstances and challenges that come with the ageing process. For many older people, attachment to place is also an outcome of time spent in a place, developing connections to people and ideas about a place as well as the material and aesthetic qualities of that place.

References

Baldacchino, G. (2012). Come visit, but don't overstay: critiquing a welcoming society. *International Journal of Culture, Tourism and Hospitality Research*, 6(2), pp. 145–153.

Bates, L., Coleman, T., Wiles, J. and Kearns, R. (2019a). Older residents' experiences of islandness, identity and precarity: ageing on Waiheke Island. *Island Studies Journal*, 14, 171–192.

Bates, L., Kearns, R., Coleman, T. and Wiles, J. (2020). 'You can't put your roots down': housing pathways, rental tenure and precarity in older age. *Housing Studies*, 35, 1442–1467.

Bates, L., Wiles, J., Kearns, R. and Coleman, T. (2019b). Precariously placed: home, housing and wellbeing for older renters. *Health and Place*, 58, p. 102152.

Blackstock, K., Innes, A., Cox, S., Smith, A. and Mason, A. (2006). Living with dementia in rural and remote Scotland: diverse experiences of people with dementia and their carers. *Journal of Rural Studies*, 22(2), pp. 161–176.

Burholt, V. (2006). 'Adref': Theoretical contexts of attachment to place for mature and older people in rural North Wales. *Environment and Planning A: Economy and Space*, 38(6), pp. 1095–1114.

Burholt, V., Scharf, T. and Walsh, K. (2013). Imagery and imaginary of islander identity: older people and migration in Irish small-island communities. *Journal of Rural Studies*, 31, pp. 1–12.

Cloutier-Fisher, D., and Joseph, A. (2000). Long-term care restructuring in rural Ontario: retrieving community service user and provider narratives. *Social Science and Medicine*, 50(7–8), pp. 1037–1045.

Coleman, T. and Kearns, R. (2015). The role of bluespaces in experiencing place, aging and wellbeing: insights from Waiheke Island, New Zealand. *Health and Place*, 35, pp. 206–217.

Collins, D. and Kearns, R. (2013). Place attachment and community activism at the coast: the case of Ngunguru, Northland. *New Zealand Geographer*, 69(1), pp. 39–51.

Eyles J., (ed.). (1985). *Senses of Place*. Warrington: Silverbrook Press.

Farmer, J., Lauder, W., Richards, H. and Sharkey, S. (2003). Dr. John has gone: assessing health professionals' contribution to remote rural community sustainability in the UK. *Social Science and Medicine*, 57(4), pp. 673–686.

Gilleard, C., Hyde, M. and Higgs, P. (2007). The impact of age, place, aging in place, and attachment to place on the well-being of the over 50s in England. *Research on Aging*, 29(6), pp. 590–604.

Gitlin, L. (2003). Conducting research on home environments: lessons learned and new directions. *The Gerontologist*, 43(5), pp. 628–637.

Golant, S., (ed.). (2015). *Aging in the Right Place*. Baltimore, MD: Health Professions Press.

Hanlon, N. and Halseth, G. (2005). The greying of resource communities in northern British Columbia: implications for health care delivery in already-underserviced communities. *The Canadian Geographer*, 49(1), pp. 1–24.

Hanlon, N. and Kearns, R. (2016). Health and rural places. In: M. Shucksmith and D. Brown eds., *Routledge International Handbook of Rural Studies*. London: Routledge, pp. 92–100.

Herron, R. and Skinner, M. (2013). The emotional overlay: older person and carer perspectives on negotiating aging and care in rural Ontario. *Social Science and Medicine*, 91, pp. 186–193.

Jivén, G. and Larkham, P. (2003). Sense of place, authenticity, and character: a commentary. *Journal of Urban Design*, 8(1), pp. 67–81.

Joseph, A. and Cloutier-Fisher, D. (2005). Ageing in rural communities: vulnerable people in vulnerable places. In: G. Andrews and D. Phillips, eds., *Ageing and Place*. London: Routledge, pp. 149–162.

Kearns, R. (1991). The place of health in the health of place: the sase of the Hokianga special medical area. *Social Science and Medicine*, 33(4), pp. 519–530.

Kearns, R. (1998). Going it alone: community resistance to health reforms in Hokianga, New Zealand. In: R. Kearns and W. Gesler, eds., *Putting Health into Place: Landscape, Identity and Well-Being*. Syracuse: Syracuse University Press, pp. 226–247.

Kearns, R. and Collins, D. (2012). Feeling for the coast: the place of emotion in resistance to residential development. *Social and Cultural Geography* 13, pp. 937–955.

Kearns, R., Lewis, N., McCreanor, T. and Witten, K. (2009). 'The status quo is not an option': community impacts of school closure in South Taranaki, New Zealand. *Journal of Rural Studies*, 25, pp. 131–140.

Kearns, R. and Reinken, J. (1994). Out for the count? Questions concerning census enumeration in the Hokianga. *New Zealand Population Review*, 20(1–2), pp. 19–30.

Keating, N., (ed.). (2008). *Rural Ageing: A Good Place to Grow Old?* Bristol: Policy Press.

Keeling, S. (2001). Relative distance: ageing in rural New Zealand. *Ageing & Society*, 21, pp. 605–619.

Kontos, P. (1998). Resisting institutionalization: constructing old age and negotiating home. *Journal of Aging Studies*, 12(2), pp. 167–184.

Lawton, M. (1985). The Elderly in context: Perspectives from environmental psychology and gerontology. *Environment and Behavior*, 17(4), pp. 501–519.

McCarthy, L. (2000). Poppies in a wheat field: exploring the lives of rural lesbians. *Journal of Homosexuality*, 39(1), pp. 75–94.

Manzo, L. and Perkins, D. (2006). Finding common ground: the importance of place attachment to community participation and planning. *Journal of Planning Literature*, 20(4), pp. 335–350.

Meijering, L., Lettinga, A., Nanninga, C. and Milligan, C. (2017). Interpreting therapeutic landscape experiences through rural stroke survivors' biographies of disruption and flow. *Journal of Rural Studies*, 51, pp. 275–283.

Oswald, F. and Wahl, H. (2004). Housing and health in later life. *Review of Environmental Health*, 19(3–4), pp. 223–252.

Parr, H. and Philo, C. (2003). Rural mental health and social geographies of caring. *Social and Cultural Geography*, 4(4), pp. 471–488.

Róin, A. (2015). The multifaceted nature of home: exploring the meaning of home among elderly people living in the Faroe Islands. *Journal of Rural Studies*, 39, pp. 22–31.

Rowles, G., (ed.). (1978). *Prisoners of Space? Exploring the Geographical Experiences of Older People*. Boulder, CO: Westview Press.

Rowles, G. (1993). Evolving images of place in aging and 'aging in place'. *Generations*, 17(2), pp. 65–70.

Rubinstein, R. (1989). The home environments of older people: a description of the psychosocial processes linking person to place. *Journal of Gerontology, Social Sciences*, 44(2), pp. S45–S53.

Rubinstein, R. (1990). Personal identity and environmental meaning in later life *Journal of Aging Studies*, 4(2), pp. 131–147.

Rubinstein, R., and Parmalee, P. (1992). Attachment to place and representation of life course by the elderly. In: I. Altman and S. Low, eds., *Human Behaviour and Environment, Vol. 12: Place Attachment*. New York: Plenum Press, pp. 139–163.

Scharf, T. and Bartlam, B. (2008). Ageing and social exclusion in rural communities. In: N. Keating, ed., *Rural Ageing: A Good Place to Grow Old?*. Bristol: Policy Press, pp. 97–108.

Sixsmith, J., Sixsmith, A., Fänge, A. M., Naumann, D., Kucsera, C., Tomsone, S., . . . Woolrych, R. (2014). Healthy ageing and home: The perspectives of very old people in five European countries. *Social Science and Medicine*, 106, pp. 1–9.

Skinner, M. and Hanlon, N., (eds). (2016). *Ageing Resource Communities: New Frontiers of Rural Population Change, Community Development and Voluntarism*. London: Routledge.

Skinner, M. and Joseph, A. (2007). The evolving role of voluntarism in ageing rural communities. *New Zealand Geographer*, 63, pp. 119–129.

Skinner, M. and Rosenberg, M. (2006). Managing competition in the countryside: non-profit and for-profit perceptions of long-term care in rural Ontario. *Social Science and Medicine*, 63(11), pp. 2864–2876.

Skinner, M., Rosenberg, M., Lovell, S., Dunn, J., Everitt, J., Hanlon, N. and Rathwell, T. (2008). Services for seniors in small town Canada: The paradox of community. *Canadian Journal of Nursing Research*, 40(1), 81–101.

Walker, J., Orpin, P., Baynes, H. and Stratford, E. (2012). Insights and principles for supporting social engagement in rural older people. *Ageing & Society*, 33(6), pp. 938–963.

Weinhold, I. and Gurtner, S. (2014). Understanding shortages of sufficient health care in rural areas. *Health Policy*, 118, pp. 201–214.

Wenger, G. (2001). Myths and realities of ageing in rural Britain. *Ageing & Society*, 21(1), pp. 117–130.

Wiles, J., Rolleston, A., Pillai, A., Broad, J., Teh, R., Gott, M. and Kerse, N. (2017). Attachment to place in advanced age: a study of the LiLACS NZ cohort. *Social Science and Medicine*, 185, p. 27.

Winterton, R. and Warburton, J. (2011). Does place matter? Reviewing the experience of disadvantage for older people in rural Australia. *Rural Society*, 20, pp. 187–197.

Winterton, R. and Warburton, J. (2012). Ageing in the bush: The role of rural places in maintaining identity for long term rural residents and retirement migrants in north-east Victoria, Australia. *Journal of Rural Studies*, 28(4), pp. 329–337.
Winterton, R. and Warburton, J. (2014). Healthy ageing in Australia's rural places: the contribution of older volunteers. *Voluntary Sector Review*, 5(2), pp. 181–201.

16 Place-bound rural community of older men

Social and autobiographical insideness of the Mill Village Boys in Finland

Marjaana Seppänen, Elisa Tiilikainen, Hanna Ojala and Ilkka Pietilä

Introduction

Older people's attachment to their physical environment and, in particular, their current home has been the topic of a considerable amount of research across international contexts. Age, it has been argued, is associated with intensified attachment, both physically and mentally, to one's place of residence (Gilleard, Hyde and Higgs, 2007; Seppänen et al., 2012). In the lives of rural older adults, place plays an especially central role: it represents something that is not only valued, but also struggled with, due to disadvantages associated with rural living (Walsh, O'Shea and Scharf, 2019).

Place itself, in more sociological studies, has been approached as one of the central elements in the creation of community (e.g., Means and Evans, 2012). In countries like Finland, where urbanisation occurred relatively late compared to elsewhere in Europe, nostalgic visions about small, intimate rural communities often present these settings as examples of well-functioning and mutually supportive communities (see Pessi and Seppänen, 2011). In general, moreover, there has been a tendency to focus on positive aspects of life in rural communities, such as the feelings of safety and security they afford to their members (McKenzie and Frencken, 2001). Attachment to the rural landscape is often seen as a provider of comfort and continuity, making up a key building block of rural older persons' identity (Gullifer and Thompson, 2006).

Even where place-based communities are understood to have disappeared, the meaning of place has shown its continuous importance, especially in connection with the phenomenon of "ageing in place". Today, the latter is globally seen as not just a worthwhile but also attainable goal, providing the guiding principle of ageing policies in many countries (see, e.g., Jolanki and Vilkko, 2015; Vasunilashorn et al., 2012; see Menec and Novek, Chapter 14 in this volume). Indeed, the majority of rural older adults still prefer to live in their own homes, or at least in the communities familiar to them. At the same time, however, many of them also feel stuck in their physical environment.

The relationship between older people and their rural places is largely subjective, embodying accumulated memories and affirmation of self-identity (Rowles, 2008; Walsh, 2015; Walsh, O'Shea and Scharf, 2019; Winterton and Warburton, 2012). Deep emotional attachments between people and places are known to emerge over time (Rubinstein and Parmelee, 1992). The experience of time is part of the dynamic behind attachment to place. It, accordingly, becomes important to understand the biographical aspects of the person – environment relationship if we are to understand the phenomenon well. Besides the current place of residence, also physical places connected to one's earlier life course strongly influence a person's identity, which thus reflects both individual and social life history (Peace, Holland and Kellaher, 2011).

Places and environments experienced earlier in the life course can play a subsequent major role in shaping older adults' identity and the way that identity becomes manifested in the current phase of the life course. The strength and prominence of such, what in this article we term "environmental identity" varies, to be sure, according to life course phase, with the geographical attachment likely at its strongest during childhood and adolescence. In any case, it is fundamentally about

> a sense of connection to . . . the nonhuman natural environment, based on history, emotional attachment, and/or similarity, that affects the ways in which we perceive and act towards the world; a belief that the environment is important to us and an important part of who we are.
>
> (Clayton 2003: 46)

Through environmental identity, personal identity reflects both individual and social life history, thus bringing up the aspect of generational experience (Peace, Holland and Kellaher, 2006). As people are becoming less dependent on spatially bound institutions hitherto sustaining communal identity, the role of place has changed. Gilleard and Higgs (2005) have thus raised an important question prompted by one key aspect of our (rural) communities of the past, namely their ability to provide support to those belonging to them: can our present symbolic communities offer the same level of support and belonging that the spatialised communities of former times?

In this chapter, we want to pay attention to just that characteristic, looking at how place-bound rural identity deriving from the earlier life course may be manifested in current life of older adults. Drawing upon an empirical study of older Finnish men, we first consider how attachment to a rural childhood place can foster a sense of communality among older persons. After that, we look at how this attachment may be manifested in their current life, in terms of the kind of social interaction and support in which they engage. As an analytical framework we use Graham Rowles' (1978, 1983) theory of "insideness", which considers the question of attachment to place from the perspective of rural older people. Our analysis is guided by the three key attributes presented by Rowles – physical, social and autobiographical insideness – with a focus on the autobiographical components of place attachment (see Wiles et al., Chapter 15).

Social and autobiographical insideness

In Rowles' (1983) framework, *physical insideness* refers to an intimate familiarity with one's environment and its barriers. This familiarity can, for instance, compensate for deficits in physiological competence, enabling old persons to continue traversing spaces that would otherwise pose challenges to them. The component of *social insideness*, on the other hand, refers to integration within the social fabric of the community, including values and behavioural norms. Integration within social networks results in both giving and receiving practical and social support. More importantly still, such integration also conveys status and a sense of belonging. The third dimension of Rowles' insideness, *autobiographical insideness*, involves a series of remembered places, in addition to the present place. The places in question are, however, not just physical settings possible for also other people to visit or view but also present themselves more subjectively as a mosaic of "incident places". Four features can be distinguished here (Rowles 1983: 304–305):

1 Autobiographical insideness is generally taken for granted and the basis for attachment in it is not easy to articulate for the actors themselves;
2 It is supported by physical participation in the spaces wherein remembered events occurred;
3 Insideness may include "incident places" from the entire space and time trajectory of older adult's life span; and
4 Autobiographical insideness represents an attachment that is self-created and thus to a certain degree fictional; often the places are idealised and remembered in a positive light.

Drawing from Rowles' theoretical framework, we examine an older men's community from the perspective of social and autobiographical insideness. Deploying the key attributes Rowles lays out, we investigate older men's place attachment and social participation in the context of the former rural community. Based on interview and observation data, we ask whether and how participation in a place-bound social community might support the social and autobiographical insideness of older rural men.

Data, methods and participants

The data for our study derived from a collective ethnographic study, in which observations were accompanied with 27 individual interviews (Ii) and four focus group interviews (Gi) with a total of 40 older Finnish men born between 1930 and 1950. The men all belonged to a local social community whose membership was based on (male) gender and the shared experience of having lived in one rural locality in Finland in the 1960s. Most men had lived in the locality all their lives, while others had moved back there in later life, typically after retirement. The youngest study participants were from the baby boomer generation born between 1945 and 1950. More than three quarters of this generation were born in rural areas. For them, traces of life-historical layers are evident through memories and connections to the traditional society (Karisto and Haapola, 2014).

Our analysis relied on a discourse analytical approach (e.g., see Nikander, 2008). In keeping with this method, we paid close attention to the interviewees' different ways of wording and structuring their answers, along with the broader, culturally distributed logics of reasoning and interpretive frameworks shaping the statements by individual interviewees.

"Mill Village Boys" as a community of older men

A group of older men living in the eastern part of Finland started to informally get together in the summer of 2006. The first meetings were initiated by a few retirees from a rural township that, in 1967, had been merged with a larger neighbouring municipality. The stated purpose of the meetings was to reminisce together about the past and the history of the community. Over the years, the initially small group grew, coming to include more than 100 men. The criteria for membership included being a retiree and having at some point lived in the area of the former rural township. The group was not formally organised, as the members did not want their character and activities to become rigid and hierarchical. Accordingly, the men termed their group simply as "the Group", or used for it the name of the former township; to preserve anonymity we call them "the Mill Village Boys", referring to the main employer in their reminisced physical home community, a paper mill.

The group met regularly once a month, with experts from various fields often invited to join in to present on or discuss current issues. In addition to the main gatherings, men organised various kinds of activities in their home community, such as fishing competitions and field trips. The main activities nevertheless consisted of collectively reminiscing on the past and maintaining "the local spirit". Even though five decades had passed since the annexation of the men's former hometown to its neighbouring city, it kept serving as a powerful source of local identity for the men.

Results

Social insideness

The idea of social insideness refers to social ties of the community, including values and behavioural norms as well as giving and receiving support. Integration within the social fabric of a community is usually created by similarity of economic position, work history and the like. In the community of the Mill Village Boys, a pre-condition for membership was that one had lived in the Mill Village at the time of its annexation to its neighbouring city in the 1960s. Even though the community was characterised by a high degree of closeness among its members and its membership criteria were clearly articulated, all of the interviewees nevertheless also participated in other groups, associations and communities that were significant to them, in terms of the support they offered them or the density of the social interactions. These other groupings, associations and communities were, however, separate from the Mill Village Boys' community having very few members in common. However, as emerged in the interviews, where one had had a close relationship with someone during childhood, contact was maintained with this person also outside the meetings of the community.

The most common way of becoming a member of the community through a network of contacts or friends, despite all the "public" membership criteria that were clearly articulated and the information the community regularly published about itself and its activities in a local newspaper, including open invitations for locals to join in. Most members' level of social involvement in the community was high, and they attended the monthly meetings virtually every time. The opportunity the meetings offered to hear interesting presentations and socialize with others like oneself was clearly attractive to the Boys.

During our fieldwork, we could observe how, at their meetings, the way the Mill Village Boys seated themselves at the tables reflected each individual's relationship to the community. One arrived half an hour before the meeting was due to start, and some of the men had their regular tables around which they went on to huddle. In the group there were, however, usually also "irregular" members; these proceeded to sit freely wherever they found space in the end. While this seating practice did not appear to constitute an exclusionary mechanism of any kind, the division into a core circle and those more loosely attached was nonetheless clearly there. One interviewee described this as follows:

> Of course there's same sort of clique behaviour as everywhere else. Guys go sit around the same table and talk their own stuff. But then you've also got these irregular cases like I, too, usually am that ain't got the kind of standard company to go socialize with.
>
> (Ii)

One of the factors creating clear lines of internal division among the men was where, exactly, they had lived within the Mill Village area in the 1960s. A certain part of the village mattered in this regard, but even the neighbourhood one came from was significant. However, also age mattered: the closest relationships were those between Boys who came from the same area of the Mill Village and were born the same year.

One of the central activities in which the Boys engaged in their meetings was shared collective reminiscence. Recollecting places and events from one's childhood and youth was of great significance to them as members of their community; indeed, reminiscence activities acted as a kind of social glue holding the community together. Humour had the same function. Especially the men-only nature of the community was responded to in jocular tones when inquired about, with no rational grounds offered for it. The horizontal homo-social bond uniting the men as a community was both expressed and cemented through three mechanisms: the community (1) sustained and reproduced a collective narrative about its history and equality as the basis of all its activities; (2) together defined the explicit rules for its full meetings (no alcohol, politics and religion); and (3) articulated the arrangements related to its operation and activities (see Pietilä et al., 2019). In both the group and individual interviews, the interviewees demonstrated a uniform understanding of the shared community-level norms and principles and an ability to clearly articulate them.

Although communities often build upon some strong aspect of commonality, such as of occupation or professional status, in the pronounced norms of the Mill Village Boys' community there was nevertheless the emphatic notion built in of every member being equal regardless of background or status. As one interviewee expressed what this meant in practice:

> No one there's gonna ask you if you've finished some school or something. We're equal. We're all the same when we're here. Even though we're engineers, a ship captain, a police officer, what have you – people in very different jobs and professions. But those are behind us now, those times. Now we're all pensioners, this whole group of us, and I find us all to be equal here; you can go talk with anyone and everyone, just the same way.
>
> (Ii)

Yet, in practice, one's professional history in the Mill Village area did matter, too. Having worked in the past as a shopkeeper, a taxi driver, or a sports coach, for instance, one came with ready-made contacts, with the background thus doing much to define the person's position and role in the community even with all of its members already retired from working life.

Another important factor maintaining social insideness in the group was the ritualisation of the entry into it. When a prospective new member came to a meeting, he was introduced and asked to tell his story***irman hich ced, asked to tell his " retired from working life.ä että tutk.kohteiden kannalta. Ref luetteloon tämä samalla tav: where he was from and how he knew the shared localities of the Mill Village Boys. After this the chairman of the meeting proposed the person for election as a new member of the community. Once this was done, the audience confirmed the election by greeting the new member by applause (confirmation was never withheld). Also, the exit from the community was associated with a ritual charged with meaning. Members having passed away were acknowledged at the monthly meeting following their death and commemorated with a moment of silence, with everyone present standing. Reflections on the end of life were notably present in the interview discussions as well, with death often addressed in them in the form of the ritual commemorations.

Our study participants were in relatively good physical condition and they had no environmental barriers to participation in the activities of their community.

Through its actions, however, also their community as such contributed to the sense of social insideness among them: those unable to drive to the meetings or otherwise make it there on their own were offered a ride with other members. What did not feature prominently in the activities of the community was concrete or social support provided to its members. Difficult or more personal matters were as a rule not taken up and discussed in the meetings. Instead, spouses and family members were cited as sources of such support for the men. Any concrete or practical help and support that the Mill Village Boys offered to one another took mainly the form of transportation provision to those with reduced mobility or no car. Yet, there could be more loosely organised, occasional forms of help and assistance given to local inhabitants in need of such help; these individuals, moreover, needed not even be members of the community nor men. Any such activities were, however, not part of the ordinary or core activities of the group.

Autobiographical insideness

In the determination autobiographical insideness, physical proximity in current life is not essential. This is due to geographical fantasy, which involves ability to project oneself in places of the past (see Rowles, 1978). Autobiographical insideness among the Mill Village Boys was to a large extent created through the meaning that sports and, in particular, a major mill that used to structure life in the Mill Village had for them. In forming part of their life history and relationship to place, these defined the Boys' sense of autobiographical insideness. To a significant extent, the collective reminiscing that functioned as the social glue of the community focused on precisely the two, as exemplified by the following interview quote:

> You know, when yours truly was a kid, there was quite a bit of rivalry between villages and neighbourhoods. Which meant that you weren't, like, exactly fond of the other guys. The sports field was then, for us, the kind of place where we'd meet, and there was lots of mistrust and suspicion there, and, you know, trash talk like 'Oh, so you're from there' and 'You know, we're from here', with some people thinking that they were better than others.
>
> (Ii)

On the other hand, autobiographical insideness also had an individual dimension to it. It was embedded in memories of places where one had, for instance, first fallen in love with one's future spouse, or where one used to play as a child. One interviewee described this dimension for his part as follows:

> Well, you know, it's like every time I drive by that place I look there, to my right, towards the mouth of the canal. That's where I used to swim as a child and ride timber rafts. Those were the childhood playgrounds for people like

> myself. And we fished there, too, especially when the wind was from the north, as then you'd catch breams, lots of them, right off the dock.
>
> (Gi)

Rowles (1983) has shown how creative "grand fiction" functions in first creating and then sustaining a sense of insideness. Its mechanisms commonly include the preservation of certain artefacts, such as photographs and small objects from the past. In the community of the Mill Village Boys, artefacts of this kind occupied a prominent position. One of the grander examples of this was a project to create a monument commemorating the Mill Village and its history. Through the large, heavy work constructed of stone and unveiled in solemn ceremony, the Boys' autobiographical insideness was permanently moored to a specific place, with also the collective character of that insideness rendered more visible.

Also books and cinematic works had this function of artefacts in the Mill Village Boys' community. With support from the Boys, several books and one film had been produced of their Village, driven by a desire to document collective history and transmit knowledge to future generations. Even the research reported in this chapter may be understood as an artefact along these lines. The initiative for it came from the Boys themselves, and the story emerging from the first round of data collected was remarkably uniform and consistent, serving for its part the purposes of autobiographical insideness. (Yet, as the research proceeded, it acquired more – and more diverse – dimensions, yielding a more nuanced and multi-faceted picture.)

As Rowles (1983) has further noted, one way to sustain the created images is through on-going participation in a familiar place – not necessarily by remaining in place but also by making trips to places and locations that bring up latent memories. This kind of autobiographical insideness creation had a prominent place in the activities of the Mill Village Boys' community. The Boys often organised walking tours and car trips to places of memory in the area, sometimes involving playful arguments about where this or that significant site had originally been and how the place had changed.

As in Rowles (1983), communication among age peers sharing the experience of the same environment was an important way to sustain the "grand fiction" also in this study. As one way of keeping alive the image of their Village, the Boys frequently gathered to collectively reminisce on it in a group setting. This image was in the first place a positive one, with the positivity of both it and the place-related communality connected to it, moreover, only reinforced by a sense of "injustice" caused by the earlier annexation of the Mill Village to its neighbouring municipality against the Mill Villagers' will.

It is the mutual correspondence between such collective and individual "fiction" (in the sense used here) that, we may expect, enables the construction of communality. Even though there were case-to-case differences in the images portrayed in individual interviews, no major discrepancies between the collective image and the individual environmental identity appeared probable in any of the cases examined. Had such a discrepancy existed, the individuals interviewed would not have been able to partake in the community the way they now did.

Attachment to place was given much emphasis in the discussions:

> Not sure if it's gonna be a flat or something else that I might move in, then. But either way, it's going to be in [the Mill Village]. Nowhere else. Nowhere.
>
> (Ii)

Also when thinking about where one wanted to be buried one day it was the Mill Village that the interviewees most often cited as their preferred final destination. That was, after all, where one had lived one's entire life.

> I've already been and picked out a place for me where I want them to lay my bones. My father's there, too, and then there are people we know, people from the village next to ours, folks like that – everyone's there.
>
> (Ii)

Conclusions

Our analysis on a place-bound community of older Finnish men born in the same rural community has been guided by the framework of Rowles, especially his notions of social and autobiographical insideness. Life in the men's community was based on an explicit code of conduct and rules regarding activities and interaction, along with specific criteria for belongingness. Entry to, and exit from, it were governed by rituals that, together with the community's rules and codes, reinforced its members' social insideness. The men's sense of communality was also cemented through the joint creation and employment of shared narratives and humour.

The community members had constructed a shared, highly uniform narrative about the community and what it stood for. This narrative was employed to protect and maintain community members' insideness, which involved, besides its collective dimension, also an individual, more personal one revealed especially in individual interviews. Reminiscing, different artefacts and trips to childhood places were means used to develop and reinforce autobiographical insideness and help construct environmental identity. The role and meaning of the community were actualised precisely here, and not, for instance, in any provision of concrete support and assistance to community members.

As research on rural ageing has highlighted, while the existence of social communities can facilitate community access and social support of older people, the same communities can also be exclusionary for others, especially those exhibiting difference and those not involved in the existing community networks (Walsh, O'Shea and Scharf, 2019; Winterton and Warburton, 2011). At-risk groups include older men in remote areas, carers and former carers (particularly men), those who have relocated to new rural areas, and those living alone (see Winterton and Warburton, 2011). These at-risk groups were underrepresented in our study, which included mostly participants living in semi-urban areas and with a partner.

Our results in this study show that spatially bound communal identity can arise from the interaction of individual and collective autobiography. However, the ability of a former common place-based rural community to provide social support for its members appeared to be limited. To conclude with Gilleard and Higgs (2005), it remains to be seen whether our present symbolic communities can offer the same level of support and belonging as the different kinds of spatialised communities of the past.

References

Clayton, S. (2003). Environmental identity: a conceptual and an operational definition. In: S. Clayton and S. Opotov, eds., *Identity and the Natural Environment: The Psychological Significance of Nature*. Cambridge, MA: MIT Press, pp. 45–65.

Gilleard, C. and Higgs, P., (eds). (2005). *Contexts of Ageing: Class, Cohort and Community*, Cambridge: Polity Press.

Gilleard, C., Hyde. M. and Higgs. P. (2007). The impact of age, place, aging in place, and attachment to place on the well-being of the over 50s in England. *Research on Aging*, 29(6), pp. 590–605.

Gullifer, J., and Thompson, A. P. (2006). Subjective realities of older male farmers: self-perceptions of ageing and work. *Rural Society*, 16(1), pp. 80–97.

Jolanki, O. and Vilkko, A. (2015). The meaning of a 'sense of community' in a Finnish senior co-housing community. *Journal of Housing for the Elderly*, 29(1–2), pp. 111–125.

Karisto, A. and Haapola, I. (2014). Generations in ageing Finland: finding your place in the demographic structure. In: K. Komp and S. Johansson, eds., *Population Ageing from a Life Course Perspective*. Bristol: Policy Press.

McKenzie F. and Frencken, M. (2001). The lively dying town? Challenging our perspectives of rural ageing. *Australian Planner*, 28(1), pp. 16–21.

Means, R. and Evans, S. (2012). Communities of place and communities of interest? An exploration of their changing role in later life. *Ageing & Society*, 32(8), pp. 1300–1318.

Nikander, P. (2008) Constructionism and discourse analysis. In: J. A. Holstein and J. F. Gubrium, eds., *Handbook of Constructionist Research*. New York: Guilford Press, pp. 413–428.

Peace, S., Holland, C. and Kellaher, L., (eds). (2006). *Environment and Identity in Later Life: Growing Older*. Maidenhead: Open University Press.

Peace, S., Holland, C. and Kellaher, L. (2011) '"Option recognition" in later life: variations in ageing in place. *Ageing & Society*, 31(5), pp. 734–757.

Pessi, A. B. and Seppänen, M. (2011) 'Yhteisöllisyys'[communiality]. In: J. Saari, ed., *Hyvinvointi – suomalaisen yhteiskunnan perusta [Well-being – The Founding of Finnish Society]*. Helsinki: Gaudeamus, pp. 288–313.

Pietilä, I., Ojala, H., Tiilikainen, E. and Seppänen, M. (2019). Horisontaalinen homososiaalisuus eläkeikäisten miesten yhteisössä [Horizontal homosociability in retired mens's community]. *Sukupuolentutkimus*, 32(1), pp. 21–33.

Rowles, G. D., (ed.). (1978). *The Prisoners of Space? Exploring the Geographical Experience of Older People*. New York: Avalon.

Rowles, G. D. (1983). Place and personal identity in old age: observations from Appalachia. *Journal of Environmental Psychology*, 3(4), pp. 299–313.

Rowles, G. D. (2008). Place in occupational science: A life course perspective on the role of environmental context in the quest for meaning. *Journal of Occupational Science*, 15(3), pp. 127–135.

Rubinstein, R. I. and Parmelee, P. A. (1992). Attachment to place and the representation of the life course by the elderly. In: I. Altman and S. M. Low, eds., *Place Attachment*. New York: Plenum Press, pp. 139–163.

Seppänen, M., Haapola, I., Puolakka, K. and Tiilikainen, E. (2012). Takaisin Liipolaan. Lähiö fyysisenä ja sosiaalisena asuinympäristönä [Back to Liipola. The suburban neighbourhood as a physical and social living environment]. Reports of the Ministry of the Environment no. 14/2012, Ministry of the Environment, Helsinki.

Vasunilashorn S., Steinman, B. A., Liebig, P. S. and Pynoos, J. (2012). Aging in place: evolution of a research topic whose time has come. *Journal of Aging Research*, 2012, pp. 1–6.

Walsh, K. (2015). Interrogating the 'age-friendly community in austerity: myths, realties and the influence of place context. In: K. Walsh, G. Carney, Á. and Ní Léime, eds., *Ageing through Austerity: Critical Perspectives from Ireland*. Bristol: Policy Press, pp. 79–95.

Walsh, K., O'Shea, E. and Scharf, T. (2019). Rural old-age social exclusion: a conceptual framework on mediators of exclusion across the lifecourse. *Ageing & Society*. Epun ahead of print 2 July. DOI: 10.1017/S0144686X19000606.

Winterton, R. and Warburton, J. (2011). Does place matter? Reviewing the experience of disadvantage for older people in rural Australia. *Rural Society*, 29(2), pp. 187–197.

Winterton, R. and Warburton, J. (2012). Ageing in the bush: The role of rural places in maintaining identity for long term rural residents and retirement migrants in north-east Victoria, Australia. *Journal of Rural Studies*, 28(4), pp. 329–337.

17 Social relations, connectivity and loneliness of older rural people

Catherine Hagan Hennessy and Anthea Innes

Introduction

Isolation, loneliness and social disconnect can occur in old age generally (Hennessy, 2015), but they are a particular challenge for certain groups of older people, for example those living with dementia (Kane and Cook, 2013), and these issues can be compounded by remote and rural geographies (Rural England, 2016). The social inclusion and participation of older people in rural communities and the impact of their social relations and connections on well-being outcomes, like loneliness, are receiving growing attention in contemporary gerontological research. This chapter will examine these issues within the evolving empirical framework of studies in this area, with special reference to dementia as a health condition that increases rural-dwelling older adults' vulnerability to social detachment and loneliness. Within this chapter "loneliness" refers to the subjective perception that one's level of social contact is inadequate, while "social isolation" is the objective measure of the number of one's social contacts (Cattan et al., 2005; Grenade and Boldy, 2008; Victor et al., 2005). Various concepts related to or indicative of a state of social isolation, e.g., "social ex(in)clusion" (Walsh et al., 2012, 2019), "social detachment" (Jivraj, Nazroo and Barnes, 2012), "social integration" (Fuller-Iglesias and Rajbhandari, 2016) and "connectivity" (Hennessy et al., 2014) are used interchangeably although they have more precise definitions.

Research approaches to social isolation and loneliness in rural older age

Historically, studies in rural gerontology on the links between social relations and well-being have focused on risk factors for exclusion, loneliness and social isolation. Early work in the United States, for example, examined factors that discriminated between rural older adults experiencing varying degrees of loneliness (Kivett, 1978, 1979) and predictors of loneliness for this geographic group (Kivett and Scott, 1979). While usefully identifying factors associated with loneliness among rural elders, these first studies tended to concentrate on factors intrinsic to the individual (e.g., marital status, self-rated health, having a confidant) with relatively little attention to extrinsic factors that would provide an understanding of older people's broader life contexts within a rural setting. Likewise, the predominance of quantitative methodologies in such early research did not allow for a more in-depth qualitative perspective on rural older people's experiences of the conditions and supports surrounding their well-being. This situation prompted Rowles (1984), for example, to argue the need for greater use of qualitative approaches in studying older people's well-being in rural environments.

Nonetheless, the epidemiologic tradition of studies on loneliness and other well-being outcomes among rural elders has continued to the present. Key developments have included greater attention to conceptually distinguishing loneliness from social isolation and on their respective predictors (Havens et al., 2004; Wenger and Burholt, 2004). The growing international scope of these studies (Guo, 2009; Theeke and Mallow, 2013; Wang et al., 2001; Wang et al., 2011, 2017; Yang and Victor, 2008) has raised obvious issues relating to their cultural contexts and the need to consider findings within an understanding of these contexts. In a study of loneliness among Chinese rural elders, for example, Wang et al. (2011) found family function (assessment of satisfaction with family relationship) to be the strongest predictor of loneliness in the study population. The authors interpret this result as reflecting the collectivistic nature of traditional Chinese society based on close-knit family relationships and expectations of filial piety. The patterning of meaning and subjective experience of loneliness among rural-dwelling older adults was explicitly explored within the Irish context in a qualitative investigation by McHugh Power et al. (2017). This research highlighted the relationship between factors such as perceived safety of the rural setting and loneliness, indicating the influence of environmental factors on feelings of loneliness.

A "contextual turn" is evident in Hinck's (2004) qualitative study of the rural oldest-old in the American Midwest in which "participants described how historical, cultural and environmental contexts shaped their everyday thoughts, activities and what was meaningful to them" (779). Although the research focused on elderly rural dwellers' adaptive strategies in general, and not on loneliness per se, these included the means individuals used to mitigate loneliness as positioned within this larger context. As Burholt and Scharf (2014) have emphasised in respect to quantitative investigations on loneliness among rural elders, however, "Although numerous studies explore the associations between sociodemographic variables and loneliness, relatively few address environmental correlates" (312). Approaches to incorporating aspects of the rural environment in this research have been advanced, for example, in Burholt, Morgan and Winter's (2018) study of individual and situational risks for social and emotional loneliness among a representative older rural population in Wales. Findings of this investigation underline the significance of factors related to community embeddedness, rural community type and area disadvantage on these respective well-being outcomes.

A more general understanding of the impact of rural community characteristics on aspects of older adults' wellness and well-being has been further developed in Winterton and colleagues' (2016) meta-synthesis of the literature on rurality, ageing and community. Results of this analysis were used to build an ecological framework encompassing influences at different levels of community context.

These include objective sociodemographic and spatial characteristics of rural communities (the "socio-spatial environment") and elements such as the service context and other features and amenities that constitute the "resource-based environment". Among the posited impacts of these rural community-level characteristics are on health and well-being including outcomes such as social connectedness and loneliness.

This ecological framework reflects related conceptual models of rural ageing developed over the past decade that provide more holistic contextualised perspectives on aspects of well-being. Keating and Phillips' (2008), critical human ecology perspective on rural ageing similarly focuses on "the ways in which contexts provide opportunities and constraints to individuals living in various settings" (9). Conceptual models of age-related social exclusion in rural places like that developed from Walsh, O'Shea and Scharf's (2012) cross-border study in Ireland include several aspects of rural settings (social relations; services infrastructure; transport and mobility; safety, security and crime; and financial and material resources) relevant to older people's risk of exclusion from rural community and consequent well-being. Reported loneliness and social isolation were examined in this study as a function of these multiple, interrelated and dynamic dimensions of rural contexts.

Similarly, in the Grey and Pleasant Land (GaPL) project conducted in several rural areas of the United Kingdom, Hennessy et al. (2014) focused on different forms of rural older people's connections to community life and place – termed "connectivity" – and their impact on social inclusion and well-being. Mixed methodologies were used to explore the nature of these connections and their contextual influences at the level of the individual, family and friends and community. A secondary analysis of the GaPL survey data (De Koning, Stathi and Richards, 2017) that specifically examined predictors of loneliness and two types of social isolation (from one's family and from one's community) underlined the positive effect of long-standing presence in the community (length of residence) on all three outcomes and of local social participation on isolation from community. The perceived salience of connectivity to rural elders' well-being is highlighted in an Australian study by Stanley et al. (2010) who conclude:

> Our findings show that loneliness is influenced by . . . whether older people feel that they have, or are seen by others as having, a sense of connectedness with the wider community. Participants expressed the importance of maintaining social contact and having a sense of connection and belonging to the community.
>
> (407)

Stewart, Browning and Sims' (2015) concept of "civic socialising" may go some way towards interpreting these results through their view of the significance of older people's presence and interactions in local neighbourhood venues as a means of "authentification of themselves as individuals and as community members" (750).

Thus, a wide range of individual characteristics have been associated with loneliness and social isolation among rural older people and there is growing recognition of the role of mediating factors – both individual and environmental – in exclusionary processes and detachment from social relations and networks (Burholt, Windle and Morgan, 2017; Walker et al., 2013; Walsh, O'Shea and Scharf, 2019). The impact of these factors and processes in the general population of rural older adults can also be illustrated for a sub-group with heightened vulnerability to social marginalisation and exclusion – those living with dementia. As with rural older adults in general (Bailey, Biggs and Buzzo, 2014; Burholt et al., 2014), the experience of those with dementia who have aged "in place" demonstrates they may have strong attachments to the rural area and the rural landscapes where they have lived and worked (Myren et al., 2017). Likewise, for people with dementia the experience of living in remote and rural areas can differ markedly in relation to whether one has migrated to a rural community in older age and therefore has had less opportunity and time to develop social networks (Blackstock et al., 2006). For example, existing social networks can fall away when a person receives a diagnosis of a particular condition, in the case of dementia this may occur due to fear, stigma and a need to "distance" from what may be difficult to understand (McParland, Kelly and Innes, 2017). However, this leads to increased social isolation for older people living with dementia and their families as their social circle and opportunities for "civic socialising" shrink, a situation which may be exacerbated in rural and remote settings (Alzheimer's Society, 2018).

Reduced mobility can compound this marginalisation when the ability to enjoy the countryside that may have been a pull factor for retirement migration becomes more restricted (Innes et al., 2005; Innes, Szymczynska and Stark, 2014) leading to isolation and loneliness. Transport, or a lack of it, is a common problem in rural areas, but when one has dementia, or another form of frailty that restricts physical movement, the ability to stay connected can rapidly decline. Although the guidelines on driving and dementia vary and are the subject of contemporary debate (Rapoport et al., 2018), when the ability to drive has been assessed as being compromised this has a direct impact on mobility and independence. For example, if an individuals' driving licence is withdrawn this has a significant effect on the person with dementia (and potentially their partner) being unable to independently travel to services, both general, such as shops, banks and libraries and health specific, such as the "local" physician's office, hospital and other care services. Accessing support and services becomes difficult and this situation can be compounded by the lack of specialist health services generally available in remote and rural areas resulting in unmet needs for older people living with dementia in rural locations (Innes et al., 2006; Morgan, Innes and Kostenieuk, 2011). The older person is often placed in the situation that they are no longer able to get out and about independently and a lack of outreach services available to support such individuals can be extremely challenging for both the person with the diagnosis and their immediate and wider family and social networks. This may result in social disconnect and feelings of isolation (Innes et al., 2005; Szymczynska et al., 2011; Innes, Szymczynska and Stark, 2014).

The previous description of the evolution of empirical approaches to loneliness, social isolation and their relationship to community connectedness and inclusion among older rural people demonstrates the increasing integration of a number of strands of research in rural gerontology around these issues. This is illustrative of the research landscape in rural gerontology more generally, in which multiple disciplinary perspectives on social, environmental and economic factors have co-existed (Scharf, Walsh and O'Shea, 2016) but until recently without the significant interdisciplinary integration that would benefit a fuller understanding of issues such as loneliness and social isolation among rural elders.

Interventions to promote social inclusion and connectivity of rural older people

Within rural contexts internationally, a variety of intervention types have been developed to reduce loneliness and isolation through addressing challenges to older people's social inclusion and connectivity to community (UNECE, 2017). Francis and Windle (2014) have classed such interventions into three main categories. The first includes one to one interventions such as befriending which pair volunteers or paid workers in a relationship with individuals in need of social and often instrumental support. Other forms of interventions in this category are mentoring schemes for personal skills development and gatekeeping initiatives to help facilitate wayfinding to support and services for vulnerable individuals. Group services are a second type of intervention and include congregate activities such as lunch clubs and social groups. Finally, schemes to stimulate older individuals' wider engagement with community through participation in local services and activities constitute a third category of interventions. In an Age UK report (2015) on approaches to reducing loneliness and isolation in later life, these modalities are categorised as "direct interventions" to support and maintain existing relationships, foster and enable new connections and assist people in thinking differently about their social connections. The effectiveness of such interventions to prevent loneliness and social isolation is further recognised as being enabled by "gateway services" like transport and technology and reflects the particular importance of such supports for older people's connectivities in rural areas (see Berg and Winterton, 2017; Hennessy, 2014).

Newer emergent forms of intervention approaches to prevent social isolation and loneliness are described as focusing on resources and assets within the community environment and characterised as "structural enablers" for creating social connections (Age UK, 2015). These are typically based in neighbourhood localities, feature volunteering roles and draw on community social capital to promote older people's participation and social inclusion. An exemplar of this approach is the Village and Community Agents Project in Gloucestershire (UK) which utilises trained local volunteers age 50 and over to provide older people in rural areas with easier access to information and services and enable them to feel part of a supportive enabling community. The project, begun in 2006, currently operates in over 90% of the county's 253 parishes, with demonstrated benefits including the support and promotion of rural older people's social networks (Wilson, Crow and Willis, 2008). Similarly in Canada, the Saskatchewan Seniors Mechanism provides a template for organising social, recreational and other activities planned and led by older people in their own rural communities. Based in residents' neighbourhoods, the project has reached vulnerable older people who would not ordinarily take part in community activities due to difficulties with access in rural locations (UNECE, 2017).

Given the acknowledged difficulties of providing support to older people with dementia and their families living in remote and rural communities (Innes et al., 2006), alternative models of support are required to tackle isolation and loneliness. Such initiatives, designed to address social isolation and loneliness while promoting social connections, have had varying degrees of success. In the United Kingdom, dementia-friendly community initiatives have emerged since the Prime Minister's dementia challenge of 2012 (Department of Health, 2012) and subsequent implementation plan (Department of Health, 2016). This was taken on board by rural communities in a range of ways (Alzheimer's Society, 2018) demonstrating the community spirit and creative ways different age and community groups across the country embraced the notion of becoming "dementia-friendly" and raising awareness of dementia in the process. The case studies in the Alzheimer's Society *Dementia-Friendly Rural Communities Guide* (2018) provides a series of examples of such initiatives in parts of the United Kingdom. In Scotland, *The Windows of the World* project has encouraged engagement with creative activities as well as accessing shopping services and health information through using tablet devices to access information. In Wales, people living with dementia in rural areas were given the opportunity to visit a farm and to meet animals and reminisce about growing up in a rural area. In the High Peak area of Derbyshire in England, a letter writing service at a local post office was provided promoting connections with family and friends at a distance and also the opportunity for those using the service to meet together for a cup of tea and a chat at the same time. In Ireland, social models have developed in response to a lack of formal policy directives and absence of formal support services. These initiatives exemplify the power of community groups to address the needs of those living with dementia in their communities (O'Shea, Cahill and Pierce, 2017; see also Herron and O'Shea, Chapter 26 in this volume).

To inform the design of interventions that effectively support social engagement and community connections for rural older people, Walker et al. (2013) emphasise the essential need to understand their perspectives on social engagement and social detachment within the rural setting. In line with the current research frameworks on loneliness and social isolation previously described, intervention approaches need to be grounded in context and "as far as possible, integrated into, facilitate, and function according to the processes of community with which the older rural participants are already familiar and comfortable" (Walker et al., 2013: 955). These insights are confirmed, for example, by the findings of an evaluation of the Upstream Healthy Living project which engages socially isolated older people in rural areas of mid-Devon (United Kingdom) in programmes of social and creative activities (Greaves, 2006). The tailoring of activities to individual needs, abilities, skills and preferences, as well as successfully overcoming barriers related to access within rural areas (e.g., transport and availability of venues) were identified as key mediators of well-being outcomes for Upstream participants. These findings reflect an evidence review on interventions to reduce loneliness and social isolation among older people (Gardiner, Geldenhuys and Gott, 2018), which concluded that the most effective were those that included adaptability, a community development approach and productive engagement. These features in turn relate to the ecological lens on loneliness and social isolation that highlights the joint influences of individual, neighbourhood and community-level factors affecting social relations and well-being outcomes for rural elders.

Conclusion

The need to systematically address the well-being, social inclusion and community connectedness of rural older adults from a strategic coordinated perspective is increasingly being realised internationally. In 2017 in the United States, for example, a multi-sector, multidisciplinary group of national stakeholders was convened for the first Connectivity Summit on Rural Aging, at which the impact of social isolation and loneliness on rural-dwelling older adults was identified as a key issue. Following this inaugural meeting, a second summit in 2018 aimed at identifying actionable solutions that could have an immediate impact on reversing social isolation among rural elders (Bipartisan Policy Center, 2018; Tivity Health, 2018). To date these convenings have produced recommendations for national awareness raising, building on existing resources and infrastructure, improving public policy and health care system reform. In Australia, the Foundation for Rural and Regional Renewal provides an example of multi-sector partnership between philanthropy, government and business which sponsors its Caring for Ageing Rural Australians (CARA) programme (see www.frrr.org.au/grants/cara). Funding available under CARA supports a wide range of efforts to improve ways and means of enabling rural older adults to stay supported by, connected to and involved in their rural communities.

As these and further efforts to reduce social isolation and loneliness among older adults in rural areas continue to develop, it is imperative that they reflect the research evidence and contextual approaches described earlier in this chapter, including the multiple levels of individual, neighbourhood and community factors that affect these outcomes. Particular health conditions, such as dementia, that can lead to further social exclusion and jeopardy are important to consider when developing initiatives designed to promote well-being and social connectivity. A one size fits all approach will not work. The diversity of rurality based on cultural, geographical and social contexts in which people are located will influence their experiences of ageing in rural areas. Addressing such diversity is a challenge that cannot be ignored if practice and policy initiatives are intended to address the needs and well-being of all older people, including those for social inclusion and connection.

References

Age UK. (2015). *Promising Approaches to Reducing Loneliness and Social Isolation in Later Life*. London: Age UK.

Alzheimer's Society. (2018). *Dementia Friendly Rural Communities Guide. A Practical Guide for Rural Communities to Support People Affected by Dementia. Prime Minister's Rural Dementia Task and Finish Group Force*. London: Alzheimer's Society. Available at: www.alzheimers.org.uk/sites/default/files/201907/AS_DF_NEW_Rural_Guide_Online_09_07_19.pdf

Bailey, J., Biggs, I. and Buzzo, D. (2014). Deep mapping and rural connectivities. In: C. H. Hennessy, R. Means and V. Burholt (eds) *Countryside Connections: Older people, Community and Place in Rural Britain*. Bristol: Policy Press, pp. 159–192.

Berg, T. and Winterton, R. (2017). 'Although we're isolated, we're not really isolated': the value of information and communication technology for older people in rural Australia. *Australasian Journal on Ageing*, 36(4), pp. 313–317.

Bipartisan Policy Center. (2018). *Rural Aging: Health and Community Policy Implications for Reversing Social Isolation*. Washington, DC: Bipartisan Policy Center.

Blackstock, K., Innes, A., Cox, S., Mason, A. and Smith, A. (2006). Living with Dementia in rural and remote Scotland. *Journal of Rural Studies*, 22(2), pp. 161–176.

Burholt, V., Curry, N., Keating, N. and Eales, J. (2014). Connecting with community: the nature of belonging among rural elders. In: C. H. Hennessy, R. Means and V. Burholt, eds., *Countryside Connections: Older People, Community and Place in Rural Britain*. Bristol: Policy Press, pp. 95–125.

Burholt, V., Morgan, D. J. and Winter, B. (2018). The role of community type, disadvantage and community embeddedness in the experience of social and emotional loneliness. *Innovation in Aging*, 2(1), pp. 631–632.

Burholt, V. and Scharf, T. (2014). Poor health and loneliness in later life: the role of depressive symptoms, social resources, and rural environments. *Journals of Gerontology, Series B: Psychological Sciences and Social Sciences*, 69(2), pp. 311–324.

Burholt, V., Windle, G. and Morgan, D. J. (2017). A social model of loneliness: the roles of disability, social resources, and cognitive impairment. *The Gerontologist*, 57(6), pp. 1020–1030.

Cattan, M., White, M., Bond, J. and Learmouth, A. (2005). Preventing social isolation and loneliness among older people: a systematic review of health promotion interventions. *Ageing & Society*, 25(1), pp. 41–67.

De Koning, J. L., Stathi, A. and Richards, S. (2017). Predictors of loneliness and different types of social isolation of rural-living older adults in the United Kingdom. *Ageing & Society*, 37(10), 2012–2043.

Department of Health (2012). *Prime Minister's Challenge on Dementia: Delivering Major Improvements in Dementia Care and Research by 2015*. London: Department of Health.

Department of Health (2016). *Prime Minister's Challenge on Dementia 2020*. London: Department of Health.

Francis, J. and Windle, K. (2014). *Preventing Social Isolation and Loneliness among Older People: In Improving Later Life: Services for Older People – What Works*. London: Age UK.

Fuller-Iglesias, H. R. and Rajbhandari, S. (2016). Development of a multidimensional scale of social integration in later life. *Research on Aging*, 38(1), pp. 3–25.

Gardiner, C., Geldenhuys, G. and Gott, M. (2018). Interventions to reduce social isolation and loneliness among older people: An integrative review. *Health and Social Care in the Community*, 26(2), 147–157.

Greaves, C. J. (2006). Effects of creative and social activity on the health and well-being of socially isolated older people: Outcomes from a multi-method observational study. *Journal of the Royal Society for the Promotion of Health*, 126(3), pp. 134–142.

Grenade, L. and Boldy, D. (2008). Social isolation and loneliness among older people: issues and future challenges in community and residential settings. *Australian Health Review*, 32(3), pp. 468–478.

Guo, Z. (2009). *Loneliness of older adults in rural China*. Master's Thesis, Georgia State University, Atlanta, GA. Available at: http://scholarworks.gsu.edu/gerontology_theses/18

Havens, B., Hall, M., Sylvestre, G. and Jivan, T. (2004). Social isolation and loneliness: differences between older rural and urban Manitobans. *Canadian Journal on Aging*, 23(2), pp. 129–140.

Hennessy, C. (2014). Promoting older people's inclusion in rural communities. In: Age UK, eds., *Improving Later Life: Services for Older People – What Works*. London: Age UK, pp. 48–50.

Hennessy, C. (2015). Supporting social participation in later life. In: *Improving Later Life: Vulnerability and Resilience in Older People*. London: Age UK, pp.18–21. Available at: https://www.ageuk.org.uk/Documents/EN-GB/For-professionals/Research/Final_Improving_Later_Life_4-Vulnerability.pdf?dtrk=true

Hennessy, C. H., Means, R. and Burholt, V., (eds). (2014). *Countryside Connections: Older People, Community and Place in Rural Britain*. Bristol: Policy Press.

Hinck, S. (2004). The lived experience of oldest-old rural adults. *Qualitative Health Research*, 14(6), pp. 779–791.

Innes, A., Blackstock, K., Mason, A., Smith, A. and Cox, S. (2005). Dementia care provision in rural Scotland: service users' and carers' experiences. *Health and Social Care in the Community*, 13(4), pp. 354–365.

Innes, A., Cox, S., Smith, A. and Mason, A. (2006). Service provision for people with dementia in rural Scotland: difficulties and innovations. *Dementia*, 5(2), pp. 249–270.

Innes, A., Szymczynska, P. and Stark, C. (2014). Dementia diagnosis and post-diagnostic support in Scottish rural communities: experiences of people with dementia and their families. *Dementia*, 13(2), pp. 233–247.

Jivraj, S., Nazroo, J. and Barnes, M. (2012). Change in social detachment in older age in England. In: J. Banks, J. Nazroo and A. Steptoe, eds., *The Dynamics of Ageing: Evidence from the English Longitudinal Study of Ageing 2002–2010 (Wave 5)*. London: Institute for Fiscal Studies, pp, 48–98.

Kane, M. and Cook, L. (2013). *Dementia 2013: The Hidden Voice of Loneliness*. London: Alzheimer's Society. Available at: www.alzheimers.org.uk/sites/default/files/migrate/downloads/dementia_2013_the_hidden_voice_of_loneliness.pdf [Accessed 31 August 2019].

Keating, N. and Phillips, J. (2008). A critical human ecology perspective on rural aging. In: N. Keating, ed., *Rural Aging: A Good Place to Grow Old?* Bristol: Policy Press, pp. 1–10.

Kivett, V. R. (1978). Loneliness and the rural widow. *Family Coordinator*, 27(4), pp. 389–394.

Kivett, V. R. (1979). Discriminators of loneliness among the rural elderly: Implications for intervention. *The Gerontologist*, 19(1), pp. 108–115.

Kivett, V. R. and Scott, J. P. (1979). *The Rural By-Passed Elderly: Perspectives on Status and Needs* (The Caswell Study, Bulletin No. 260). Raleigh, NC: Agricultural Research Service.

McHugh Power, J. E., Hannigan, C., Carney, S. and Lawlor, B. A. (2017). Exploring the meaning of loneliness among socially isolated older adults in rural Ireland: a qualitative investigation. *Qualitative Research in Psychology*, 14(4), pp. 394–414.

McParland, P., Kelly, F. and Innes, A. (2017). Dichotomising dementia: is there another way? *Sociology of Health and Illness*, 39(2), pp. 258–269.

Morgan, D., Innes, A. and Kostenieuk, J. (2011) Formal dementia care in rural and remote settings: A systematic review. *Maturitas*, 68(1), pp. 17–33.

Myren, G. E. S., Enmarker, I. C., Hellzén, O. and Saur, E. (2017). The influence of place on everyday life: observations of persons with dementia in regular day care and at the green care farm. *Health*, 9(2), pp. 261–278.

O'Shea E., Cahill, S., Pierce, M. (2017). *Developing and Implementing Dementia Policy in Ireland*. Galway: National University of Ireland, Galway.

Rapoport, M. J., Chee, J. N., Carr, D. B., Molnar, F., Naglie, G., Dow, J., Marottoli, R., Mitchell, S., Tant, M., Herrmann, N., Lanctôt, K.L., Taylor, J. P., Donaghy, P. C., Classen, S. and O'Neill, D. (2018). An international approach to enhancing a national guideline on driving and dementia. *Current Psychiatry Reports*, 20(3), p. 16.

Rowles, G. D. (1984). Aging in rural environments. In: I. Altman, M. P. Lawton and J. F. Wohlwill, eds., *Elderly People and the Environment*. New York: Springer, pp. 129–157.

Rural England (2016). *Older People in Rural Areas: Vulnerability Due to Loneliness and Isolation Paper*. Available at: www.ruralengland.org/wp-content/uploads/2016/04/Final-report-Loneliness-and-Isolation.pdf [Accessed 31st August 2019].

Scharf, T., Walsh, K. and O'Shea, E. (2016). Ageing in rural places. In: M. Shucksmith and D. L. Brown, eds., *Routledge International Handbook of Rural Studies*. London: Routledge, pp. 80–91.

Stanley, M., Moyle, W., Ballantyne, A., Jaworski, A., Corlis, M., Oxlade, E., Stoll, A. and Young, B. (2010). 'Nowadays you don't even see your neighbours': loneliness in the everyday lives of older Australians. *Health and Social Care in the Community*, 18(4), pp. 407–414.

Stewart, J., Browning, C. and Sims, J. (2015). Civic socialising: a revealing new theory about older people's social relationships. *Ageing & Society*, 35(4), pp. 750–764.

Szymczynska, P., Innes, A., Stark, C. and Mason, A. (2011) Identification and post-diagnostic support of people with dementia: thematic review of experience in rural and urban areas. *Journal of Primary Care and Community Health*, 13(2), pp. 233–247.

Theeke, L. A. and Mallow, J. (2013). Loneliness and quality of life in chronically ill rural older adults: findings from a pilot study. *American Journal of Nursing*, 113(9), pp. 23–28.

Tivity Health. (2018). *The Power of Connection: Reversing Social Isolation in Rural America. Highlights of the 2018 Connectivity Summit on Rural Aging*. Portland, Maine, 7–8 August. Available at: www.ruralage.com/the-power-of-connection-reversing-social-isolation-in-rural-america/ [Accessed 31st August 2019].

UNECE (United Nations Economic Commission for Europe). (2017). *Older Persons in Rural and Remote Areas* (UNECE Policy Brief on Ageing, no. 18). Available at: www.unece.org/fileadmin/DAM/pau/age/Policy_briefs/ECE-WG1-25.pdf [Accessed 13 January 2020].

Victor C, Scambler S, Bowling A, and Bond J. (2005). The prevalence of, and risk factors for, loneliness in later life: a survey of older people in Great Britain. *Ageing & Society*, 25(6), pp. 357–375.

Walker, J., Orpin, P., Baynes, H., Stradford, E., Boyer, K., Mahjouri, N., Patterson, C., Robinson, A. and Carty, J. (2013). Insights and principles for supporting social engagement in rural older people. *Ageing & Society*, 33(6), pp. 938–963.

Walsh, K., O'Shea, E. and Scharf, T., (eds). (2012). *Social Exclusion and Ageing in Diverse Rural Communities: Findings of a Cross-Border Study in Ireland and Northern Ireland.* Galway: Irish Centre for Social Gerontology, National University of Ireland Galway.

Walsh, K., O'Shea, E. and Scharf, T. (2019). Rural old-age social exclusion: a conceptual framework on mediators of social exclusion across the lifecourse. *Ageing & Society.* Epub ahead of print 2 July. DOI: 10.1017/S0144686X19000606.

Wang, G., Hu, M., Xiao, S. and Zho, L. (2017). Loneliness and depression among rural empty-nest elderly adults in Liuyang, China: a cross-sectional study. *British Medical Journal Open*, 7, p. e016091.

Wang, G., Zhang, X., Wang, K., Li, Y., Shen, Q., Ge, X. and Hang, W. (2011). Loneliness among rural older people in Anhui, China: prevalence and associated factors. *International Journal of Geriatric Psychiatry*, 26, pp. 1162–1168.

Wang, J. J., Synder, M. and Kaas, M. (2001). Stress, loneliness, and depression in Taiwanese rural community-dwelling elders. *International Journal of Nursing Studies*, 38, pp. 339–347.

Wenger, G. C. and Burholt, V. (2004). Changes in levels of social isolation and loneliness among older people in a rural area: a twenty-year longitudinal study. *Canadian Journal of Aging*, 23(3), pp. 115–127.

Wilson L., Crow A., Willis M. (2008). *LinkAge Plus Project: Village Agents: Gloucestershire County Council in Partnership with Gloucestershire Rural Community Council: Overall Evaluation Report.* Birmingham: INLOGOV, School of Government and Society, the University of Birmingham.

Winterton, R., Warburton, J., Keating, N. Petersen, M., Berg, T. and Wilson, J. (2016). Understanding the influence of community characteristics on wellness for rural older adults: a meta-synthesis. *Journal of Rural Studies*, 45, pp. 320–327.

Yang, K. and Victor, C. R. (2008). The prevalence of and risk factors for loneliness among older people in China. *Ageing & Society*, 28, pp. 305–327.

18 Understanding and performing care in rural contexts in Central Europe

Anna Urbaniak

Introduction

In rural areas, the historical reliance on family (especially wives and daughters) for caregiving, as well as the high costs associated with institutionalisation are very visible. Therefore measures to reinforce or stimulate informal care, and thereby allow people to age in place for as long as possible, are seen as both culturally acceptable and cost-effective by policy makers and families alike (Bigonnesse and Chaudhury, 2020). However, this prioritisation of informal care result from a combination of unexamined cultural assumptions about what constitutes appropriate care in rural settings, financial concerns and/or existing power relations between interest groups (Broese van Groenou, and Boer 2016; Oudijk, Woittiez and de Boer, 2011). Therefore, an investigation of meanings assigned to care practices in a rural environment is much needed. Specifically, an examination of how care is defined and performed will assist in advancing an understanding of care practices in rural environments (see Glasgow and Doebler, Chapter 10 in this volume).

In this chapter, I aim to investigate the multi-dimensional connections between rural context and understanding and performing care for older people. I examine the role of spatial and sociocultural factors in influencing how care is provided and received by members of care convoys in rural environments in the central European context, using Kemp's convoys of care as a model (Kemp et al., 2013).

I begin by reviewing the gerontological literature on care and contextualise the meaning of care for different members of care convoys in rural environments. I then explore the understanding of care and care practices performed across different types of care convoys in rural Poland. I draw on interview data that are derived from a broader mixed methods study on informal carers' needs related to caring for people aged 60+. In doing so, I show that the hierarchical – compensatory model of care (Cantor, 1979) is embedded in understanding and performing care in the central European context. The chapter concludes by highlighting implications for our understanding of care practices in global ageing rural contexts.

Rural contexts of care practices

A growing body of literature, across domains of environmental gerontology, geographical gerontology and social gerontology highlight the particular circumstances created by rural contexts for the processes of ageing (e.g., Keating, 2008; Skinner, Andrews and Cutchin, 2018). Older people in rural environments are recognised as a rapidly growing population (UNECE, 2017; USDA, 2018) that are facing complex challenges and in some circumstances marginalisation (Walsh, O'Shea and Scharf, 2012). Among rural-dwelling older people, care-dependent older adults are a particularly vulnerable group. Their vulnerability originates from the intersection of micro, mezzo and macro-level factors that encompass individual circumstances (e.g., socio-demographic factors), lack of service infrastructure in rural environments as well as migration processes that all together transform not only the meaning of ageing in place in rural environments (e.g., Milne, Hatzidimitriadou and Wiseman, 2007; Hogan and Young, 2013) but also influence care practices in these settings. There is a growing body of research evidence that suggests that as a result of these factors, individual experiences of ageing and providing/receiving care vary across different places (Manthorpe et al., 2008; Milligan and Wiles, 2010).

There is a longstanding focus on relationships between places and processes of care. This is related to the fact that in older age people tend to spend more time in their spatial location (Dahlin-Ivanoff et al., 2007; Oswald and Wahl, 2005) and at the same time, might be more vulnerable to environmental factors (Golant, 2011; Wahl, Iwarsson and Oswald, 2012). Within rural gerontology, there is also an emphasis on how environmental factors impact on well-being of rural older people and their access to care services such as hospitals or home care. As highlighted in the literature, with an age-related decline, the need for accessible and appropriate services becomes more apparent, especially in rural environments where time, distance and transport required to access them needs to be taken in account (Innes et al., 2006).

On one hand, the evidence base that rural environments might be more at risk of lacking care service infrastructure (due to spatial factors, lack of market etc.) is well established within the literature (Joseph and Chalmers, 1996; Skinner and Rosenberg, 2002). However, on the other hand, some studies show that tight-knit rural communities are sometimes capable to compensate for the shortcomings or lack of formal services in place (Walsh, O'Shea and Scharf, 2012). This accounts for the fact that rural communities represent complex, interrelated systems of formal and informal socio-political units, where cultural context impacts care practices in a complex manner across diverse environments (Heenan, 2010; Long, Campbell and Nishimura, 2009).

Cultural context of care practices for older people in Central Europe

The impact of cultural and legal context on care arrangements is widely recognised. As suggested by Saraceno (2010) and Saraceno and Keck (2010), patterns of division of responsibilities between state and family in providing care can be described as (1) familialism by default, or unsupported familialism, when the responsibility for providing care is assigned mainly to the family (most often to women) because there are neither publicly provided alternatives to the family nor explicit financial provisions for family care; (2) supported familialism, when the family is supported with care-related leave, payments for care or tax relief; (3) defamilialisation, when there is a high level of services for the frail elderly (publicly financed services and/or market provision) and the individualisation of social rights reduces family responsibility (along gender and generational lines).

In many central European countries, the role of the family in providing care is strongly embedded in the culture. This is evident in both: everyday care practices and policy formulation of the long-term care system (Bouget, Spasova and Vanhercke, 2016; Mestheneos and Triantafillou, 2005). As research has established, the role of female care providers (especially daughters and wives) in providing care for older people in central Europe remains crucial (Naldini, Pavolini and Solera, 2016).

Determinants of informal care provision at the individual level consist of (1) the physical and mental health of care-dependent older person, (2) the caregiver's dispositions (attitudes and affections, norms of solidarity and reciprocity and perceived restrictions in terms of money, competences and distance) and (3) social network (family, community) (Broese van Groenou and Boer, 2016). Depending on individual circumstances, the care arrangements might differ. Whole spectrums of "care networks" (Keating et al., 2003) might be observed in rural environments.

Kemp et al. (2013) have proposed a care convoy model that emphasises the intersection between formal and informal care and the influence of community-level factors and macro-level factors (such as culture and policy). This model also takes into account changes and negotiations of care among all individuals "who may or may not have close personal connections to the recipient or to one another, but who provide care, including help with activities of daily living (IADLs), socio-emotional care, skilled health care, monitoring and advocacy" (Kemp et al., 2013: 18). As noted in empirical research (Byrd, Spencer and Goins, 2011; Page et al., 2018), care convoys have outcomes for all members involved, including the well-being of care recipient, level of care burden for the informal carer and job satisfaction for formal caregivers. This chapter examines understandings of caring for older people and care practices in central European rural environments using Kemp's convoys of care model.

Methods

The present study is a qualitative analysis of five case studies of care convoys of rural-dwelling older adults. Data in this chapter are drawn from a broad scope mixed-method study on informal care in Poland, which aimed to explore the lived experiences of members of care convoys and needs of informal caregivers providing care to people aged over 60 years.[1]

In the original study, 33 cases were selected using a purposive sampling procedure. One hundred twenty-one in-depth interviews with care-dependent older people, informal caregivers, formal caregivers and local stakeholders were collected. Care dependent older people were included based on their age (60 and more) and need of care (score 1–4 on the ADL scale). Exclusion criteria were hospitalisation and the person's incapacity to participate in the study. Informal caregivers were included based on the time and frequency on care provision (at least eight hours per week). Formal caregivers and local stakeholders were included based on information provided by the older person and informal caregiver(s) (if they were identified as members of the given care convoy).

Eligible participants received written information about the study before participation and expressed consent to participate. All interviews were digitally recorded and transcribed ad verbatim. All data were anonymised and coded using the software program MAXQDA.

The present study is focused on the care convoys of (5) older people living in rural environments and used qualitative interviews (N = 19) with older people, formal and informal caregivers (e.g., family member, friend or neighbour).

A sample overview is presented in Table 18.1.

A qualitative analysis using thematic (content) techniques was conducted on the data and incorporated both deductive, concept-driven coding and inductive, data-driven coding (Fereday and Muir-Cochrane, 2006). Results presented in this chapter focuses on presenting the understandings of caring for older people and care practices in rural environments.

Understanding of caring for older people

Collected data suggest that in creating the meaning of caring for older people, the following aspects seemed to be most important: internalised values, normative beliefs and perceived obstacles to providing care. Internalised values are most general and influence normative beliefs on the care that again might be (re)negotiated and (re)shaped by perceived obstacles to providing care.

Internalised values, in this case, refer to standards for discerning what is good and what is bad in relation to providing care. At the most general level and as recognised in all analysed cases, the culturally defined goal was to provide care to older people who might need it. This might be linked to ethical and/ or religious values internalised during the socialisation process. Participants often referred to religious or ethical obligations in defining care: "For me, this is natural, normal, I can't imagine not taking care of her (mother-in-law), and I think it's the same for my wife. This comes from our beliefs, the principles that we trust in" (Case 3_FM). The value of providing care to older people was perceived as undisputable and natural what often was translated into normative beliefs regarding care.

Normative beliefs on care provision on the most general level impact how strongly one adheres to the general norm that family is responsible for helping others in times of need (Cooney and Dykstra, 2011). In the Polish cultural context, the family is perceived to be the primary source of support. The preferred

Table 18.1 Sample overview from rural care convoy research

Case	*Number of interviews within case*	*Characteristic of a care-dependent person*	*Age of primary informal caregiver*	*Relationship with care-dependent person*	*Structure of the care convoy*	*Living arrangement*	*Duration of care*
1	2 (OP, IC)	71 female, stroke	46	Daughter	Only informal caregiver provides care	Living together	3 months
2	3 (OP, IC, FM)	92 female	68	Son	Informal carer and other keens	Living together	14 years
3	4 (OP, IC, FM*, SW)	78 male	57	Daughter	Informal carer, other keens, social support	Living together	5 years
4	5 (OP, IC, FM, SW, FC)	83 female	64	Daughter-in law	Informal carer, other keens, social support, formal caregiver	Living in separate households, sharing one house	9 years
5	5 (OP, IC, FM*, SW, FC)	80 male	54	Son	Informal carer, other keens, social support, formal caregiver	Living separately	2 years

OP- older person, IC-informal caregiver, FM- family member (*secondary carer), SW- social worker, FC- formal caregiver.

hierarchy assumes the following pattern: in the first circle of support is spouse and children, followed by friends and neighbours, while formal organisations are at the external circle. This selection process was evident across narratives from different members of care convoys, starting from care-dependent older people themselves: "I don't want anybody else, just my [daughter], we wouldn't like any outsiders" (Case 2_OP); to informal caregivers: "We think that this is somehow on us . . . that it's the best when we do everything for him" (Case 3_IC); to stakeholders: "it's a family responsibility to provide support, we can only assist the family, but primarily it's family responsibility" (Case 5_SW). Additionally, in context of the care provision for older people, internalised norms not only on caregiving as such but also on gender roles, reciprocity and (intergenerational) solidarity might come into play as highlighted by one of the family caregivers: "For me, it's absolutely obvious that at the beginning there are children, parents take care of them and then those kids take care of their parents" (Case 1_IC).

Normative beliefs regarding care might be (re)negotiated and (re)shaped depending on perceived obstacles in providing care. Across analysed cases, the following barriers were recognised as most often changing the meaning of care: the geographical distance between older person and caregiver, poor health of the caregiver, poor economic situation of caregiver, weak emotional relationship between caregiver and older person, caregivers' responsibilities towards family of procreation and caregivers' professional duties. Norms regarding reciprocity were renegotiated within care convoys depending on physical proximity between an older person and potential caregivers, as highlighted by one of the women taking care of her mother in-law: "Her daughter is in England, other son moved as well so they are not here. Therefore, we took mother in law to live with us as we are here, in place" (Case 4_IC).

Poor health might also influence the understanding of care as a process that might be transferred to more distant relatives if a closer family member is not able to provide care, as highlighted by one of the stakeholders: "I think that his daughter would have helped but she is unable to walk so how can you expect her to be a caregiver?" (Case 5_SW). Similarly, in the case of potential caregivers, who were focused on professional career or family of procreation, norm of reciprocity and solidarity was renegotiated as described by one of the formal caregivers: "You know, her (dependent older person) daughter has her own family, (family caregiver) is single so he might focus on caring for his mom . . . he has more time" (Case 3_SW).

The earlier-mentioned quote represents also the renegotiation of norm ascribing the care responsibilities in the first line to female carers. It reflects the dynamic aspects of defining care in relation to convoy structure and availability of culturally most preferred carers. When material situation didn't allow informal caregiver to provide sufficient care, care convoys were expanded and included formal institutions as highlighted by one of the stakeholders: "The fact that they turn to us for help indicates that they can't cope on their own; to pay for services privately it's very expensive" (Case 4_SW).

Care practices in rural environments

The analysed data highlight that informal care may vary in intensity, type of help provided, location and duration. Sometimes remoteness of the place caused issues for home care services, as formal carers were unable to get to dwellings of care-dependent older persons, as highlighted by one of the interviewees: "They (formal carers) just don't come. Only one called and said that it's just too far for her to come" (Case 5_FM). Stakeholders also pointed to difficulties in organising care in remote rural areas not only due to transportation issues but also because of depopulation of areas caused by migration of younger generations.

> It's very difficult to make it all work somehow, we try to plan routes with clients in the same region, but they are scattered across and if the carer comes from (name of large city) and relies on public transport there is no chance to make it in assigned time . . . it's impossible to recruit caregivers in some areas, there are just not enough young people left there.
>
> (Case 4_SW)

Difficulties in accessing formal care were compensated sometimes by strong social networks in place, as highlighted by one participant taking care of her mother: "So my neighbours come and help. It isn't their responsibility but they do it anyway, just because they are kind-hearted" (Case 2_IC). This presented quote points also to the role of normative beliefs on caregiving where care is perceived as the responsibility of family and not neighbours.

The normative beliefs and perceived limitations in providing care were mirrored in care practices across diverse care convoys. Within analysed cases, three types of strategies in performing care were identified: care conformism, care innovativeness and care withdrawal. These strategies reflect the variety across approaches to the culturally defined goals and normatively defined means in the area of taking care of the older people.

Care conformism is the strategy that is characterised by a high acceptance of norm defining care provision as primarily family responsibility. The main mean of delivering care was through informal care provision. This was often linked with ignoring costs (both material and nonmaterial) of providing care by both: older person and informal carer(s) and a reluctance to expand care convoy. This strategy was identified in case 1 and 2. Here it's important to note that rural communities might be characterised by a high level of social control and lack of anonymity, which impacts care practices disregarding the internalisation of norms and beliefs regarding care: "People would be wondering if I'm so young why I don't take care of mum, so I would be thinking what people would say?" (Case 1_IC). Narratives of interviewees suggested that this strategy was common in the early stages of providing care, before extending care convoys.

Care convoys that were more extended employed strategies that could be labelled as care innovativeness. This strategy of performing care is characterised by a high acceptance of norm defining care provision as primarily family responsibility but at the same time lower acceptance of the belief that it is primarily family who should provide the care. Therefore, the main difference compared to the previous strategy is recognising other parties as possible care providers. In this strategy, caregiver(s) and care-dependent older person are aware of the cost of providing care and limited possibilities of providing care only within the family. This strategy was identified in cases 3 and 5. The division of task in extended care convoys runs along with gender roles as highlighted by one of the carers: "He is a man and I am a man so it's obvious that it is my wife and a nurse who are helping him with all hygienic procedures" (Case 5_ IC). The division of tasks within extended care convoys was negotiated on the basis of norms and obstacles in providing care.

When obstacles in providing care are perceived as too high and make it impossible to deliver care, even within the extended care convoy structure care withdrawal can be observed. In this strategy, caregivers are unable to meet the needs of care-dependent older people. This is linked with a too high level of needs of an older person or with very low possibilities of the carer(s), resulting from the financial situation, health condition, carer burden or a combination of these factors. In this situation, caregivers strive to cease care provision and delegate care responsibilities elsewhere. In most cases this would mean delegating care to public institutions as described by one caregiver (who had significant functional limitations).

> Well, my father did not raise a blind, handicapped one so I tried to repay him. But now it's really hard. I can't take it anymore, I have to place him in the hospice because there is no other option.
>
> (Case 4 _IC)

Conclusions

In this chapter, I have used Kemp's model (Kemp, Ball and Perkins, 2013) to explore how care is understood and performed by different actors within care convoys. Poland is a good example of the central European country, where the cultural and formal context of care provision is strongly family oriented which means that care is provided mostly by informal, unpaid family caregivers (Kurpas et al., 2018; Leszko and Bugajska, 2015).

In this context, providing care for an older, fragile family member is perceived as a legitimate cultural value and delivery of care by a family member as a culturally accepted norm (Connidis and Barnett, 2018). Similar patterns were observed in the Czech Republic (Dudová, 2018), Slovenia (Hlebec et al., 2014), Romania and Bulgaria (Kulcsár and Brădățan, 2014).

In rural environments across central Europe impact of norms on the care-provision is intertwined with post-socialist heritage (Galčanová and Staveník, 2019) that translates into the lack of services and underdeveloped civil society institutions that could participate in the long-term care provision system.

In this chapter, I have argued that understanding and providing care for older people in central European rural context is not individual experience but rather process shared within the context of care convoy and community. Effort in disentangling meanings attached to care practices in a rural environment is much needed to avoid oversimplifications that on one hand focus on weaknesses of service infrastructure in rural areas and on the other might be too optimistic in perceiving rural environment as somewhat idyllic settings where norms of reciprocity and solidarity overcome weaknesses of the long-term care system (see Hanlon and Poulin, Chapter 4). Caring for older people takes place and it's enacted with and through spatial infrastructures and cultural practices. Without further analysis, we might overlook important factors and processes influencing understanding and performing care in global ageing rural contexts.

Acknowledgements

In this chapter, I draw on previously published work (Urbaniak and Szukalski, 2018). I thank the participants for their willingness to participating in this research.

Note

1 The project was founded by Regional Centre for Social Policy in Łódź. The principal investigator was A. Urbaniak and co-investigators were P.Mielczrek, A. Szczerbik, A.Bujwicka, M.Beczkowska, M.Błaszczyk, K.Brzezińska-Krakowiak, M.Tomczak and E. Żmurkow-Poteralska.

References

Bigonnesse, C. and Chaudhury, H. (2020). The landscape of "aging in place" in gerontology literature: Emergence, theoretical perspectives, and influencing factors. *Journal of Housing For the Elderly*, 34, pp. 233–251.

Bouget, D., Spasova, S. and Vanhercke, B. (2016). *Work-Life Balance Measures for Persons of Working Age with Dependent Relatives in Europe: A Study of National Policies*. Brussels: European Comission.

Broese van Groenou, M. and Boer, A. (2016). Providing informal care in a changing society. *European Journal of Ageing*, 13(3), pp. 271–279.

Byrd, J., Spencer, S. M. and Goins, R. T. (2011). Differences in caregiving: does residence matter? *Journal of Applied Gerontology*, 30, pp. 407–421.

Cantor, M. H. (1979). Neighbors and friends: an overlooked resource in the informal support system. *Research on aging*, 1(4), pp. 434–463.

Cooney, T. M. and Dykstra, P. A. (2011). Family obligations and support behaviour: a United States – Netherlands comparison. *Aging and Society*, 33, 1026–1050.

Connidis, I. A. and Barnett, A. E., (eds)., (2018). *Family Ties and Aging*. London: SAGE Publications.

Dahlin-Ivanoff, S., Haak, M., Fänge, A. and Iwarsson, S. (2007). The multiple meaning of home as experienced by very old Swedish people. *Scandinavian Journal of Occupational Therapy*, 14(1), pp. 25–32.

Dudová, R. (2018). Doing gender and age: the case of informal elderly care in the Czech Republic. *International Journal of Ageing and Later Life*, 12(1), pp. 41–73.

Fereday, J., Muir-Cochrane, E. (2006). Demonstrating rigor using thematic analysis: a hybrid approach of inductive and deductive coding and theme development. *International Journal of Qualitative Methods*, 5, pp. 80–92.

Galčanová, L. and Staveník, A. (2019). Ageing in the changing countryside: the experience of rural Czechs. *Sociologia Ruralis*, 60(2), pp. 329–356.

Golant S. M. (2011). The quest for residential normalcy by older adults: relocation but one pathway. *Journal of Aging Studies*, 25, pp. 193–205.

Heenan, D. (2010). Social capital and older people in farming communities. *Journal of Aging Studies*, 24(1), pp. 40–46.

Hlebec, V., Mali, J. and Filipovič Hrast, M. (2014). Community care for older people in Slovenia. *Anthropological Notebooks*, 20(1), pp. 5–20.

Hogan A. and Young M. (2013). Visioning a future for rural and regional Australia, Cambridge *Journal of Regions, Economy and Society*, 6(2), 319–330.

Innes, A., Cox, S., Smith, A. and Mason, A. (2006). Service provision for people with dementia in rural Scotland: difficulties and innovations. *Dementia*, 5(2), pp. 249–270.

Joseph, A. E. and Chalmers, A. I. (1996). Restructuring long-term care and the geography of ageing: a view from rural New Zealand. *Social Science and Medicine*, 42(6), pp. 887–896.

Keating, N., (ed.). (2008). *Rural Ageing: A Good Place to Grown Old?* Bristol: Policy Press.

Keating, N., Otfinowski, P., Wenger, C., Fast, J. and Derksen, L. (2003). Understanding the caring capacity of informal networks of frail seniors: a case for care networks. *Ageing & Society*, 23, pp. 115–127.

Kemp, C. L., Ball, M. M., and Perkins, M. M. (2013). Convoys of care: theorizing intersections of formal and informal care. *Journal of Aging Studies*, 27(1), pp. 15–29.

Kulcsár, L. J., and Brădăţan, C. (2014). The greying periphery – Ageing and community development in rural Romania and Bulgaria. *Europe-Asia Studies*, 66(5), 794–810.

Kurpas, D., Gwyther, H., Szwamel, K., Shaw, R. L., D'Avanzo, B., Holland, C. A., and Bujnowska-Fedak, M. M. (2018). Patient-centred access to health care: a framework analysis of the care interface for frail older adults. *BMC geriatrics*, 18(1), p. 273.

Leszko, M. and Bugajska, B. (2015). Towards creating a comprehensive care system for elders: an overview of long-term systems across the developed countries. *Social Policy and Models of Services for the Elderly International Perspective*, 2(10), pp. 13–24.

Long, S. O., Campbell, R. and Nishimura, C. (2009). Does it matter who cares? A comparison of daughters versus daughters-in-law in Japanese eldercare. *Social Science Japan Journal*, 12(1), pp. 1–21.

Manthorpe, J., Iliffe, S., Clough, R., Cornes, M., Bright, L., Moriarty, J., and Older People Researching Social Issues. (2008). Elderly people's perspectives on health and well-being in rural communities in England: Findings from the evaluation of the National Service Framework for Older People. *Health and Social Care in the Community*, 16(5), pp. 460–468.

Mestheneos E. and Triantafillou J. (2005). *Supporting Family Carers of Older People in Europe, The Pan-European Background Report. Supporting Family Carers of Older People in Europe, Empirical Evidence, Policy Trends and Future Perspectives* (Vol. 1, eds Döhner, H., Kofahl, C). Münster: LIT-Verlag. Available at: www.lit-verlag.de/isbn/3-8258-9121-6

Milligan C. and Wiles J. (2010). Landscapes of care. *Progress in Human Geography*, 34, pp. 736–754.

Milne, A., Hatzidimitriadou, E. and Wiseman, J. (2007). Health and quality of life among older people in rural England: exploring the impact and efficacy of policy. *Journal of Social Policy*, 36(3), pp. 477–495.

Naldini, M., Pavolini, E. and Solera, C. (2016). Female employment and elderly care: the role of care policies and culture in 21 European countries. *Work, Employment and Society*, 30(4), pp. 607–630.

Oudijk, D., Woittiez, I. and de Boer, A. (2011). More family responsibility, more informal care? The effect of motivation on the giving of informal care by people aged over 50 in The Netherlands compared to other European countries. *Health Policy*, 101(3), pp. 228–235.

Oswald, F. and Wahl, H-W. (2005). Dimensions of the meaning of home in later life. In: G. D. Rowles and H. Chaudhury, eds., *Home and Identity in Late Life*. New York: Springer, pp. 21–46.

Page, K. J., Robles, Z., Rospenda, K. M. and Mazzola, J. J. (2018). Understanding the correlates between care-recipient age and caregiver burden, work-family conflict, job satisfaction, and turnover intentions. *Occupational Health Science*, 2(m4), pp. 409–435.

Saraceno, C. (2010). Social inequalities in facing old-age dependency: a bi-generational perspective. *Journal of European Social Policy*, 20(1), pp. 32–44.

Saraceno, C. and Keck, W. (2010). Can we identify intergenerational policy regimes. *European Societies* 12(5), pp. 675–96.

Skinner, M. W., Andrews, G. J. and Cutchin, M. P., (eds). (2018). *Geographical Gerontology: Perspectives, Concepts, Approaches*. Oxon: Routledge.

Skinner, M. and Rosenberg, M. (2002). *Health care in rural communities: Exploring the development of informal and voluntary care*. Working paper, McMaster University, Social and Economic Dimensions of an Ageing Population Research Papers. Available at: http://socserv2.socsci.mcmaster.ca/~sedap/p/sedap148.pdf

UNECE. (2017). Older persons in rural and remote rural area (UNCE policy brief on ageing no. 18). Available at: www.unece.org/fileadmin/DAM/pau/age/Policy_briefs/ECE-WG1-25.pdf

Urbaniak, A. and Szukalski, P. (2018). *Sytuacja opiekunów rodzinnych w kontekście sprawowania opieki nad osobami w wieku 60+ z terenu województwa łódzkiego*. Raport końcowy. Available at: http://dspace.uni.lodz.pl/xmlui/bitstream/handle/11089/25534/sytuacja-opiekunow-rodzinnych---7-08-2018%20ZAAKCEPTOWANE.pdf?sequence=1&isAllowed=y

USDA. (2018). *Rural America at a Glance*. United States Department of Agriculture, Available at: www.ers.usda.gov/webdocs/publications/90556/eib-200.pdf

Wahl, H-W., Iwarsson, S. and Oswald, F. (2012). Aging well and the environment: toward an integrative model and research agenda for the future. *The Gerontologist*, 52, pp. 306–316.

Walsh, K., O'Shea, E. and Scharf, T. (2012). *Social Exclusion and Ageing in Diverse Rural Communities: Findings from a Cross-Border Study in Ireland and Northern Ireland*. Galway: Irish Centre for Social Gerontology.

Part IV

Emerging critical perspectives

19 Postcolonial perspectives on rural ageing in (South) Africa

Gendered vulnerabilities and intergenerational ambiguities of older African women

Jaco Hoffman and Vera Roos

Introduction

This chapter critically reflects on contemporary rural ageing in South Africa, historically situated within the postcolonial and post-Apartheid eras. A critical exploration of this kind however, ventures beyond the historical moment and draws on contemporary postcoloniality where the *subaltern* (Spivak 1988) (in this case, older Black women and younger generations in rural settings) speak for themselves to produce a complex and heterogeneous range "of contradictions, of half-finished processes, of confusions, of hybridity and of liminalities" (Lee and Lam, 1998: 970). "Postcoloniality" is used here to signify "an attitude of critical engagement with colonialism's aftereffects and its constructions of knowledge" (Radcliffe, 1997: 1331) and is applied to convey the legalistic continuation of the economic, cultural and linguistic power relationships that control the generation, production and distribution of knowledge and conventions. Concurrently, it also provides a broad conceptual frame to *destabilize* some of the dominant discourses/categories/divisions to challenge inherent assumptions and to critique the material and discursive legacies of colonialism (Crush, 1994; Perera, 1998). Against the backdrop of postcoloniality and post-Apartheid we acknowledge as two white South African social gerontologists our limited (outsider) perspective on the reality of the multiplicity of issues impacting on older Black women in rural areas. However, by adopting empathy and perspective taking (Vorster, Roos and Beukes, 2013), as well as drawing on the notion of crystallisation (Ellingson, 2009), we consciously moved towards a deeper understanding of the perspective of older Black women.

To this end, we first contextualise the contemporary South African reality broadly, driven by five interrelated trends, namely poverty exacerbated by rural-urban migration and HIV/AIDS; the increase in non-communicable diseases (NCDs) and changing family dynamics. We next qualitatively explore how a lack of essential basic and municipal service delivery in rural settings contributes to older persons' vulnerability; the generational dynamics between older and younger generations; and the resilience of these older women. In conclusion, we relate the rurality of older women to postcoloniality, capturing Africa's present realities and location and contribute to a more nuanced understanding as well as reconstruction of a contemporary African space.

Contextualisation

Despite profound socio-political transformations and diversities in postcolonial Africa and post-1994 South Africa, the majority of older adults in Sub-Saharan Africa (SSA) share the same interrelated trends, namely poverty exacerbated by rural-urban migration and HIV/AIDS; the increase in NCDs and changing family dynamics.

Poverty

Most prominently, entrenched poverty characterises SSA. Of the 38 countries with the lowest human development ranking listed in the 2018 *Human Development Report* of the UN Human Development Index (UNDP, 2018), 32 are situated in SSA. Although South Africa ranks 113th out of 189 positions and is categorised as a medium development country, it has one of the highest levels of economic inequality in the world, with a Gini coefficient of 0.68 in 2015 (StatsSA, 2017). As a legacy of policies implemented under colonialism and Apartheid, the group most affected by this inequality is Black Africans (categorised as such by Statistics South Africa). South Africa's unemployment rate reached 29.0% in the second quarter of 2019, with youth unemployment more than 56% (StatsSA, 2019).

With an expected increase in social grant beneficiaries from 17.9 million (2019/20) to 18.6 million in 2020/21 (StatsSA, 2019), the South African government presently oversees one of the most rapidly expanding social welfare systems in the developing world to address issues of poverty and inequalities. Over 3.1 million individuals aged 60 years and older received the means-tested Old Age Pension (OAP) in 2015 compared with 2.7 million in 2011 – this is expected to increase to around 4 million by 2021/22. These OAPs (R1 800 [£98] per month) are used to support own households (Knight, Hosegood and Timaeus, 2016) as well as grandchildren (e.g., paying school fees), children orphaned through HIV/AIDS, ill household members (Hosegood et al., 2007) and unemployed individuals (Klasen and Woolard, 2009). This asymmetrical dependency of younger generations on older generations is increasingly relevant in view of the continuous impact of rural/urban migration and the HIV/AIDS epidemics specifically.

Migration

SSA is regarded as the world's fastest urbanising region. Urban areas currently comprise a population of 472 million which is set to double over the next 25 years. The global share of urban residents in Africa is projected to grow from 11.3% in 2010 to 20.2% by 2050 (United Nations, 2017). The South African case is unique in that under Apartheid with the Group Areas Act, Black and Coloured citizens' movements between geographic areas were restricted by legislation to keep them separate in rural, underdeveloped areas and to prevent their mobility to urban areas (see Durrheim, Mtose and Brown, 2011). The Group Areas Act was revoked in 1988, followed by massive urbanisation. In certain provinces, the exodus of working age individuals from rural areas to cities in other provinces has increased the proportion of older persons in the province of origin.

Over the past decades three trends of rural/urban dynamics evolved: Older parents no longer coping alone in a rural area, may follow their migrant children to the urban area where they often have to contend with poor and overcrowded housing in informal settlements, an unfamiliar environment and social ills, at risk of displacement and alienation (e.g., see Nxusani, 2004). A strong circulatory migration pattern furthermore exists whereby older rural dwellers visit an urban area and join their kin for a number of months in order to obtain health care and other services. The older migrants return to their rural homestead over an extended period during which they sow and reap crops, tend livestock, visit family and perform traditional rituals. Another trend noted is that of young in-migrants to an urban area who having contracted AIDS return to their ancestral rural home to be cared for by an older parent (Møller and Ferreira, 2003; Nxusani, 2004).

Impact of HIV/AIDS

It is estimated that around 7.7 million people in South Africa are currently HIV-positive (UNAIDS, 2019). The pandemic's long-term generational momentum affects both ascending and descending generations, illustrated by the high proportion of affected grandchildren who live in households headed by older persons (mostly grandmothers) (Schatz and Seeley, 2015).

A substantial body of research has been undertaken on the effects of the AIDS epidemics on older persons in SSA countries. These studies have focused mainly on associated morbidity and mortality rates and the resulting dysfunction in affected households (Oramasionwu et al., 2011; Small et al., 2017). Localised qualitative studies in South Africa have assessed the perceived needs of older carers in impoverished settings (Singo et al., 2015). Gómez-Olivé et al.'s (2010) research is one of the few quantitative studies assessing the well-being of older carers. These studies highlight the burden of care (financial, emotional and physical) and the multiple responsibilities of older carers but also evidence the significant contributions older persons (particularly women) make to the welfare and development of younger generations and their communities generally (Hosegood and Timaeus, 2006). Many neglect their own health because of the time and resources they devote to giving care to others (Small et al., 2017). Older adults furthermore suffer the burden of NCDs.

Rise in non-communicable diseases

Prevailing NCDs are high blood pressure, diabetes and arthritis, all of which require life-long management. This places a strain on ageing-related resources and the availability of good formal long-term care (Hajat and Stein, 2018). Antiretroviral therapy is likely to increase the prevalence of NCDs as a result of older adults' increased survival and/or through association of drug regimens with NCD risk (Negin et al., 2012).

There is considerable functional impairment and disability among older adults across SSA (Aboderin and Beard, 2014). In South Africa age is strongly correlated with functional disability at the population level from age 70, with those aged 80 and older reporting a threefold increased risk of poor functionality (Gómez-Olivé et al., 2010).

Families and communities thus experience mounting pressure to provide almost all long-term care for older adults in the region. There are profound inadequacies in informal (familial) care provision due to economic and infrastructural strain exacerbated by migration, mental and physical caregiver strain (Aboderin and Hoffman, 2017) and comprehensive social and demographic change caused by the HIV/AIDS epidemic, particularly in southern Africa (Schatz and Gilbert, 2014).

Changing family dynamics

Against a backdrop of poverty, the impact of HIV/AIDS, the rise in NCDs and migration, the position of dominant policy instruments is that the African family will continue to care for their older persons. This draws on the ideal and power of "the African family" and regard family values as a major moral asset in dealing with challenges associated with ageing.

Two interrelated trends challenge the familist ideology (Aboderin and Hoffman, 2015): first, continuous ascending family support patterns with indications of growing inadequacies in support for older persons, and second descending patterns of care whereby older persons are increasingly looked upon as continuous providers of family support to their own financial, social and health disadvantage.

Locating the "postcolonial" in South African rurality

From a spatial perspective, some 18 million of South Africa's 56 million people are estimated to be living in rural areas under tribal leadership.

Ageing under tribal leadership in rural South Africa

Rural communities in South Africa are organised under the leadership of a *kgosikgolo* (king) who, with traditional structures such as hereditary headmanship and advisers (male family members or influential people in the royal family), lead and govern the community; arguably seek its best interests; act as custodians; and foster a sense of community (Nkosi et al., 1994). Building on precolonial structures and as a result of exclusionary Apartheid policies, rural communities in South Africa today are generally under-resourced with limited or no economic development. Despite major political changes after 1994, whereby the institution, status and influence of traditional leadership were recognised in the Constitution (Act 108 of 1996) of South Africa (Chapter 12), rural communities have been slow to recover from a backlog of nearly five decades of oppression or to develop infrastructure and provide required services for its inhabitants, with particularly dire consequences for older people (South African Government, 2018/19).

Women under tribal authority

In general, African society is patriarchal and has deep gender divides which are most pronounced in rural areas where they are reinforced by tribal structures and traditional authority (Maitse and Majake, 2005). Although the Constitution (specifically the Bill of Rights) and legislation in South Africa provide for the protection of the rights of all citizens, application of the legislation in practice may be less systematic in the case of older persons in rural areas (see Ferreira, 2004). The recently signed Traditional Khoi-San Leadership Act (3 of 2019) again enacted precolonial and Apartheid powers of traditional authorities, such as kings, queens, chiefs and headmen and is heavily opposed by civil society: "These traditional authorities will now have governmental, law-making, judicial, custom-making and land administration powers, all at the same time. The people who will suffer the most are women" (Naki, 2019). In rural South Africa, Black older women may lack the power to assert their rights and to protect themselves from exploitation and abuse regarding land rights, security of tenure, inheritance and succession (Theron, 2014).

Research settings and methods of data collection

Against this backdrop, we draw on data collected over the past decade in rural parts of the North West Province of South Africa: Lokaleng (Ngaka Modiri Molema District Municipality), the Vaalharts region and Ganyesa (Dr Ruth Segomotsi Mopati (previously Bophirima) District Municipality). People aged 60 years (n = 152) and older, and younger people between 12 and 35 years (n = 51), participated in semi-structured interviews, focus group discussions and the Mmogo-method®, an indigenous, visual, data-collection method (see Roos, 2016 for detail). Research was conducted on topics related to: intergenerational care and respect experiences; perceptions of community life in rural settings; coping with drought in rural South Africa; a baseline assessment of the status quo of service delivery by local government to older citizens; and relational experiences around mobile phone use by older persons.

Themes

Three broad, interrelated themes are identified in the analysis: (1) structural exclusion; (2) intergenerational dynamics; and (3) precarity.

Structural exclusion: (pre-1994) inequalities and post-Apartheid service delivery realities

People who have aged in rural South Africa have been subjected to lifelong structural exclusion and systemic discriminatory practices implemented during colonialism/Apartheid. These left them vulnerable in their old age in undeveloped areas, lacking services and with ill-equipped infrastructure.

From 1994 onwards, the rights and duties of local government in relation to redressing service delivery to all citizens (including older persons) have been explicitly stated in various laws and numerous policies (Roos, Du Plessis and Hoffman, in progress). However, an assessment of the *de facto* municipal service delivery to older citizens in 2013 indicated a lack of, poor or inappropriate service offerings, and a negative attitude of local government officials towards older persons (Roos et al., in progress). Consequently, many of the basic and municipal service needs of many older persons in rural South Africa, have remained unmet: "I need running water, geyser, a house and accessible public transport". Services are non-existent or offered sporadically or are age-inappropriate, as the following quotes illustrate: "We don't have ambulances and our clinic is quite a distance"; and "sometimes the [mobile] clinic . . . comes to us. They refuse to come to my side of the neighbourhood though. I haven't seen them". Water, for example, is provided but in some instances only from a communal tap to which older persons have to walk great distances, with buckets on uneven ground: "The community taps are too far".

If ill-health prevents older people from addressing their own basic and municipal service needs, they are by default excluded: "I can't go to the clinic due to my legs being painful".

Even in an emergency, community members have to mobilise their own resources to obtain help, which could mean a death sentence for some if they do not succeed in time. "We called an ambulance. But the ambulance took [too] long . . . we went to the veld [savannah grasslands] and fetched the donkeys . . . when we get to our clinic gate, she died".

Much needed information and support from local government officials are often not provided or are unreliable: "Most of the time, they do not answer our calls when we call, or they are too busy to help. Sometimes they say they will come, but they never do". Younger people also confirmed the abusive behaviour of health workers towards older persons: "Some of the older community members are afraid to go to the clinic because the nurses don't appreciate their problems. They shout at them". Repetitive failures to access services eventually result in responses of learned helplessness, emphasising the precarity of many older persons: "When I'm sick and don't know who to call, I just sit at home". Consequently, and despite major efforts in postcolonial rural South Africa, older persons still do not receive the services they were promised, compromising their already vulnerable position and emphasising their reliance on related and unrelated younger people.

Figure 19.1 An older woman with a walking frame

Intergenerational dynamics: strained and supportive relationships

Close interpersonal relationships within multigenerational households fluctuate between being supportive, discordant or frustrating. Generally, members of multigenerational households reciprocally offer physical, instrumental and emotional care: "Then I give them food" and "they help me", with an explicit expectation that younger people should assume their moral obligation to care for older persons: "You, the youth, you are our last hope". Some older persons however,

Figure 19.2 Donkey cart used as transport in a rural community

realise that the care they bestow on younger people may not be reciprocated: "There are some children who do not care about you . . . [are] still not caring".

Intergenerational discord seems to be fuelled by misaligned demonstrations and expectations of care, as illustrated by middle adolescents' care expressions in relation to older persons. They perform physical tasks, assist with household chores, acquire an education and respect older persons, while in return older persons provide food and teach and discipline younger people (Roos, Silvestre and De Jager, 2017). However, older persons' care needs are frustrated by younger people who do not fulfil their side of the bargain. In this illustration, an older woman complains about her grandson's disobedience when asked to do household chores. Her grandson would say: "Mama, I will come just now', but you will not see him until eight or nine o'clock at night". From the perspective of the younger people, older persons are rigid and not open to dialogue:

> When you need to advise, some old man or old lady would just say: 'You are young . . . you can't tell me nothing!' That is why the youth are out of control and they take the law in their hand and show some bad attitude.

Failing to adopt the perspective of the generational other can result in the breakdown of relationships, with severe implications for older persons' care and survival:

"I live in the same house with my child but she wants her place back. Her father died long time ago. She always chases me out of her house. I don't have a place to stay".

Precarity: precarious positions and living courageously

The precarious position of many older Black South Africans in rural settings resulted from structural exclusion in the past but also from increasing uncertainty about upward support from younger generations. In pre-1994 South Africa, the majority Black citizens were subjected to inferior education, excluded from obtaining decent work or were being forcibly relocated to rural areas with few opportunities to make enough money to support them in their later years (Durrheim, Mtose and Brown, 2011). "We used to be cow watchers [herders], that's why today we don't even know how to write our names. In the past most of us were deprived [of] the rights to get proper education".

The precariousness of older persons also persists in the close interpersonal context of the family or household. For example, older persons and younger people confirmed how younger people use older persons' mobile phone airtime without the latter's knowledge and consent. In response to a question how a younger person experienced interacting with her grandmother around mobile phone use, the younger person replied: "Well for me it's fun, because I know that I can use her airtime. She will not know . . . she will ask me to buy airtime and then she will call only for a little while. I'll just use it to call my friends" (Roos and Robertson, 2019: 993).

However, despite limited economic development and insufficient resources in rural settings, women succeed in navigating the environment innovatively. They secure food and water by rearing chickens, growing vegetables for their own consumption as well as selling the products and making clay pots to store water.

Juxtaposed against the deep depletion of resources, the women have constructed a protective layer of communal compassion. They are pooling their collective resources, augmenting individual monetary deficits, buffering families against hunger and gifting the deceased with a dignified burial:

> We formed a women's club where we share ideas of how we can sustain our families. We also thought of saving our money and lending money to each other. The burial society is also helping us a lot as you all know that death comes without warning.
>
> (Roos, Chigeza and Van Niekerk, 2010: 8)

Older women are daring collectively to obtain the help they need. Local government officials' reluctance to respond to requests for help has prompted older women to take matters into their own hands, illustrated here:

> I usually go there myself because they ignore our phone calls. Most of the time they say they are busy or they will come and fix the problem [which they don't]. So going there and being persistent allows us to get the help we need immediately.

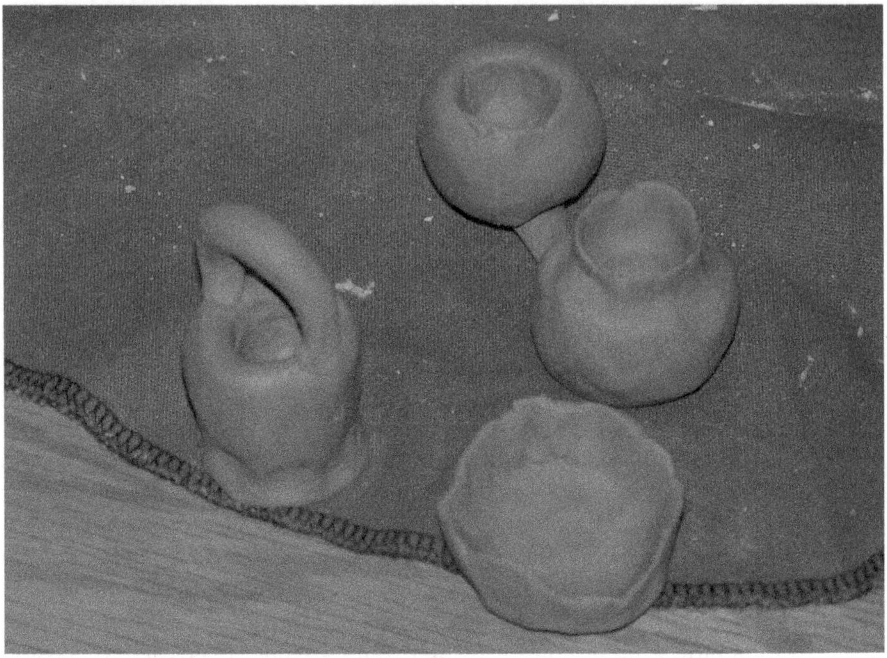

Figure 19.3 Clay posts: Kitchen utensils used for cooking and storing water

Discussion

Three interrelated arguments about rurality and ageing from a postcolonial perspective are raised. Conceptually, the notion of uniformity or homogeneity in old-age and experiences of rurality should be sensitively problematised. It is contentious to assume distinctive homogeneous categories of rural versus urban and mutually exclusive categories of traditionalism and modernity. We rather propose that migratory flows between rural and urban settings – in all its many manifestations – are continuously "changing, yielding to new ways, partially disappearing, coming up again and adapting" aligned with the notion of Mphahlele (1982).

The "traditional" extended family as often "imagined" within rural settings has been impacted and shaped by processes of colonialism, migratory dynamics and urbanisation, creating an increasingly hybrid family form. Traditional values of family life are not only being retained but are actually being integrated with modern family characteristics with a resulting synthesis of diverse systems of family life. "African family life is neither purely traditional nor purely Western." (Kayongo-Male and Onyango, 1984: 105). The support and long-term care management implications for rural older people of such rural/urban family dynamics should be prominent on the rurality research agenda.

Despite contemporary older women's (often) precolonial status within rural settings and being on the receiving end of colonial/Apartheid effects, they are people with a strong pensioners' identity who are supporting entire households, raising and educating grandchildren, caring for sick and disabled family members, engaging in many ways to augment household income, volunteering and participating in community activities. Their resilience against the odds often initiate and sustain some agency to the location about conditions which may be considered problematic. Even to the extent that developmental solutions may be proposed to the authorities by these older women, questioning hegemony.

Conclusion

Although rural poverty is a hard reality and the burden on older women in rural areas substantial, the urbanisation of rurality implores us to problematise the neocolonial binary of homogeneous distinctions between rural and urban settings. It also challenges us methodologically in terms of scope (inclusion/exclusion criteria) and practically in terms of interventions. The postcolonial critique aims to methodologically unmask the over-inclusiveness (homogeneity) as well as the exclusivity of the rural/urban binary discourse.

Ageing in rural South Africa consists of a series of cyclical, nonlinear iterations of living with contradictions. It is about being abandoned and respected; being exposed yet situated in multilayered caring community structures; living precariously but also courageously. The stark contemporary reality of rural stagnation, strained intergenerational relations and age-unfriendly resource-vulnerable settings leave many older persons neglected and at risk but also illustrates their resilience against the odds. A postcolonial reading of rural ageing offers a critical position to elucidate the complexity of rural/urban spaces, debunking hegemony and recognising agency of the subaltern.

References

Aboderin, I. and Beard, J. (2014). Older people's health in sub-Saharan Africa. *The Lancet*, 385(9968), pp. e9–e11.

Aboderin, I. and Hoffman, J. (2015). Families, intergenerational bonds, and aging in sub-Saharan Africa. *Canadian Journal of Aging*, 34(3), pp. 282–289.

Aboderin, I. and Hoffman J. (2017). Research debate on 'older carers and work' in sub-Saharan Africa? Current gaps and future frames. *Journal of Cross-Cultural Gerontology*, 32(3), pp. 387–393.

Crush, J. (1994). Post-colonialism, decolonization and geography. In: A. Godlewska and N. Smith, eds., *Geography and Empire*. Oxford: Blackwell, pp. 333–350.

Durrheim, K., Mtose, X. and Brown, L., (eds). (2011). *Race Trouble: Race, Identity and Inequality in Post-Apartheid South Africa*. Plymouth: Lexington Books.

Ellingson, L. L., (ed.). (2009). *Engaging Crystallization in Qualitative Research*. Thousand Oaks: SAGE Publications.

Ferreira, M. (2004). Born in the eastern Cape and now a social pensioner. In: M. Ferreira and E. van Dongen, eds., *Untold Stories: Giving Voice to the Lives of Older Persons in New South African society: An Anthology*. Cape Town, South Africa: UCT Press, pp. 25–41.

Gómez-Olivé, F. X., Thorogood, M., Clark, B.D., Kahn, K. and Tollman, S.M. (2010). Assessing health and well-being among older people in rural South Africa. *Global Health Action Supplement*, 2, pp. 23–35.

Hajat, C. and Stein, E. (2018). The global burden of multiple chronic conditions: a narrative review. *Preventative Medicine Reports*, 12(2018), pp. 284–293.

Hosegood, V., Preston-Whyte, E., Busza, J., Moitse, S. and Timaeus, I. M. (2007). Revealing the full extent of households' experiences of HIV and AIDS in rural South Africa. *Social Science & Medicine*, 65(6), pp. 1249–1259.

Hosegood, V. and Timaeus, I. M. (2006). HIV/AIDS and Older People in South Africa. In: B. Cohen, B. and L. Menken, eds., *Aging in Sub-Saharan Africa: Recommendation for Furthering Research*. Washington, DC: National Academies Press, pp. 250–275.

Kayongo-Male, D. and Onyango, P. (1984). *The Sociology of the African family*. New York: Longman.

Klasen, S. and Woolard, I. (2009). Surviving unemployment without state support: unemployment and household formation in South Africa. *Journal of African Economies*, 18(1), pp. 1–51.

Knight, L., Hosegood, V. and Timaeus, I. M. (2016). Obligation to family during times of transition: care, support and the response to HIV and AIDS in rural South Africa. *AIDS Care*, 28(Suppl. 4), pp. 18–29.

Lee, G. B. and Lam, S. S. K. (1998). Wicked cities: cyberculture and the reimagining of identity in the 'non-Western' metropolis. *Futures*, 30(10), pp. 967–79.

Maitse, T. and Majake, C., (eds). (2005). *Enquiry into the Gendered Lived Experience of Older Persons Living in Conditions of Poverty*. Johannesburg, South Africa: Commission on Gender Equality.

Møller V. and Ferreira M. (2003). *Getting By: Benefits of Noncontributory Pension Income for Older South African households*. Cape Town, South Africa: Institute of Ageing in Africa, University of Cape Town.

Mphahlele, E. (1982). *Afrikan Affirmations*. Available at: www.eskia.org.za/

Naki, E. (2019). Group slams traditional leadership bill for 'oppression of rural women'. *The Citizen*, 4 December. Available at: https://citizen.co.za/news/south-africa/society/2214347/group-slams-traditional-leadership-bill-for-oppression-of-rural-women/

Negin, J., Mills, E. J. and Bärnighausen, T. (2012). Aging with HIV in Africa: the challenges of living longer. *AIDS*, 26, S1–S5.

Nkosi, S. A., Kirsten, J. F., Bhembe, S. M. and Sartorius von Bach, H. J. (1994). The role of traditional institutions in rural development: the case Bogoši (Chieftainship). *Agrekon*, 33(4), pp. 282–286.

Nxusani, N. C. (2004). Late-life migration and adjustment of older persons: between the eastern Cape and the western Cape. In: M. Ferreira and E. van Dongen, eds., *Untold Stories: Giving Voice to the Lives of Older Persons in New South African society: An Anthology*. Cape Town, South Africa: UCT Press, pp. 13–24.

Oramasionwu, C. U., Daniels, K. R., Labreche, M. J. and Frei, C. R. (2011). The environmental and social influences of HIV/AIDS in sub-Saharan Africa: a focus on rural communities. *International Journal of Environmental Research and Public Health*, 8(7), pp. 2967–2979.

Perera, N., (ed.). (1998). *Society and Space: Colonialism, Nationalism, and Postcolonial Identity in Sri Lanka*. Boulder, CO: Westview Press.

Radcliffe, S. A. (1997): Different heroes: genealogies of postcolonial geographies. *Environment and Planning A*, 29(8), pp. 1331–1333.

Roos, V. (2016). Conducting the Mmogo-method. In: V. Roos, ed., *Understanding Relational and Group Experiences Through the Mmogo-method®*. Springer, pp. 19–31.

Roos, V., Chigeza, S. and Van Niekerk, D. (2010). Coping with drought: Indigenous Knowledge application in rural South Africa. *Indigilinga: African Journal of Indigenous Knowledge Systems*, 9(1), pp. 1–11.

Roos, V., Du Plessis, A. and Hoffman, J. (in progress). Municipal service delivery for older persons: understanding the contexts and gaps towards ICT interventions. In: V. Roos, ed., *Age-Inclusive ICT Innovation for Service Delivery: A Developing Country Perspective*. Cham: Springer.

Roos, V. and Robertson, C. (2019). Young adults' experiences regarding mobile phone use in relation to older persons: implications for care. *Qualitative Social Work*, 18(6), pp. 981–1001.

Roos, V., Silvestre, S. and De Jager, T. (2017). Intergenerational care perceptions of older women and middle adolescents in a resource-constrained community in South Africa. *Journal of Gerontological Social Work*, 60(2), pp. 104–119.

Schatz, E. and Gilbert, L. (2014). "My legs affect me a lot . . . . I can no longer walk to the forest to fetch firewood": challenges related to health and the performance of daily tasks for older women in a high HIV context. *Health Care for Women International*, 35(7–9), pp. 771–788.

Schatz, E. and Seeley, J. (2015). Gender, ageing and carework in East and Southern Africa: a review. *Global Public Health*, 10(10), pp. 1185–1200.

Singo, V. J., Lebese, R. T., Maluleke, T. X. and Nemathaga, L. H. (2015) The views of the elderly on the impact that HIV and AIDS has on their lives in the Thulamela Municipality, Vhembe District, Limpopo Province. *Curationis*, 38(1), pp. 1–8.

Small, J., Kowal, P. and Ralston, M. (2017). Context and culture: The impact of AIDS on the health of older persons in Sub-Saharan Africa. *Innovation in Aging*, 1(Suppl_1), p. 621.

South African Government (2018/19). *Traditional Affairs*. Available at: www.gov.za/about-government/government-system/traditional-leadership

Spivak, G. C. (1988). Can the subaltern speak? In: C. Nelson and L. Grossberg, eds., *Marxism and the Interpretation of Culture*. Urbana, IL: University of Illinois Press, pp. 271–313.

StatsSA (2017) *Poverty Trends in South Africa: An Examination of Absolute Poverty between 2006 and 2015*. Available at: www.statssa.gov.za/?p=10341.

StatsSA (2019). *Department of Statistics South Africa, Republic of South Africa*. Available at: www.statssa.gov.za/?m=2019

Theron, P. M. (2014). Being treated like "waste" during the "golden years": practical-theological perspectives. *HTS Teologiese Studies/ Theological Studies*, 70(2), Art. #2636.

UNAIDS (2019). *AIDSinfo*. Available at: www.unaids.org/en/regionscountries/countries/southafrica [Accessed September 2019].

United Nations Development Programme (UNDP). (2018). *Human Development Indicators and Indices: 2018 Statistical Update*. Available at: http://hdr.undp.org/sites/default/files/2018_human_development_statistical_update.pdf

United Nations (2017). *Drivers of migration and urbanization in Africa: key trends and issues*. Expert group meeting on sustainable cities, human mobility and international migration, 7–8 September, New York. Available at: www.un.org/en/development/desa/population/events/pdf/expert/27/presentations/III/presentation-Awunbila-final.pdf

Vorster, C., Roos, V. and Beukes, M. (2013). A psycho-diagnostic tool for psychotherapy: interactional pattern analyses (IPA). *Journal of Psychology in Africa*, 23(3), pp. 163–169.

20 Posthumanist traditions and their possibilities for rural gerontology

Andrew S. Maclaren and Gavin J. Andrews

Introduction

In this chapter, we discuss the ways in which a posthumanistic gerontology is emerging and what posthumanistic thinking can bring to the study of rural gerontological enquiry. The advent of posthumanism within geographical gerontology and wider disciplines concerned with the geographical dimensions of ageing have been discussed elsewhere (Andrews et al., 2013; Skinner et al., 2015; Andrews and Duff, 2019). However, this chapter builds from these calls and considers what engagement there has been to date between rural ageing and posthumanistic thought, but more usefully, we argue for this book, what a posthumanistic rural gerontological research can do. To do this we identify what we describe as certain posthumanistic orientations that engage with more-than-human aspects of ageing. We consider where rural ageing research has engaged with these orientations and how rural gerontological scholarship could develop further through these posthumanistic approaches.

However, we begin by considering humanistic gerontology. To date, humanistic conceptualisations of space and place have dominated geographical gerontological research. Rowles' work (1978a, 1978b, 1980, 1983a, 1983b, 1986), along with the work of Harper (1987a, 1987b, 1987c) and Rollinson (1990a, 1990b), is widely considered as the foundation of this tradition of humanistic geographical gerontology (see, Harper and Laws, 1995; Andrews et al., 2007; Andrews et al., 2009; Skinner et al., 2015) and humanistic thought in geographical scholarship more generally (Cresswell, 2013). Humanistic work in gerontological thinking has developed from these original endeavours. This conceptual lens has been expanded to include poststructuralist thinking, articulating how power within social and institutional contexts and processes affect older people, as well as acknowledging how human agency is operational within, and despite, contexts and social structures of ageing (Skinner et al., 2015). Indeed, the contexts of ageing within gerontological thought have expanded most notably here, where spatial considerations of difference in older populations that consider the rural environment have been explored (Rowles, 1988; Chalmers and Joseph, 1998). Despite this critical expansion and consideration in humanistic work, the human has remained at the centre of most inquiry.

Burgeoning research in the social sciences, arts and humanities, however, has presented new and diverse ways of thinking that challenge the centrality of the human in research (Ginn, 2017). Most notably in (human) geography and sociology where a "posthuman turn" is defined (see, Franklin, 2007; Panelli, 2010), such developments have spread to sub disciplines and beyond. The emergence, thus, of geographical gerontology as an expanding field of study, encompassing a broad swathe of diverse disciplinary traditions that focus on the relationship of ageing and older populations to space and place, has led to an abundance of work that calls for (Andrews, Evans and Wiles, 2013; Hanlon and Skinner, 2016; Skinner et al., 2015, 2018) and presents (e.g., Emmerson, 2018; Maclaren, 2018; Barron, 2019) a range of contemporary theoretical perspectives that can be drawn together as being "posthumanist".

Posthumanism has been occasioned by a wider "posthuman condition" emerging in the developed world; a particular historical moment in advanced human development (Castree and Nash, 2006). While recognising that since the establishment of human civilisation there has never been a pure human era/condition free from nonhuman materials and technologies, scholars have argued that in the twenty-first century the idea of human alone no longer accurately reflects the conditions of lived experience; there being a more level and dependent rather than tiered and distinct, relationship between the human and nonhuman. A key claim is that under advanced capitalism, humans are increasingly part of disparate networks of algorithmic calculation, virtual worlds (including through electronic media diffusion) and engineering proliferation (including bio-technologies). These developments, it is posited, have produced a situation whereby material forces are inescapable and influential; this creating both opportunities and threats for the advancement and well-being of humans. At issue for posthumanist gerontologists is hence the particular relationships older people have to this emerging posthuman world, positive or negative, inside it or outside it.

Posthumanism presents a challenge to humanistic modes of knowing the world, and, although associated with postructuralism, is distinct through the collective aim to "dissolve binary distinctions that characterize humanism, most notably culture/nature and self/world" (Ginn, 2017: 5269). The human then is reconceptualised from a fully conscious, complete actor in the world to an actor that is itself incomplete, open, dependent and less-than-fully conscious. Such perspectives ally themes closely with philosophies that emphasise engagements with the material, embodied and vital nature of lifeworlds. Work that can be described as posthumanist draws liberally and differentially on the pathbreaking work of such eminent philosophers and theorists as Gilles Deleuze, Bruno Latour, Manuel DeLanda, Donna Haraway, Karen Barad, Jane Bennett, Rosi Braidotti, Brian Massumi and Nigel Thrift, with a number of traditions constituting the posthuman turn in gerontology. For scholars interested in ageing, and indeed *rural* ageing, we argue this has involved moving beyond understanding the rural in rural ageing, or rural older people's lives, as being more than things held knowingly, logically and represented by humans. It moves towards understanding them as being produced by the interaction of diverse processes involving human and non-human actors in particular spaces and places, whether biological, natural, material or technological, that exist through the vital forces that emerge both within and between such spaces and places. In considering rural space, a posthumanistic approach is a challenge to the contention that the "'rural' . . . [is] viewed more as a research setting rather than being seen as an ever-changing context that can potentially shape experiences and outcomes for older people" (Scharf, Walsh and O'Shea, 2016: 50). In decentring the subject/human, posthumanism provides a view of rural ageing that shows how all rural biological and material entities emerge together in the same moments. It considers older people's shared spaces and their common futures with the material worlds they inhabit, providing a truly holistic rural gerontology.

Ageing in the rural as performed and experienced

Posthumanism involves decentring the human and drawing together heterogenous actors, human or non-human and appreciating how they relate. Within ageing studies this has meant appreciating more than the biological or physiological changes that potentially come with ageing, or subjective humanistic judgments and or opinions of a body as it changes through time, to explore "social, affective and material expressions of bodily ageing" (Andrews and Duff, 2019: 49). To do this, scholars have taken an interest in ageing as being enacted and performed and the vital energies they afford one and other whether human or non-human, animate or inanimate.

A critique of the humanistic approach to geographical gerontology, and one that posthumanist thought addresses, is the inherent static and unrelated picture that previous work in geographical gerontology has presented. The diversity, complexity and richness of places of ageing are appreciated, but spaces and places are still treated as discrete and static, with space in some way isolated and the meanings of specific places fixed. Non-representational theories (NRT) are an approach developed in human geography to remedy the perceived deadening effect of humanistic preoccupations in the discipline with the explanation (representation) of meaning-laden (representational) phenomenon – often forms of language and text (Thrift, 2008). NRT's particular remedy is hence to energise research inquiry. To focus instead on the ever-present base practices and immediacies of life and, instead of explaining them, to show them (Anderson and Harrison, 2010). These critiques and significant departures arguably make NRT the most adventurous theoretical tradition of the broader posthumanist turn. Non-representational theories have emerged as a way to "better cope with our self-evidently more-than-human, more-than-textual, multisensual worlds" (Lorimer, 2005: 83). Our bodies and emotions are fundamental to our experience of being in the world. Ignoring these means ignoring a fundamental and important aspect of our everyday lives. Older people are no different, and with an increasingly older population, particularly in rural areas, their embodied practices in rural spaces and places are of increasing importance.

As Carolan articulates: "consciousness is corporeal; thinking is sensuous. In short, our understanding of space is more-than-representational. It is a lived process. To ignore how understandings of rural spaces are embodied, is to ignore a major source of our understanding" (Carolan, 2008: 409). Non-representational theories afford a mode of thinking to address this concern and to engage with the complex interdependencies older people have with the spaces and places of their lives. With older people forming an increasingly larger percentage of rural populations in developed countries through complex interactions between outmigration of younger people and in-migration of retirement and pre-retirement rural migrants (see Berry, Chapter 2; Cheng et al., Chapter 9 in this volume), as well as falling fertility rates, the embodied experience of older people will be an increasingly important aspect of rural lives.

There has, however, been little non-representational research engaged with specific aspects of rural ageing. Age may play a part in research that takes forward the importance of the rural as performed and experienced immediately, pre-personally and more-than-representationally in rural gentrification, migration and ageing studies, of young and old, (see Maclaren, 2019 for an overview of rural studies engaged with non-representational theories), but methodologically the emergence of non-representational theories in particular, we argue, offers the most notable way forward to explore the performed and experienced aspects of older people's rural lives.

Carolan's (2008) work on the experience of rural people in Iowa, USA presents an example of such work. Of the older respondents a number described of having "lost a step or two over the years" or only being able to get around using their tractor on their farm. These differences in physical mobility shaped their experiences and practices in space such as being less able to cope outdoors with the long hot summers, limiting their desire to leave the house. However, physical limitations brought by old age were not always seen limiting, indeed, with respondents noting that such changes in mobility meant they dwelt more in the spaces and places of their lives as one respondent remarked "If anything, it was when I was younger – that's when I was missing stuff. Always rushing but never paying attention" (420). Indeed, this temporal layer is important in McHugh's (2009) excursion into non-representational thought where he explores the entanglement of (rural) landscapes with the memories ageing affords. Maclaren (2018) has built on this rural geographical work drawing on contemporary understandings of affective and atmospheric experiences as part of a consideration of the rural in rural ageing. In particular he notes the importance of encounters with the more-than-human actors of rural spaces, whether flora and fauna, the quality of a landscape or other materialities and with people, to the everyday lives of older people. Such encounters, however, are not solely theoretical in exploring the experiences of older people.

Research such as this that draws on non-representational theories has the potential to widen these conceptual studies and provide useful and timely insights into how ageing intersects with rural spaces and places, but in particular specific geographies of rural spaces and places under study, through the experience of the countryside. If scholars are to respond to criticisms of the lack of embodied considerations of the countryside, there also needs to be a question of intersectionality, and difference with regard to ageing, and how space and place affects ageing but also ageing affects spaces and places.

Affect theory

Thinking further about the everyday experience of spaces and places, scholars have turned to the interrelated concepts of affect and emotion to consider these embodied, corporeal experiences of being in the world. To summarise swathes of debate in human geography, affect is an infectious inter-body process in that it involves affecting and being affected, resulting in increases or decreases in individual and collective bodily energy (Thrift, 2004). It is a less-than-fully consciously felt intensity of involvement and intensity of the environment. Somatically registered as a "feeling state", affect is not a typical emotional category and experience (such as fear, love, empathy and others), as these are often more personal and internal. Affect, however, is certainly not immune from being exposed to forms and moments of consciousness. Affect can be very briefly realised in fleeting conscious breaks when the mind "surfaces" from being immersed in participation. Affect can also be reflected upon afterwards as a simple but powerful thing that happened and that could happen again. This being a form of reawakening – an affective bodily pre-emption of what certain places or situations might feel like if encountered (Anderson, 2014). Understanding affect relies upon the manifest ways of the body to move, walk, rest, touch, gesture, sense, feel and perceive the world around us (Latham et al., 2009). The body can thus be considered political, as thinking with bodies involves dealing with how bodies register difference but also thinking with bodies depends on a foregrounding of the affective dimensions of space.

Affect and emotion can be understood in a three-part structure consisting of Affect-Feeling-Emotion (Ahmed, 2004; Anderson, 2006, 2014), where *affect* is the

> pre-personal intensity of relation between bodies, where bodies do not necessarily need to be human . . . *feeling* can be understood as the sensed registering of this intensity in a body . . . *emotion* can be understood as sensed intensity articulated and expressed in a socially recognisable form of expression.
>
> (Latham et al., 2009: 112, emphasis added)

Little work in rural gerontology has drawn on affect theory directly (although see Maclaren, 2018 as engagements with non-representational theories routinely engage with specific understandings of affect through a focus on practices). Gerontological work, more widely, that draws on affect has thus far considered how affective energies exist in spaces of care such as in nursing homes (Emmerson, 2018) or in care practices for those with dementia (Kontos et al., 2017; Kontos and Grigorovich, 2018). Such spaces of care might be usefully considered, and how a rural setting affords different affective encounters in specific spaces of care. Even everyday spaces, such as where Maclaren (2018) draws on the everyday encounters of a village, produce atmospheres of care. This wider consideration of affect and affective encounters older people have in rural areas may be useful in thinking of rural spaces as being "therapeutic landscapes", where the affective and emotional aspect of space and place, of being, is valued. Such research into therapeutic landscapes is burgeoning in wider geographical discipline where Emmerson (2019) has noted their importance in "attuning ourselves to the more ordinary ways in which certain other things, activities, and events come to matter within [spaces of care]" (600). Indeed, expanding, as Maclaren (2018) articulated, to concepts of everyday spaces of a village, country lane or otherwise, we can think of the rural spaces as being important in this consideration. As Conradson (2005) has argued, individuals' specific places within rural landscapes can have beneficial impacts on physical and mental well-being and health. Such work would add theoretical depth to emerging engagements with affect as part of the posthuman turn within gerontological studies but also add further insights into the importance of affect in understanding how "nature . . . adds value to culture" (Cruickshank, 2009: 104). This is especially the case when engaging with questions behind people's migration choices and the affective connection older people have to specific rural spaces and places. It can also be why older people, despite experiencing a lack of services in their rural area, and perhaps with increasing pressures from changes in their own or family members' health or care needs, still choose to stay in more inaccessible and/or remote rural places.

Ageing emerging through relational material rural assemblages

Posthumanism recognises that all social realities emerge through the generative capacities of "assemblages" composed of human and nonhuman entities. According to scholars (see DeLanda, 2016; Duff, 2014), this working involves (1) entities as the "components" of assemblages that might stay in-situ, or move into or out of them; (2) the "mechanisms" of assemblages (such as events and relations) that operate internally and within wider networks; (3) the internal "processes" of assemblages (such as territorialisation, homogenisation) continually in operation and (4) the "excessive" expressive vital outcomes of assemblages (such as affects) that emerge from them. The understanding is that new social realities are created through impermanent unities being reached within assemblages (through particular components possessing time-limited "symbiosis"). Yet, as Müller (2015) suggests, all this happens without assemblages disclosing transcendental origins or all-embracing organising principles (see Deleuze and Guattari, 1988). Overall, as a concept, assemblage has given posthumanist research a way to explain social realities that tells us a lot about their content, order and how they are produced through the ways in which the world is arranged. Yet it is an approach that is not deterministic like others that look to establish single or multiple "causality" for phenomenon.

Posthumanism takes forward the notion that human and nonhuman bodies and materialities are responsible for the realities we experience. Each actor has agency, and so evolve concurrently together and follow what can be seen as a flat ontological relation. Within geographical gerontology the development of this area of thought has focused on how the lived realities and experiences of ageing grow out of assemblages of these human and non-human bodies and materialities (Delanda, 2016; Deleuze and Parnet, 2006). In the case of rural ageing, this would suggest that experiences of older people in rural areas emerge and shift over time and are the product of an individual's varied life experience and the environment in which they currently live and have previously lived in.

Andrews and Duff (2019) have argued that this view of posthumanism for gerontology affords a "way to explain ageing that is not structuralist or constructionist yet still speak to some degree about the content and order involved" (48). Such assemblage theory has influenced and underpinned, even if not always credited as such, recent critical engagements in gerontological thinking where "New Materialism" work in gerontology addresses these relational materialities of ageing. For example, Buse, Martin and Nettleton (2018) consider the "materialities of care", conceptualising the multiple spaces of care and the relations between care and the materials that contribute to older people's ongoing lives. Indeed, such a view can be developed to think more about not just the environment an older person experiences but also how such environmental assemblages and interaction with materialities, clothes, objects of and in the spaces of their lives, affect their feeling of self and illuminate intimate social meanings that develop and exist in relation to the process of ageing older and across people's life course (Chapman, 2006; Buse and Twigg, 2014, 2015, 2016, 2018). Expanding the ageing process beyond spaces of care or caring, independence in later life has also come into consideration through work focused on the importance of assemblages of home or private and community or public spaces in maintaining such independences.

There has, however, yet to be real engagement with rural spaces, ageing and assemblage. Yet, Cloke (2013) has articulated that the rurality is "gazed on, lived in, performed, and experienced in so many different ways. It combines and narrates the human and the non-human in diverse assemblages and in both representative and quotidian registers" (225), in particular related to notions of the rural idyll. With particular emphasis in developed nations on independence in later life, as part of broader neoliberal policies, there is significant scope to see how the rural in rural ageing assembles in everyday rural spaces and places. Indeed, longer trajectories of older people's lives in rural places bring differing views to the ageing process for someone who has only lived in a rural place compared to someone who has lived in contrasting spaces over the course of their lives.

Moving forward methodologically and with posthumanism in rural geographical gerontology

Rowles (1988) was ahead of his time when he described the need to consider a phenomenological aspect of rural ageing. Posthumanism presents differing modes of thought; we have outlined but three themes of posthumanism, that we believe would allow for academics to expand their "conceptualization of rural from an objective circumstance "outside" the individual into an internal creation, an element of the way in which individuals organize and construe their life-worlds" (Rowles, 1988: 121). Posthumanism pushes beyond social constructivist understandings to see the lifeworlds of older people in rural spaces and places as emergent through heterogenous actors. We end here with some constructive points towards where such modes of thoughts can be taken forward methodologically and why such modes of thought are important overall in the study of rural ageing.

Methodologically such modes of thought have often been reticent or unclear for those "keen to learn but foxed by venturing out alone" (see Lorimer, 2015: 184, on non-representational theories). Those engaged in posthumanist thought, in all its diversity, focus on the situatedness of knowledge. That knowledge produced is partial and situated to its locality, there are very few methodological textbooks that offer ways forward. And so the challenge is "to develop methods or methodological strategies that are sensitive to human and nonhuman agencies, entanglements, and thresholds, which confound and unsettle humanist and sometimes posthumanist beliefs and sentiments" (Blackman, 2015: 26).

But like any research, we advocate starting with the questions at hand that a researcher wants answered. With this, then the most appropriate way forward must be considered as how will such a posthumanist perspective help. Most often, those drawing on posthumanist modes of thought are engaged with participatory and/or ethnographic methods where the researcher is always "in relation" to others (Ginn, 2017: 5277), through ethnography, for example, that is "participant observation *plus* any other appropriate methods/techniques/etc . . . *if they are appropriate for the topic*" (Crang and Cook, 2007: 35, emphasis in original). This openness we see as an encouragement for scholars to explore and expand and to "get out of the armchair" as has been advocated so much in rural studies most widely (Bunce, 1994), to get out there into the lifeworlds of their research participants in their rural spaces and places.

Future attention is needed to understand assemblages of rural ageing, what is happening in them and their networks. This involves efforts to establish and describe causative pathways in the emergence of successful or unsuccessful rural ageing in places. A particular focus would be on the key components of assemblages that disclose as such in the course of these outcomes arising. General questions include: Which types of assemblages provide the optimal ageing experiences and outcomes? Which types of political-economic, institutional, family and socio-cultural processes aid or obstruct successful assemblages? Are there optimal setting, family, community or practice assemblages, and what do they look like? How might social assemblages be (re)organised or augmented to create circumstances whereby material and human actors are in fullest possession of their ability to act (hence which components and processes within a given assemblage could be utilised, which could be assisted, which could be countered and which overlooked)? What methodological innovations might help answer the aforementioned questions and more generally unpack the processual aspects of rural ageing assemblages, witness and animate their vital immediacy and even intervene acting into and "nudging" them in particular positive ways? As we articulated in our introduction (after Scharf, Walsh and O'Shea, 2016) rural spaces and places are more than research settings, they are ever-changing contexts. Work that considers ageing as being performed and experienced, and emergent from relational material *rural* assemblages, we argue, has the potential to establish a number of fruitful, pertinent and pressing concerns through an engagement with posthumanism.

References

Ahmed, S. (2004). Collective feelings: or the impressions left by others. *Theory, Culture and Society*, 21, pp. 25–42.

Anderson, B. (2006). Becoming and being hopeful: towards a theory of affect. *Environment and Planning D: Society and Space*, 24, pp. 733–752.

Anderson, B., (ed.). (2014). *Encountering Affect: Capacities, Apparatuses, Conditions.* Aldershot: Ashgate.

Anderson, B. and Harrison, P. (2010). The promise of non-representational theories. In: B. Anderson and P. Harrison, eds., *Taking-Place: Non-Representational Theories and Geography*. Farnham: Ashgate, pp. 1–34.

Andrews, G. J., Cutchin, M., McCracken, K., Phillips, D. R. and Wiles, J. (2007). Geographical gerontology: the constitution of a discipline. *Social Science and Medicine*, 65(1), pp. 151–168.

Andrews, G. and Duff, C. (2019). Understanding the vital emergence and expression of aging: how matter comes to matter in gerontology's posthumanist turn. *Journal of Aging Studies*, 49, pp. 46–55.

Andrews, G. J., Evans, J. and Wiles, J. L. (2013). Re-spacing and re-placing gerontology: relationality and affect. *Ageing & Society*, 33(8), pp. 1339–1373.

Andrews, G. J., Milligan, C., Phillips, D. R. and Skinner, M. W. (2009). Geographical gerontology: mapping a disciplinary intersection. *Geography Compass*, 3(5), 1641–1659.

Barron, A. (2019). More-than-representational approaches to the life-course. *Social and Cultural Geography*. Epub ahead of print 14 May. DOI: 10.1080/14649365.2019.1610486.

Blackman, L. (2015). Researching affect and embodied hauntologies: exploring an analytics of experimentation. In: B. T. Knudsen and C. Stage, eds., *Affective Methodologies: Developing Cultural Research Strategies for the Study of Affect*. Basingstoke: Palgrave Macmillan, pp. 25–44.

Bunce, M., (ed.). (1994). *The Countryside Ideal: Anglo-American Images of Landscape*. Abingdon: Routledge.

Buse, C., Martin, D. and Nettleton, S. (2018). Conceptualising 'materialities of care': making visible mundane material culture in health and social care contexts. *Sociology of Health and Illness*, 40(2), pp. 243–255.

Buse, C. and Twigg, J. (2014). Women with dementia and their handbags: negotiating identity, privacy and 'home' through material culture. *Journal of Aging Studies*, 30, pp. 14–22.

Buse, C. E. and Twigg, J. (2015). Clothing, embodied identity and dementia: maintaining the self through dress. *Age, Culture, Humanities*, (2).Available at: https://kar.kent.ac.uk/53618/

Buse, C. and Twigg, J. (2016). Materialising memories: exploring the stories of people with dementia through dress. *Ageing & Society*, 36(6), pp. 1115–1135.

Buse, C. and Twigg, J. (2018). Dressing disrupted: negotiating care through the materiality of dress in the context of dementia. *Sociology of Health and Illness*, 40(2), pp. 340–352.

Carolan, M. S. (2008). More-than-representational knowledge/s of the countryside: how we think as bodies. *Sociologia Ruralis*, 48, pp. 408–422.

Castree, N, and Nash, C. (2006) Posthuman geographies. *Social and Cultural Geography*, 7(4), pp. 501–504.

Chalmers, A. I. and Joseph, A. E. (1998). Rural change and the elderly in rural places: commentaries from New Zealand. *Journal of Rural Studies*, 14(2), pp. 155–165.

Chapman, S. A. (2006). A 'new materialist' lens on aging well: special things in later life. *Journal of Aging Studies*, 20(3), pp. 207–216.

Cloke, P. (2013). Rural landscapes. In: N. C. Johnson, R. H. Schein, and J. Winders, eds., *The Wiley-Blackwell Companion to Cultural Geography*. London: John Wiley & Sons, pp. 225–237.

Conradson, D. (2005). Landscape, care and the relational self: therapeutic encounters in rural England. *Health and Place*, 11, pp. 337–348.

Crang, M., and Cook, I. G., (eds). (2007). *Doing Ethnographies*. London: SAGE Publications.

Cresswell, T., (ed.). (2013) *Geographic Thought: A Critical Introduction*. Oxford: John Wiley & Sons.

Cruickshank, J. (2009). A play for rurality: modernization versus local autonomy. *Journal of Rural Studies*, 25, pp. 98–107.

DeLanda, M., (ed.). (2016). *Assemblage Theory*. Edinburgh: Edinburgh University Press.

Deleuze, G. and Guattari, F., (eds). (1988). *A Thousand Plateaus: Capitalism and Schizophrenia*. London: Bloomsbury Publishing.

Deleuze, G. and Parnet, C. (2006). *Dialogues II: 1977*. New York: Continuum International Publishing Group.

Duff, C., (ed.). (2014). *Assemblages of Health: Deleuze's Empiricism and the Ethology of Life*. Cham: Springer.

Emmerson, P. (2018). From coping to carrying on: a pragmatic laughter between life and death. *Transactions of the Institute of British Geographers*, 44(1), pp. 141–154.

Emmerson, P. (2019). More-than-therapeutic landscapes. *Area*, 51(3), pp. 595–602.

Franklin, A. (2007). Posthumanism. In: G. Ritzer, ed., *The Blackwell Encyclopedia of Sociology*. Oxford: Blackwell, pp. 3548–3550.

Ginn, F. (2017). Posthumanism. In: D. Richardson, N. Castree, M. F. Goodchild, A. Kobayashi, W. Liu, and R. A. Marston, eds., *International Encyclopaedia of Geography*. Oxford: Wiley-Blackwell.

Hanlon, N. and Skinner, M. W. (2016). Towards new rural ageing futures. In: M. W. Skinner and N. Hanlon, eds., *Ageing Resource Communities: New Frontiers of Rural Population Change, Community Development and Voluntarism*. Abingdon: Routledge, pp. 206–212.

Harper, S. (1987a). The kinship network of the rural aged: a comparison of the indigenous elderly and the retired immigrant. *Ageing & Society*, 7(3), pp. 303–327.

Harper, S. (1987b). The rural-urban interface in England: a framework of analysis. *Transactions of the Institute of British Geographers*, 12(3), pp. 284–302.

Harper, S. (1987c). A humanistic approach to the study of rural populations. *Journal of Rural Studies*, 3(4), pp. 309–319.

Harper, S. and Laws, G. (1995). Rethinking the geography of ageing. *Progress in Human Geography*, 19(2), 199–221.

Kontos, P. and Grigorovich, A. (2018). Rethinking musicality in dementia as embodied and relational. *Journal of Aging Studies*, 45, pp. 39–48.

Kontos, P., Miller, K.-L., Mitchell, G. J. and Stirling-Twist, J. (2017). Presence redefined: the reciprocal nature of engagement between elder-clowns and persons with dementia. *Dementia*, 16(1), pp. 46–66.

Latham, A., McCormack, D. P., McNamara, K. and McNeill, D., (eds). (2009). *Key Concepts in Urban Geography*. London: SAGE Publications.

Lorimer, H. (2005). Cultural Geography: the busyness of being "more-than-representational." *Progress in Human Geography*, 29(1), pp. 83–94.

Lorimer, H. (2015). Afterword: non-representational theory and me too. In: P. Vannini, ed., *Non-Representational Methodologies: Re-Envisioning Research*. New York; London: Routledge, pp. 177–187.

Maclaren, A. S. (2018). Affective lives of rural ageing. *Sociologia Ruralis*, 58(1), pp. 213–234.

Maclaren, A. S. (2019). Rural geographies in the wake of non-representational theories. *Geography Compass*, 13(8), p. e12446.

McHugh, K. E. (2009). Movement, memory, landscape: an excursion in non-representational thought. *GeoJournal*, 74, pp. 209–218.

Müller, M. (2015). Assemblages and actor-networks: rethinking socio-material power, politics and space. *Geography Compass*, 9(1), pp. 27–41.

Panelli, R. (2010). More-than-human social geographies: posthuman and other possibilities. *Progress in Human Geography*, 34(1), pp. 79–87.

Rollinson, P. (1990a) The story of Edward: the everyday geography of elderly single room occupancy (SRO) hotel tenants. *Journal of Contemporary Ethnography*, 19(2), pp. 188–206.

Rollinson, P. (1990b). The everyday geography of poor elderly hotel tenants in Chicago. *Geografiska Annaler B*, 72(2–3), pp. 47–57.

Rowles, G. D., (ed.). (1978a). *Prisoners of Space? Exploring the Geographic Experience of Older People*. Boulder, CO: Westview Press.

Rowles, G. D. (1978b). Reflections on experiential fieldwork. In: D. Ley and M. Samules, eds., *Humanistic Geography: Prospects and problems*. Chicago, IL: Maaroufa Press, pp. 173–193.

Rowles, G. D. (1980). Towards a geography of growing old. In: A. Buttimer and D. Seamon, eds., *The Human Experience of Space and Place*. London: Croom Helm, pp. 55–72.

Rowles, G. D. (1983a). Between worlds: a relocation dilemma for the Appalachian elderly. *International Journal of Aging and Human Development*. 17(4), pp. 301–314.

Rowles, G. D. (1983b). Place and personal identity in old age: observations from Appalachia. *Journal of Environmental Psychology*, 3(4), pp. 299–313.

Rowles, G. D. (1986). The geography of ageing and the aged: towards an integrated perspective. *Progress in Human Geography*, 10, pp. 511–539.

Rowles, G. D. (1988). What's rural about rural aging? An Appalachian Perspective. *Journal of Rural Studies*, 4(2), pp. 115–124.

Scharf, T., Walsh, K., and O'Shea, E. (2016). Ageing in rural places. In: M. Shucksmith and D. L. Brown, eds., *Routledge International Handbook of Rural Studies*. Abingdon: Routledge, pp. 50–61.

Skinner, M. W., Andrews, G. J., and Cutchin, M. P. (2018). Introducing geographical gerontology. In: M. W. Skinner, G. J. Andrews, and M. P. Cutchin, eds., *Geographical*

Gerontology: Perspectives, Concepts, Approaches. London; New York: Routledge, pp. 3–10.

Skinner, M. W., Cloutier, D., and Andrews, G. J. (2015). Geographies of ageing: progress and possibilities after two decades of change. *Progress in Human Geography*, 39(6), pp. 776–799.

Thrift, N. (2004). Intensities of feeling: towards a spatial politics of affect. *Geografiska Annaler. Series B, Human Geography*, 86(1), pp. 57–78.

Thrift, N., (ed.). (2008). *Non-Representational Theory: Space, Politics, Affect*. London: Routledge.

21 A Deweyan pragmatist perspective on rural gerontology

Graham D. Rowles and Malcolm P. Cutchin

Introduction

Rural ageing as we have conceived of it in much of the gerontological literature of the past 30 years, no longer exists, if it ever did. Bucolic romanticised images of agrarian bliss with older adults gently rocking on their porch and staring out over the family farm as they wile away their final years, are just that – images. Instead, the contemporary experience of growing old in rural places is far more diverse and dynamic. The majority of older adults in rural areas have few ties to farming and forestry and many of those that do are beset by emotional anxiety about the economic challenges of contemporary family farming. For many, a working life may have been centered in a city to and from which they engaged in a lengthy daily commute. For others, retirement to the country represents an aspiration for rural living based on Rockwellian images of what it might be like to escape the bustle, tensions and alienation of life in the metropolis. For yet others, ageing in rural America is simply an extension of a life spent working in the service sector of a village or employed in one of a diverse array of commercial and service enterprises that are increasingly located in rural settings (Fitchen, 1991).[1]

Even at the time of the publication of "What's rural about rural aging?" (Rowles, 1988),[2] it was acknowledged that rural America was an environment of change and instability, a milieu constantly evolving into myriad different manifestations of rurality determined by local circumstances. Since then, the pace of change has accelerated and the diversity of scenarios for rural ageing has increased. With the transformation of space accompanying the internet age, the potential is for even the most remote rural environments to become integrated within a more universally homogeneous culture. But the adoption of technology and its effects on communities and the lives of older residents has been varied. Older adults in some rural communities are fully embraced within regional telemedicine, communication and social media networks. In other settings, through choice or lack of opportunity, older adults are estranged from the digital world (see Kosurko et al., Chapter 27 in this volume). Isolated rural communities still have limited internet access, but this characteristic will gradually disappear with the trend toward universal coverage. And as this occurs, the fabric of life for rural older adults will continue to evolve.

In this chapter, we propose a reframing of the discourse on rural ageing. First, we critique established views of rural ageing as an ecological, socio-cultural or phenomenological experience. Within the framework of understanding how rural older adults' relationship with their environment is a function of their ability to sustain a supportive sense of "being in place" that is to some degree shared with their neighbours, we argue that each of these perspectives is limited in its ability to encapsulate the essence of rural living and community. Our critique provides a context for introducing a more dynamic perspective on rural ageing that embraces the situational uniqueness of each rural environment. Here, we introduce a pragmatist theoretical perspective grounded in situationally defined manifestations of place integration within an ever-changing milieu (Cutchin 1997, 2016). Place integration is the continual process of social inquiry to meliorate problematic situations and create a more harmonious and meaningful experience of place. Central to this idea is the notion that place integration in rural areas involves commitment to and caring for place and community. The place integration process nurtures a sense of belonging that becomes manifest in individual and group behaviour and actions focused on support and betterment of the community. In the final portion of the chapter we explore implications of this perspective, not only for critical research but also, more importantly, for establishing a participatory action lens through which older adults in rural areas are intimately and dynamically engaged in creating and nurturing environments and an ageing experience attuned to their needs and aspirations for a good life in old age.

The need to re-conceptualize rural ageing

More than 30 years ago, in synthesising extant conceptualisations of rural ageing, Rowles (1988) identified three perspectives. The first, and by far the most prevalent, was an ecological perspective on *ageing in rural environments*. Here, the focus was on the characteristics of rural environments and their ageing populations. Many studies were based on demographic characterisations of rurality determined on the basis of population density (Krout and Hash, 2015, Table 1.1: 10–12).

With respect to ageing populations, distinctions emerged among older adults ageing in place (having grown up and grown old in these places), older adults who were new migrants to rural environments (first time residents moving to these areas in search of amenities and/or an idealised rural way of life) and return migrants (persons who left during their youth and chose to return "home" to retire). Each of these subpopulations had different needs and expectations (Rowles and Watkins, 1993). Contrasting expectations on occasion led to the potential for conflict.

The ecological perspective facilitated documenting changes in the structural fabric of rural America. Farming became progressively less important. By 1990, less than 4% of rural residents were involved in farming as their primary activity; many farmers transitioned to working part time. Residential mobility, especially young people leaving in search of employment, resulted in the transformation of rural community social structures. Many rural communities became ghost towns

(Norris-Baker and Scheidt, 1994) or bedroom communities with their populations not only working elsewhere but also purchasing groceries and supplies in the cities to which they commuted. Different economic activities, prisons, noxious and hazardous waste sites, factory outlet complexes, recreational sites and retirement communities, transformed rural landscapes. Responding to economies of scale, rural resources became concentrated within fewer settlements which became growth centres as schools were consolidated and as higher order retail services and health care resources became concentrated in selected communities. The socio-economic divergence of rural and urban places over the last few decades has been reflected in lagging rural employment and income. It has caused distress and hardships for rural life in many areas of North America and has resulted in many communities that can be characterised as "left-behind places" (Hendrickson, Muro and Galston, 2018).

While an ecological perspective can interpret the dynamic circumstances of rural ageing, it does not capture the social experience of growing old in a rural setting. For this, it is necessary to consider the shared socio-cultural context, the *environment of rural ageing*. This focus moves beyond the life circumstances of rural older adults. It shifts towards rural ways of life and lifestyles and the way in which the diverse cultures of rural places influence well-being and help to explain a paradox of inferior objective life circumstances but either equal or higher levels of subjective well-being than urban older adults that was first reported in the early 1980s (Grams and Fengler, 1981; Lee and Lassey, 1980). A slower pace, the rhythm and routine of personal and community activities, the smaller scale of life, intimate lifelong affinity with the land and a sense of belonging or being physically, socially and autobiographically "inside" a rural setting help to sustain older adults as they age (Rowles, 1983a). Adherence to shared community values of self-resilience and yet mutual care and social support amidst community-wide recognition of the vulnerability of ageing neighbours who are known rather than anonymous also distinguishes many rural settings (Rowles, 1983b).

A final perspective focuses directly on phenomenological aspects of lived experience in rural settings. Here, the focus is on the *rural environment of the ageing*, how older adults themselves perceive and identify with their rurality and proactively envision and reinforce the essence of rural life. Narratives of older adults in rural areas reveal how their identity is tied up with personal community history and recollections of the essential features of rural community life as well as a sense of social connectedness (Buys et al., 2015; Dorfman et al., 2004; Shenk et al., 2002).

Ecological, socio-cultural and phenomenological perspectives provide important complementary lenses on rural ageing. But, by themselves, these perspectives are not sufficient for understanding growing old in contemporary rural environments. The challenge for rural gerontology becomes even more complex when we acknowledge health and well-being disparities and diverse manifestations of disadvantage and social exclusion (Walsh, O'Shea and Scharf, 2019). Such threats to the viability of rural life in old age are reinforced by the demise of traditional social and cultural structures that historically provided social support and a buffer to the challenges of ageing. It is necessary to integrate diverse perspectives within a comprehensive model of evolving political and social processes of contemporary life for rural older adults. Such a model must acknowledge constantly evolving combinations of individual and community decisions within the political economy of each rural situation.

Moving in this direction, we begin by assembling the building blocks of a general model of the manner in which individuals and social groups in rural settings create and strive to sustain and evolve a unique mode of being in a place. This involves ongoing processes of reconciling individual and group needs and aspirations with opportunities available in the local setting.

On being in place

On the most fundamental level, our model acknowledges the reciprocity of two elements: *place identification* and the *identity of place* in establishing for each person a sense of *being in place* (Rowles, 2018) (Figure 21.1). The first element, place identification, is an inherent individual characteristic. Moving through their life, each person develops preferences for place that condition a visceral level of comfort. For example, a person may eschew the hustle and bustle and claustrophobic intensity of a central city metropolitan environment. Alternatively, he or she may feel alone and uncomfortable in an isolated rural setting. Over time, a shared sense of place identification may be experienced by groups with similar histories of habitation; we engage in reinforcing (or changing) the manner in which we experience our community as our community becomes part of us and we become part of our community in a perpetually iterative process.

In contrast, the second element, the identity of place is a characteristic of a setting. Rural places develop an identity reflecting their physical attributes, resource and service infrastructure and social and cultural history. They become places

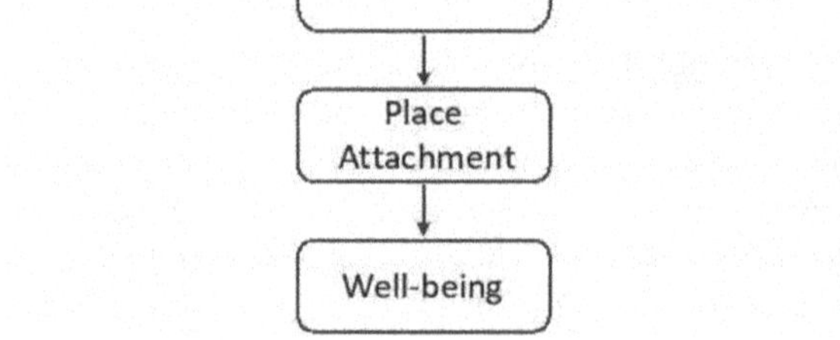

Figure 21.1 Being in place

with their own personality. Each rural community is different as it assumes an identity apparent both to residents and outsiders; one is a dying isolated ghost town, another a thriving retirement destination.

Over time, interaction between people's place identification and the identity of the places they inhabit evolves through critical community engagement as each place shapes and, in turn, is shaped by its residents. The dynamic outcome for each resident is a sense of being in place or, if the processes of engagement are unsuccessful, a sense of alienation or being out of place. We suggest that being in place creates or enhances place attachment and associated levels of investment in or estrangement from the community. It is also closely linked with well-being and provides motivation for action. Building on this conceptualisation, we next present a dynamic framework for understanding rural ageing that is consistent with an evolving sense of being in place as people actively engage with their milieu through ongoing processes of *community engagement*.

Facilitating being in place and place integration through processes of social inquiry

We view rural places and the experience of people ageing in those places as situationally unique, dynamic and ever-changing. Yet, as we have noted, rural gerontology has tended to categorize places and problems of rurality in ways that focus on sociodemographic conditions and traits of rural life rather than the evolving dynamics that underpin life in particular places. Recent work suggests a shift in focus.

The economic stress on much of rural America provides evidence of increasing *precarity* – "life worlds characterized by uncertainty and insecurity" that may be conceptualised both as a condition and as a point for mobilisation (Waite, 2009: 412). Grenier et al. (2017) have developed the concept of precarity as it applies to ageing and later life. These authors focus on decline of the welfare state and associated austerity policies as prime movers of increased precarity. They conclude that "An extended consideration of precarity renders visible the relationship between structures, life events, and everyday experiences of ageing; highlights the shared experiences of risk and insecurity with regards to disadvantage and care; and underscores the political imperative for addressing inequality" (Grenier et al., 2017: 13).

Since Waite's work on precarity, interest in the concept has grown dramatically with a primary focus on labour markets and migrants. But gerontologists such as Grenier and colleagues pay limited attention to the socio-spatial contexts of precarity, a notable exception being Bates et al. (2019). As far as we know, there has been little work on the precarity of rural communities and the way in which this translates into the precarious situations of many rural older adults. Here, we extend existing perspectives on precarity and suggest a framework through which rural communities and their older residents can mobilize to improve the situation and experience of rural ageing. This framework is grounded in ideas both acknowledging the precarious nature of rural life and showing how we can creatively and intelligently respond to associated challenges and improve the quality of life in rural communities. The philosopher John Dewey offered such a collection of well-articulated ideas which remain of great interest to social scientists and rural development scholars (e.g., see Skinner and Hanlon, 2016; Wills and Lake, 2020).

Dewey's philosophy is based, in part, on recognition of the precarious – and the uncertain futures of our life worlds – as central to community life and human action (Dewey, 1980, 1989). Uncertainty arises not from problems in human cognition; it is integral to the life worlds with which we have to continually respond, coordinate and shape. Dewey stressed the need to fully acknowledge contingency and uncertainty. He intended his philosophy to drive both theory and practice by communities and individuals in uncertain situations. Uncertainty, in the Deweyan view, is not an issue of knowledge *per se*. Dewey argued that uncertainty is always integral to experiencing: a community, an evolving situation and constantly changing relationships of the entities that comprise them. Dewey's philosophy moves beyond the view of precarity in the current social science literature because, in addition to the need for criticism, he suggests a method of trying to reduce the precariousness of any situation. Dewey was emphatic that we drop the pretence of certainty and stability and replace these notions with "a demand for imagination" to deal with uncertain situations (Rorty, 1999: 34; Cutchin, 2004).

Dewey explained that situations evolve, shift, or otherwise change in a problematic way and force us to attend to them (Dewey, 1957). Situations are uncertain in what they will become and what that change means for us. For Dewey, this reoccurring process of situational uncertainty, our attention and its expansion into thought and action was the process of *inquiry* (Dewey, 1960). Rural communities can be considered as Deweyan situations necessitating and benefitting from inquiry (Cutchin, 2016). Communities facing precarious and uncertain conditions need a way to reconstruct relationships, repair processes and reconfigure institutions and fundamental socio-economic bases in ways that reduce precarity and foster ongoing positive dynamics of change. For rural residents, such dynamics result in an enhanced experience of being in place, reinforcement of place attachment and improved well-being, all parts of a larger process we have termed "place integration" (Cutchin, 1997, 2016). Figure 21.2 expands the model of being in place (discussed earlier) by incorporating a Deweyan approach.

Rural communities and their older adult populations have an opportunity to engage in place integration utilising what Dewey described as social inquiry for the purpose of "social reconstruction" (Cutchin, 2016). Dewey offered a framework for social inquiry into community problems that allows citizens to collectively discuss and devise what they deem to be the best path forward to address problems and enhance place integration (Campbell, 1995; Cutchin, 2016, 2020). Dewey acknowledged that community problems could be complex and seemingly intransigent, and that proposed or enacted solutions should always be considered provisional and subject to change. He argued that without an understanding of rigorous processes of social inquiry and reconstruction, community life – as experienced in challenging rural settings – would not improve. We therefore need to sketch what Dewey meant by social inquiry and social reconstruction because it

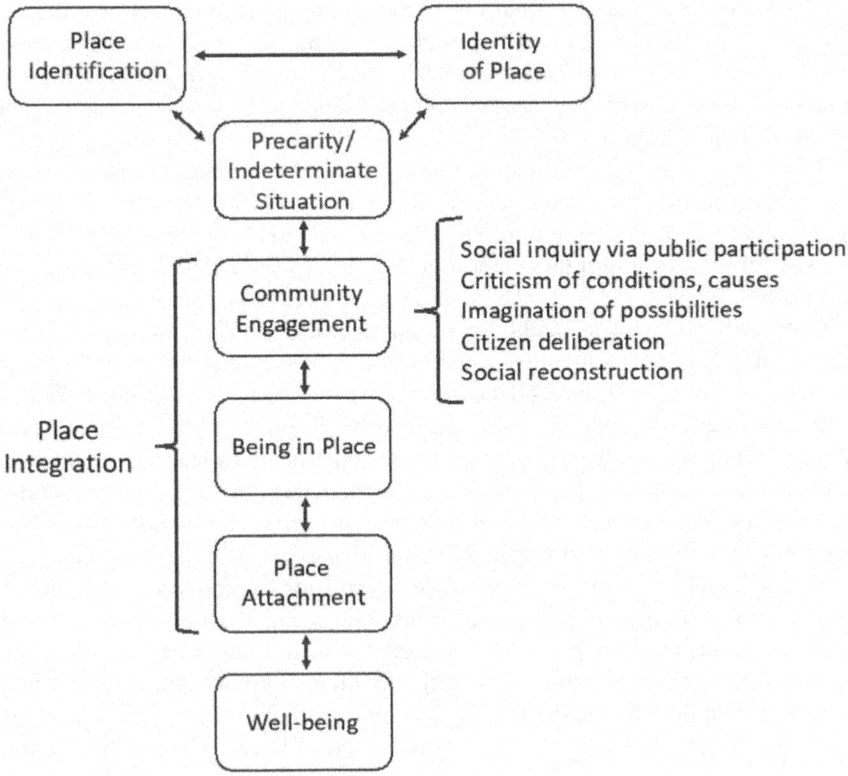

Figure 21.2 Facilitating place integration through community engagement

has potential to empower rural communities, including older members of those communities, to stimulate community and personal development as well as enhance quality of life.

Dewey's rethinking of social and political theory some 90 years ago shifted the definition of democratic engagement away from elections and the state and towards the cooperative activity of "publics" and affected citizens often co-located in communities. Dewey recognised the American system's failure to promote participation of publics to address their own issues. He therefore put forth a concept of democracy as community life itself, and his view was grounded in his belief that community members can cooperate, discuss issues, imagine potential futures, make intelligent judgments and act together to improve situations with proper supports and frameworks (Campbell, 1998). America in Dewey's time (and our own) was plagued by dysfunctional politics and social institutions that prevented publics from forming and collaborating to understand and address problematic situations. This raises critical questions regarding the role of government and external constituencies in shaping or nurturing rural community development. In many respects, the precarity that characterises rural communities and the precariousness of residents' lives within such communities is determined by external forces beyond their control. The issue becomes one of the degree to which the processes of community engagement can be enabled and flourish within the constraints of larger social forces.

Dewey clearly stated, "the actual problem is one of reconstruction of the ways and forms in which men [sic] unite in associated activity" (Dewey, 1954: 192). Cooperative inquiry in communities is the approach Dewey developed in order to foster "improvement of the methods and conditions of debate, discussion, and persuasion" (Dewey, 1954: 208). Such inquiry is central to place integration.

We argue that Dewey's "method of social reconstruction" (Campbell, 1992) – a process of collaborative intellectual inquiry and deliberation followed by practical enactment of the consensus decisions and plans – has great relevance for rural communities and their ageing residents. Briefly, the method suggests that a community's citizens work together at naming and framing their shared problems and collecting information and explanations. This includes the ability and freedom to not only criticize current conditions and their potential causes but also identify and build on community strengths (Cooperrider et al., 2008). The goals are to clarify the situation and develop possible options for, and consequences of, collective action. Experts, such as social critics and scientists, may augment community citizens' efforts through input of materials and ideas, but experts should not dominate those efforts. Dewey argued that publics in communities should be in control of such practices because they have the ability to evaluate the knowledge of others and how it applies to their shared issues (Dewey, 1954). This includes sharing and discussing imaginative and creative possible solutions and outcomes. The final step in the method is to suggest and enact changes in social institutions (social reconstruction) which have failed to adjust to evolving conditions, including the needs of older adults. Older members of rural community publics have a wealth of experience from which to draw and share in order to help improve the method of social reconstruction and its outcomes. As illustrated by Norris-Baker and Scheidt's (1994) observations on the manner in which older residents of a dying small town in Kansas town assumed key roles in order to sustain the community, this level of engagement and participation is potentially more meaningful and contextually important than typical volunteerism by older adults.

To summarize, we conceptualize being in place, in parallel with community engagement (through community-engaged social inquiry), and attachment to place, as complementary dynamic processes within the larger process of place integration. The outcomes of these processes are always provisional; they are at risk of failure and are constantly in need of revision and reconstruction as circumstances change.

Research and action in creating a new rural gerontology

Our proposed model of nurturing being in place through community engagement and place integration suggests processes at the community level that may be leveraged to address negative dimensions of rural places and rural ageing. Precarity and social exclusion are always potentially countervailing forces to community engagement. In this final part, we briefly sketch implications for rural ageing and rural gerontology that address the challenge of precarity and social exclusion within the context of community processes of social inquiry and place integration. Such processes entail a coming to awareness through community self-education to a consciousness and critique of the conditions and causes of current circumstances (Freire, 1970). Such awareness provides a baseline for imagining possibilities through citizen deliberation as part of the process of social reconstruction.

As important as the inclusion of older community members in such focused work, is the realisation that "successful community development involves a commitment [attachment] to place through cooperative action and deep engagement that is reflective of the responsibility inherent in Dewey's philosophy" (Cutchin, 2016: 33). Some older community members, often those with lengthy residence, have such commitments and attachments and are therefore essential assets and potential leaders for addressing the challenges involved in social reconstruction of rural communities. It is well recognised that community development efforts are underpinned by strong local leadership and that older community members offer a ready and often untapped resource of leadership to be utilised in community reconstruction and place integration (Scheidt and Norris-Baker, 1999).

More than leadership is needed to build a sustainable process of ongoing inquiry and place integration. What can be a valuable approach for rural community development and inclusion of older people in community problem-solving, the method of social inquiry and reconstruction also develops habits of inquiry that serve a community over the long term and build community capacity. This approach is consistent with a growing tradition of participatory action research over the past few decades (Blair and Minkler, 2009; Boog, 2003; Minkler, 2000). As it evolves through repeated practice, participation in the local situation through participatory action research can become habitual in setting supportive precedent for participation. This includes an emphasis on diversity of participants through conscious and explicit efforts to ensure the inclusion of constituencies previously excluded from the deliberative process in rural communities. If inclusive of older residents as well as persons with disability or chronic health conditions, and fully using their talents, commitments, and attachments, the process of addressing situated challenges of rural communities from the ground up and by the community itself can transmit wisdom and induce community learning that serves succeeding generations (Blair and Minkler, 2009). As we have suggested,

> our shared experience in places or communities makes us inescapably connected to each other through shared habits; and this is not the case just because we interact, but because the values and ideas of a place or community also are shared.
>
> (Cutchin, 2016: 30)

Through inclusion and practice, positive habits of participation in community social inquiry and reconstruction can be developed and sustained. We acknowledge that inclusion of all constituencies in community engagement throughout the process of social inquiry is easily advocated for but often difficult to achieve. One of the challenges of implementing our model is developing and employing practical strategies and procedures to ensure such inclusion both at the outset and as part of the ongoing process of social inquiry.

Leadership, learning and participatory habits are therefore the means of building and sustaining rural community capacity to enhance place integration and well-being for older community members. Over time, such development fosters community resilience and response to precarity (Norris-Baker and Scheidt, 1994).

What does this imply for the practice of a more critical rural gerontology? First, it means shifting focus to the dynamic processes of evolving rural places, life and consequences for older adults. Here, we have suggested a focus on being in place (Rowles, 2018) and the forms and phases of community engagement that Dewey suggested were critical in ongoing inquiry and reconstruction of shared situations – in this case, rural communities (Cutchin, 2016, 2020). While we do not suggest exclusion of other types of work, we do argue that an emphasis on the processes we have described is an essential part of moving forward critically in rural gerontology. That means that rural gerontologists must play a role in teaching, evaluating and generating in-depth understandings of how social inquiry, reconstruction and place integration can work in rural places for the betterment of rural ageing and well-being in place.

The key practical issue becomes how the process is initiated and what the appropriate relationship is between the need for inherent motivation within a rural community and stimulation and support from outside (Minkler, 2004). Informing and educating community leaders to the potential of a Deweyan approach may provide an initial stimulus. Technical support from outside the community (for example with respect to advice and assistance in the optimal conduct of inquiry) may also be helpful. Such contributions must be invited; they cannot be forced. Attention to inclusion of older adults and others who are often left out of community decision making is important. Ultimately, rural communities shape their own destiny through the actions of community leaders, decision makers and citizens (including older adults) who through their actions and processes of place identification forge the ongoing identity of each rural place and, in doing so, determine the lived experience of growing old within their community.

Throughout this chapter, we have emphasised the uniqueness of each rural community, the influence of local circumstances and the role of citizens in sustaining and nurturing the identity of place. Some rural communities are resilient; they are endowed with natural resources, economically viable, blessed with strong and creative leadership, populated by an engaged citizenry (including older adults) committed to fostering and/or preserving a sense of community. The identity of such places is pervaded by an aura of hope and willingness to explore new options even though these options may sometimes fail. Other rural communities have limited or declining resources, are economically unsustainable or vulnerable, lack leadership and have a citizenry that is disempowered, disengaged and apathetic. The identity of such places reflects their struggle with overcoming a sense of futility and entrapment in an inexorable march towards a dismal future. Of course, most rural communities fall between these extremes. All rural communities trace individual trajectories; they face periodic or ongoing challenges through which an engagement and place integration perspective can be helpful.

A Deweyan approach to social reconstruction, based primarily on processes of community engagement, operationalised through participatory action research or related methods, offers the possibility for rural communities to evolve successfully in ways consistent with local conditions and potential. We do not contend that this is a panacea. Rather, it provides a basis for reflection and experimentation as rural places and their ageing residents create their future and evolve as settings in which to grow old. Inevitably, some communities will fail. They will become ghost towns and eventually disappear because they lack the economic and human capital to succeed. A Deweyan approach to social reconstruction and enhanced place integration has the potential to open the door to a brighter future. But only a community's leaders and residents can walk through.

Notes

1 This chapter focuses entirely on ageing in rural areas of contemporary Western and primarily North American societies. It does not address the nature of the environmental context, processes of growing old or the experience of ageing rural areas of non-western societies. Our insights are drawn largely from North America, but we believe that our observations may be pertinent to other developed countries and, perhaps, even beyond.

2 In using 1988 as a baseline for this chapter, we do not wish to imply that there was no significant work on ageing in rural environments before this time.

References

Bates, L., Wiles, J., Kearns, R. and Coleman, T. (2019). Precariously placed: home, housing and wellbeing for older renters. *Health & Place*, 58, pp. 102–152.

Blair, T. and Minkler, M. (2009). Participatory action research with older adults: key principles in practice. *The Gerontologist*, 49(5), pp. 651–662.

Boog, B. W. M. (2003). The emancipatory character of action research, its history and the present state of the art. *Journal of Community and Applied Social Psychology*, 13, pp. 426–438.

Buys, L., Burton, L., Cuthill, M., Hogan, A., Wilson, B. and Baker, D. (2015). Establishing and maintaining social connectivity: an understanding of the lived experiences of older adults residing in regional and rural communities. *Australian Journal of Rural Health*, 23, pp. 291–294.

Campbell, J., (ed.). (1992). *The Community Reconstructs: The Meaning of Pragmatic Social Thought*. Urbana, IL: University of Illinois Press.

Campbell, J., (ed.). (1995). *Understanding John Dewey: Nature and Cooperative Intelligence*. Chicago, IL: Open Court.

Campbell, J. (1998). Dewey's conception of community. In: L. A. Hickman, ed., *Reading Dewey: Interpretations for a Postmodern Generation*. Bloomington, IN: Indiana University Press, pp. 23–42.

Cooperrider, D., Whitney, D. D., Stavros, J. M., and Stavros, J., (eds). (2008). *The Appreciative Inquiry Handbook: For Leaders of Change*. Oakland, CA: Berrett-Koehler Publishers.

Cutchin, M. P. (1997). Physician retention in rural communities: the perspective of experiential place integration. *Health & Place*, 3(1), pp. 25–41.

Cutchin, M. P. (2004). A Deweyan case for the study of uncertainty in health geography. *Health & Place*, 10(3), pp. 203–213.

Cutchin, M. P. (2016). Place integration: notes on a Deweyan framework for community development. In: M.W. Skinner and N. Hanlon, eds., *Ageing Resource Communities: New Frontiers of Rural Population Change, Community Development and Voluntarism*. London: Routledge, pp. 24–37.

Cutchin, M. P. (2020). Habits of social inquiry and reconstruction: a Deweyan vision of democracy and social research. In: J. Wills and R.W. Lake, eds., *The Power of Pragmatism: Knowledge Production and Social Inquiry*. Manchester: Manchester University Press, pp. 55–68.

Dewey, J., (ed.). (1954). *The Public and Its Problems*. Athens, OH: Swallow Press; Ohio University Press. (Original work published in 1927)

Dewey, J., (ed.). (1957). *Human Nature and Conduct: An Introduction to Social Psychology*. New York: The Modern Library. (Original work published in 1922)

Dewey, J., (ed.). (1960). *Logic: The Theory of Inquiry*. New York: Holt, Rinehart & Winston. (Original work published in 1938)

Dewey, J., (ed.). (1980). *The Quest for Certainty: A Study of the Relation of Knowledge and Action*. New York: Perigree Books. (Original work published in 1929)

Dewey, J., (ed.). (1989). *Experience and Nature*, 2nd ed. La Salle, IL: Open Court. (Original work published in 1929)

Dorfman, L. T., Murty, S. A., Evans, R. J., Ingram, J. G. and Power, J. R. (2004). History and identity in the narratives of rural elders. *Journal of Aging Studies*, 18, pp. 187–203.

Fitchen, J. M., (ed.). (1991). *Endangered Spaces, Enduring Places: Change, Identity and Survival in Rural America*. Boulder, CO: Westview Press.

Freire, P. (1970). *Pedagogy of the Oppressed*. New York: Continuum.

Grams, A. and Fengler, A. P. (1981). Vermont elders: no sense of deprivation. *Perspective on Aging*, 10, pp. 12–15.

Grenier, A., Phillipson, C., Rudman, D. L., Hatzifilalithis, S., Kobayashi, K., and Marier, P. (2017). Precarity in late life: understanding new forms of risk and insecurity. *Journal of Aging Studies*, 43, pp. 9–14.

Hendrickson, C., Muro, M. and Galston, W. A., (eds). (2018). *Countering the Geography of Discontent: Strategies for Left-Behind Places*. Washington, DC: Brookings Institution.

Krout, J. A. and Hash, K. M. (2015). What is rural? Introduction to aging in rural places. In: K. M. Hash, E. T. Jurkowski and J. A. Krout, eds., *Aging in Rural Places: Policies, Programs and Professional Practice*. New York: Springer, pp. 3–22.

Lee, G. R. and Lassey, M. L. (1980). Rural-urban differences among the elderly: economic, social and subjective factors. *Journal of Social Issues*, 36, pp. 62–74.

Minkler, M. (2000). Using participatory action research to build healthy communities. *Public Health Reports*, 115, pp. 191–197.

Minkler, M. (2004). Ethical challenges for the "outside" researcher in community-based participatory research. *Health Education and Behavior*, 31(6), pp. 684–697.

Norris-Baker, C., and Scheidt, R. J. (1994). From 'our town' to 'ghost town'? The changing context of home for rural elders. *The International Journal of Aging and Human Development*, 38(3), pp. 181–202.

Rorty, R., (ed.). (1999). *Philosophy and Social Hope*. London: Penguin Books.

Rowles, G. D. (1983a). Place and personal identity: observations from Appalachia. *Journal of Environmental Psychology*, 3, pp. 299–313.

Rowles, G. D. (1983b). Geographical dimensions of social support in rural Appalachia. In: G. D. Rowles and R. J. Ohta, eds., *Aging and Milieu: Environmental Perspectives on Growing Old*. New York: Academic Press, pp. 111–130.

Rowles, G. D. (1988). What's rural about rural aging? An Appalachian perspective. *Journal of Rural Studies*, 4(2), pp. 115–124.

Rowles, G. D. (2018). Being in place, identity and place attachment in late life. In: M. W. Skinner, G. J. Andrews and M. P. Cutchin, eds., *Geographical Gerontology: Concepts and Approaches*. New York: Routledge, pp. 203–215.

Rowles, G. D. and Watkins, J. F. (1993). Elderly migration and development in small communities. *Growth and Change*, 24(4), pp. 509–538.

Scheidt, R. J. and Norris-Baker, C. (1999). Place therapies for older adults: conceptual and interventive approaches. *International Journal of Aging and Human Development*, 48(1), pp. 1–15.

Shenk, D., Davis, B., Peacock, J. R. and Moore, L. (2002). Narratives and self-identity in later life: two rural older American women. *Journal of Aging Studies*, 16, pp. 401–413.

Skinner, M. and Hanlon, N., (eds). (2016). *Ageing Resource Communities: New Frontiers of Rural Population Change, Community Development and Voluntarism*. London: Routledge.

Waite, L. (2009). A place and space for a critical geography of precarity? *Geography Compass*, 3, pp. 412–433.

Walsh, K., O'Shea, E. and Scharf, T. (2019). Rural old age social exclusion: a conceptual framework on mediators of exclusion across the lifecourse. *Ageing & Society*. Epub ahead of print 22 July. DOI: 10.1017/S0144686X19000606.

Wills, J. and Lake, R. W., (eds). (2020). *The Power of Pragmatism: Knowledge Production and Social Inquiry*. Manchester: Manchester University Press.

22 Interrogating the nature and meaning of social exclusion for rural dwelling older people

Kieran Walsh, Sinéad Keogh and Brídín Carroll

Introduction

Social exclusion among rural-dwelling older people is a product of the multi-faceted nature of rural life, the complexity of the individual ageing process, and its social constitution, and the interlinkage between different scalar and political contexts. But reflecting the construct's partial origins as a descriptor of urban disadvantage (Silver, 2019), and its relatively low-level of application to the circumstances of older people, work on rural ageing and exclusion can still be regarded as underdeveloped. This presents an opportunity for a volume such as this that is interested in perspectives that can nurture a critical focus. With its consideration of comprehensive forms of disadvantage that vary over the life course, social exclusion may deliver considerable value in this regard (Scharf et al., 2016). It possesses four conceptual attributes which enhances its explanatory power: a multidimensional structure; a dynamic processual nature; a relative construction; and a focus on agency, or the actual act of exclusion (Atkinson, 1998). Together these attributes may help encapsulate and address contemporary challenges of rural ageing – such as those outlined in the previous section of this book. They may also help not only to unearth the role of structural mechanisms and the social construction of ageing but how older people experience and interpret inequalities while living in rural institutional and cultural contexts, illuminating lines of critical inquiry at the interface of critical and humanist perspectives (Barrs and Phillipson, 2013). However, work on social exclusion can in itself be descriptive and lack a critical engagement with these conceptual attributes for diverse groups and contexts. Scholars, policy makers and practitioners, therefore, continue to be challenged by rural exclusion and the task of identifying, measuring and reducing its prevalence for heterogeneous older populations, living in heterogeneous rural places.

This chapter aims to interrogate the nature and meaning of these four attributes within the lived experience of diverse older adults, in diverse rural settings. In doing so, it will explore how these attributes can inform a critical knowledge of rural ageing. The chapter begins by introducing the four features and reviewing the conceptual underpinnings of exclusion relevant to rural-dwelling older people. A set of illustrative cases are then presented to situate the multidimensional, the dynamic, the relative and the agentic attributes of exclusion within the narratives of rural ageing. The chapter closes with a consideration of what is necessary to operationalise social exclusion to nurture critical inquiry.

Conceptual understandings and considerations

We draw on the work of Atkinson (1998) and Scharf and Keating (2012) to first delineate the four attributes. Multidimensionality refers to the non-binary manifestation of exclusion where rural-dwelling older people can be excluded across multiple different areas of life or can experience exclusion in one area and not in others. Multidimensionality helps to broaden traditionally narrow views of what inequality might involve for older rural lives (Scharf et al., 2016). Dynamic aspects refer to the ways in which older adults can drift in and out of exclusion and experience various forms of disadvantage at different points of the life course. Whereas, relativity alludes to how levels of exclusion are relative to a specific population base and the institutions, norms and values of a particular society. Together these two attributes offer a means of understanding the positioning of individual and cohort experiences in the context of historical and social structure (Barrs et al., 2006). Finally, the involvement of agency points to how older people can be excluded against their will, can lack the capacity to integrate or, whether consciously or subconsciously, can choose to exclude themselves in certain circumstances. As such it helps attribute power, powerlessness and responsibility across the various actors implicated in the exclusionary processes. Facets of each attribute are embedded within the majority of contemporary understandings of old-age exclusion, and are evident, through both implicit and explicit references within the following definition. Social exclusion in later life is described as involving:

> interchanges between multi-level risk factors, processes and outcomes. Varying in form and degree across the older adult life course, its complexity, impact and prevalence are amplified by old-age vulnerabilities, accumulated disadvantage for some groups, and constrained opportunities to ameliorate exclusion. Old-age exclusion leads to inequities in choice and control, resources and relationships, and power and rights in key life domains. . . . Old-age exclusion implicates states, societies, communities and individuals.
>
> (adapted from Walsh et al., 2017: p. 93)

However, formal theoretical elaboration in relation to rural-dwelling older people has been limited within the international literature. While a detailed review of conceptual frameworks on exclusion in later life has been published elsewhere (Van Regenmortel et al. 2016; Walsh et al., 2017), here we focus on the three frameworks that consider the specific circumstances of rural ageing. Two formally conceptualise older adult exclusion in a rural context (Scharf and Bartlam, 2008; Walsh et al., 2019), while one focuses on rural "connectivity" and its barriers and facilitators (Curry, Burholt Hagan-Hennessy, 2014). In addition to testifying to the multidimensional, dynamic, relative and agentic elements, these frameworks usefully illustrate a set of important considerations that can nurture critical inquiry with respect to the manifestations of exclusion for rural-dwelling older people, namely: rural related drivers of exclusion; the heterogeneity of rural older populations; and the diversity of these rural areas.

Drivers of exclusion for rural-dwelling older people

Focusing on prominent structural mechanisms, consideration here is given to three key interconnected drivers. First, the repositioning of rural societies in global economies has seen a significant shift in power and resources to the institutions, industries and populations that are centred within urban regions (Giarchi, 2006; Skinner and Winterton, 2018). With material, and symbolic consequences, there has been a socio-political re-construction of rural life that has been suggested to relegate rural settings to secondary positions within society and the needs and identities of their residents to the margins of decision-making fora (Shucksmith, 2018). Rural communities can be positioned within discourse and policy as being "backward" and underdeveloped, with the only rationale future goal being to become urban (Shucksmith, 2012). For rural ageing societies, it can be argued that these associations are even more pronounced with demographic ageing linked to mounting needs and a lack of innovation.

Second, due to resource sector contraction and population decline, rural infrastructures can be subject to extensive retrenchment and restructuring. There can be difficulties in accessing essential day-to-day services, social outlets and opportunities and routine and specialised care supports (Dwyer and Hardill, 2011; Warburton et al., 2014). However, such a de-coupling of infrastructure is not just about what rural communities are struggling to maintain but also what they may never be able to secure. For instance, a systemic underdevelopment of infrastructure has led some to argue that ambitions around digital connectivity and health are currently ill-feted (Ali et al., 2020) (see Kosurko et al., Chapter 27 in this volume for a counter example). But, weak rural structures are not a recent development, and older people are likely to have experienced a lifetime of under provision (Neville et al., 2020), giving rise to an accumulation of education, employment and health inequalities.

Third, rural settings are subject to significant forces of change that can dislocate some longer-term rural residents from their localities. These forces can be rooted in broader macroeconomic and globalised social patterns, as much as locally bounded place-based factors (Phillipson, 2007). Transformations in the economic, social and cultural collective practices that characterised earlier life experiences of rural residents can weaken the relevance and meaning of local environments for older people, resulting in cultural exclusion (Winter and Burholt, 2018). While depopulation may be a part of these changing milieus, other processes such as counter-urbanisation and, in some cases, population churn – where there is a consistent turnover in the composition of localities – are increasingly evident. In these instances, new older adult residents may be a part of such changes.

These short descriptions, however, do not consider the differential impact on different groups of older people and different communities, nor the documented capacity of both to adapt, manipulate and resist these drivers (Skinner and Winterton, 2018; Walsh et al., 2019).

Diversity of rural-dwelling older people

Comprehending the heterogeneity of rural-dwelling older populations is essential to understanding the pathways, and meanings ascribed, to social exclusion. As espoused by critical political economy of ageing, class, wealth, gender and ethnicity have the capacity to alter later-life experiences (Hagan-Hennessy and Means, 2018). A growing body of work demonstrates the need to consider the social location of diverse older subgroups, not just because their perspectives are rarely considered, but because the nature of exclusion can be quintessentially different for such populations. Grant and Walker (2020), for instance, found that in a study of older lesbians ageing in place in rural Tasmania that a heteronormative culture dominated social activities and care provision, rendering healthy ageing lifestyles irrelevant, and even exclusionary. But different combinations of life events and transitions have also been shown to influence later life outcomes. Scharf and Bartlam (2008) and Curry, Burholt Hagan-Hennessy (2014) found that earlier life experiences and transitions (e.g., difficult relations; bereavement; ill-health) can influence capacities for exclusion/connectivity in place. Relocation has in particular been highlighted in this regard. While fragmented life experiences are generally considered as a function of an individual trajectory, they can also reflect sets of cohort experiences – whether this is early life poverty, subsistence living or high-levels of mortality.

Diversity of rural contexts

At the broadest scalar level, and as demonstrated in a number of chapters in this volume (see Keating et al., Chapter 5 and Hoffman and Roos, Chapter 19), drivers of exclusion in later life can reflect the social, cultural and economic contexts of specific rural world regions. Bernard et al. (2019) highlight that the consideration of regional characteristics (e.g., post-socialist rural transformation; austerity in southern European countries; population shrinkage in central European nations) is essential to avoid an oversimplification derived from Anglo-Saxon perspectives, pointing to the relevance of *unequal spatial opportunity structures*. In the post-Soviet space of Russia and the Ukraine, rapid socio-economic changes since the dissolution of the Soviet Union means that rural-dwelling older people are more likely to experience lower incomes and difficulties in accessing care (Grigoryeva, Parfenova and Dmitrieva, 2020). Exclusionary mechanisms can also vary across different kinds of rural places. Walsh et al. (2019), in a study of village, dispersed, remote, near-urban and island rural communities, found clear impacts as a result of differences in infrastructure, location and development trajectories: individuals in remote and island settings unsurprisingly received less services than those in near-urban settings. Cholat and Daconto (2020) in a study of mountainous regions in France and Italy highlighted how geo-spatial characteristics both restricted access to services but also gave rise to adaptive practices through *reverse mobilities*, where services and assistance travelled to older residents. However, the need to attend to community diversity also extends to the cultural/historical milieu, and plain vibrancy, of particular settings, which can influence the quality of people's lives and the capacity of places to address adverse events. As such, Walsh et al. (2019) identify rural places and older people's interactional relationship with place, as a key mediator of exclusion.

To some extent each of these considerations are evident within the set of contemporary and diverse narratives of rural ageing presented in the next section.

Contemporary narratives of rural ageing

Illustrative cases are derived from a series of studies conducted at the Irish Centre for Social Gerontology on life-course constructions of exclusion and place. While Ireland has been long-established as a site of rural transformation, local settings have in more recent times been implicated in and influenced by macro policy issues that reflect contemporary international societal challenges. These include retirement migration and counter-urbanisation, homelessness, forced migration and re-settlement and the rights and status of ethnic groups. The illustrative cases employed in this chapter will pick up on some of these issues.

Multidimensionality of rural exclusion – Case 1: Tess

Existing conceptual frameworks all testify to the multidimensionality of disadvantage among rural dwelling older people, with broad agreement as to the domains of exclusion. For example, both Scharf and Bartlam (2008) and Walsh et al. (2019) identify material and financial resources, basic services and transport, social relations, civic activities and crime and safety as being principal forms of exclusion for older people in rural settings. The simultaneous occurrence of multiple disadvantages can be particularly prevalent in these communities, with empirical evidence indicating that, despite diversity in structure and scale, many rural sites are underserved and susceptible to significant rates of change (Warburton et al., 2014). It is when inequities amass across multiple life domains that people's lives can be especially impacted. However, it is the non-binary manifestation of exclusion that demonstrates the need to consider not just multiple exclusions but the relative importance given to different life domains. It is these features of multidimensionality that are evident within Tess's story.

Reflecting the growing prevalence of retirement migration to rural sites in many developed nations (Stockdale, 2017; and Berry, Chapter 2), Tess moved with her husband, John, to Gleannacroi, a small dispersed rural community on the West Coast of Ireland 17 years ago when aged 63 years. Having relocated from the UK 15 years before that to a large Irish urban town, Tess and John began exploring rural Ireland when they discovered Gleannacroi. Located in the hinterland of a small rural village, Gleannacroi offered the scenic location and small-tight knit community that the couple always wanted to retire to. When John passed away in 2014, things began to change for Tess, and some of the everyday challenges that she had come to expect from rural residence became more acute with exclusions emerging across multiple domains. Although Tess, now aged 80 years, still drives a car, the lack of public transport and other service infrastructure, meant service exclusion promised to be more of a concern into the future, as it was for many of Tess's neighbours:

> There will probably come a time when I won't be able to drive and what'll happen then, you know? God only knows. It's like with a lot [older people], they're trapped in their homes because there is no transport.

Tess also has become more aware of the lack of opportunities for social contact and participation within the broader area. Like many remote rural settings, available social activities can be heavily gendered and revolve around the pub and local sports clubs, or else involve secondary social outlets like local shops. With little in the way of amenity infrastructure in Gleannacroi, even these opportunities are rare. This has meant that Tess's relational ties to the community are not as developed as she would have liked, and a sense of exclusion from social relations can sometimes be an issue. Tess contrasts these circumstances with what life might have been like if she lived in Galway city, an hour's drive away:

> Because you know, if I was in the city, if I was in Galway I'd have no problems because there are so many social activities go on there, so many . . . But there's nothing out here.

Even though social connectedness is a challenge, it is Tess's continuing sense of being an outsider and a "blow-in" that really threatens her feelings of being a part of the community and her overall attachment to Gleannacroi as a place.

> [I am a blow-in]. It just means that you are not, you're not, you weren't born, and you weren't raised here. And you have no longevity in family ties to here. So your parents and your grandparents weren't born or lived here so therefore you have no real understanding of a situation, of if something is happening. That could be raised, and has been raised, yeah.

But despite this form of community/neighbourhood exclusion, and despite challenges regarding exclusion from services and social relations, Tess still has a strong feeling of being at home in Gleannacroi. For Tess, it is this perceived inclusion that trumps all of the other challenges.

The dynamics of rural exclusion – Case 2: Imelda

As outlined earlier, exclusion in later life is not static and this dynamic nature is particularly evident among older populations. For rural dwelling older adults, and as testified to by a number of the cases presented in this chapter, individual circumstances can alter quickly due to the diversity of individual experiences that rural older people can encounter. For example, Walsh et al. (2019) found that life-course transitions (bereavement; ill-health onset; ageing) could serve as mediators of risk and disadvantage in older people's lives. It is this diversity of exclusionary/inclusionary experiences across the life course that is evident within Imelda's story of homelessness in rural Ireland.

Older adult homelessness is a growing phenomenon across developed nations, and while it is only beginning to become a recognised public issue in rural contexts (see Petersen, Chapter 11), its impact can be immense. Imelda, who is now 60 years old, spent most of her life living in a large regional town, Rathtanion, in Ireland's rural mid-lands. She describes a series of turning points that have punctuated her life with exclusionary and inclusionary mechanisms of different forms. Imelda describes a childhood that was happy with loving parents. But as one of 14 children, she also recalls early experiences of profound deprivation that was reflective of the conditions of the time: "times were bad that time now . . . you'd be hungry, but you go begging for it [food] like". As Imelda entered adult life, she met her husband and she had six children when still in her 20s. Soon after, she started working in the local hospital and describes the positive contributions that her job made to her life:

> I'd a good auld job you know. I was a cleaner [in a hospital]. You know, I loved it . . . Well I tell you the truth now I was respected. . . . And all the sisters [of the religious order running the hospital] respected me and trusted me and that's a good thing to have.

Then Imelda's home situation began to deteriorate and her relationship with her husband became abusive, with Imelda experiencing frequent incidences of domestic violence. In recounting one of her experiences of becoming temporarily homeless, Imelda describes the relative speed at which her circumstances changed and her reluctance to seek help:

> And then as they say, your life is up and then it goes down fast. . . . Not downhill with my health but with you know [domestic violence], so I didn't go back into the hospital again. . . . I'd never ask anyone. Id sooner do without before I'd ask anyone. See, that was my downfall. . . . When things got tough you know, when things started, I brought my children to a hostel.

Later in life Imelda ended up permanently homeless and, while she is unsure about the exact length of time, she lived in precarious circumstances for over ten years, characterised by transitory accommodation experiences. Imelda reflects on her reluctance, once again, to seek help and indicates the stigmatised identity of being a homeless person in her community:

> Couch to couch. And you know when you're independent and you say to yourself: 'Well, I won't tell anyone to make them worry', you know what

> I'm saying. . . . And I wouldn't tell anyone but I mean it's not really, it's not a shame is it? It's not a shame to be homeless.

But as Imelda began to enter older age she started to engage with a local branch of a charity organisation. Imelda is now living in supported accommodation for the last three years and credits this organisation as giving her a second chance.

> I was in [organisation name] in Rathtanion and I did let them know that I was kind of [living with domestic abuse]. You see when you're in a situation that there's [violent] people around you. . . . And you're kind of ashamed to say it . . . but the women down there [in the organisation] . . . they gave me a chance in life, you know. They gave me hope.

While Imelda's story indicates the sort of invisibility that characterises some forms of exclusion in rural contexts, it illustrates the fluid nature of disadvantage and the diversity of exclusionary mechanisms that an individual can encounter over their life course.

Relative aspects of rural exclusion – Case 3: Mary

Atkinson's (1998) observation that social exclusion cannot be assessed in absolute terms emphasises the need to compare individual circumstances to those experienced by the majority of a population within a particular society. For older people, this has led to questions as to what population base exclusion in later life should be measured against (Scharf and Keating, 2012), For rural ageing, these questions become a little more complicated and relate to which group (older or general population) and in which society (rural or urban, or both) integration and exclusion should be benchmarked. As will be demonstrated in Mary's story, the meaning of exclusion is also relative to sets of cohort experiences linked to context-specific cultural, socio-economic and geo-political forces (Bernard et al., 2019). Understanding how cohort experiences inform present-day expectations is essential to determining how best to identify and intervene in exclusionary circumstances.

Mary lives three miles outside Ballydrum, a village on the border of Northern Ireland. Like many of Mary's generation in Irish rural communities, she has during her lifetime witnessed substantial economic and social change within her local area. Together with a decline in collective agricultural and social practices, these changes have brought a general increase in standards of living. This has in turn meant improvements in material well-being and a shift away from subsistence style agriculture that characterised the early lives of many rural dwelling older people. Having raised nine children with her late husband on a small farm, Mary now reflects on her life in Ballydrum and her attachment to the locality:

> [I am] attached to it as I go about it every day. . . . There's a lot of beauty about it. There's a lot of nice people around you, and there's a freedom about it.

Residing one mile away from the public road, Mary emphasised a need for independence, even if that meant sometimes living a somewhat isolated life. In Mary's case, solitude was superseded by the desire for a personal sense of autonomy and space:

> Some people want to have company at all times, you know. . . . I'm quite happy with my own company and I can chat to people when they do come. . . . You have the space. You have the time. You're your own boss.

But Mary also did not have a telephone or mains electricity and used a tractor battery wired to room lamps to light her home. While Mary admitted to sometimes being short of money to do the things she would like to, she asserted that: "My needs are small. I never wanted a lot. . . . I don't like ornaments, mirrors. I have most I want." Mary spoke about her standard of living and how, because of the nature of her upbringing, excessive material wealth was not a priority. Consequently, she anticipated having to cope with shortages and deprivation. Mary talked about not focusing on, or complaining about, the negatives and this is inherent in how she assessed her personal satisfaction and happiness.

> I am contented yah. . . . D'you know taking the rough with the smooth. Life was never that bad. You see, some people say 'Oh God, aye it was." Not really, ya know. You can either get into a routine or a tummy rut. The routine is: 'take it as it comes, make the best of it.' That was my attitude. Tummy rut is: 'I have no money to go anywhere'. . . . So, the tummy rut part is no good.

However, Mary also acknowledged the role that her own expectations played in her assessment of life satisfaction. In the following quote Mary reflects on this and hints at her awareness of how her expectations, and her satisfaction, were socially informed and constructed.

> Yes, I am satisfied. But no one did say to me: 'You could improve' . . . . So, therefore I took it for granted that I was happy enough.

Agency and the acts of rural exclusion – Case 4: Farid

Recalling the Walsh et al. (2017) definition, social exclusion in later life can implicate the power of states, societies, communities and individuals in the construction of inequities. As referenced earlier, exclusion can be perpetrated, intentionally or unintentionally, can be a result of a lack of restricted individual capacity, or can reflect choice. Consequently, there are different forms of acts, and different sources, or agents, of those acts layered within an individual's life course trajectory. This can have a particular relevance for rural communities, given the strength of the interlinkage between scalar levels, and the mobility across them, that is evident in some rural settings. And it is this layering that can be discerned from Farid's story of forced migration from Syria and resettlement in rural Ireland.

Like many of the 5.6 million Syrian refugees (UNHCR, 2018), Farid and his family fled the civil war within his homeland amidst significant regional conflict, threat to life and personal loss. Aged now 55 years, Farid began his migration journey in 2015 before entering a refugee programme, obtaining refugee status and relocation to Ireland in 2018. Farid had already accumulated multiple exclusions and traumas prior to arrival due to the conflict. In Ireland, Farid and his family were relocated to an Emergency Reception and Reorientation Centre (EROC), which, reflecting a broader national trend to avoid large urban areas, is located in a privately run former hotel in a small regional rural town. Farid has been at this Centre for five months. Now, despite the support of an orientation programme, Farid describes a sense of profound exclusion across multiple spheres as a result of the settlement process in Ireland. This betrays the uncertainty of continuing to be in a transitional period and space:

> We're not fully into this country and we don't have the right to vote yet because have to fill out papers and apply and so on. [Our experiences of exclusion] include such things as income, material goods and [lack of] good health.

For Farid, these exclusions have been exacerbated by being resident in the hotel, which was not purpose built to accommodate the longer-term residency of individuals and families. Reflecting publicised complaints of overcrowding, poor food quality and a lack of transport for essential appointments, Farid conveys the isolating structures of his environment and the stress of not knowing when he will move to more suitable housing and begin a more permanent resettlement process. This illustrates again the involvement of the state programme, and its local manifestation, in his exclusion, and his own sense of powerlessness in this process.

> And the stress of living in a hotel and the food, [and] waiting for a house [is exclusion]. We've been waiting for a house for three months now and they constantly tell us 'Next time, next time'. And now, well we just met the Department of Justice now and they just told us 'This house is not going to work for you' so they just cancelled the house for us.

Despite high unemployment and public service cuts in the town, a lack of state consultation around establishing the EROC in the community, and concerns about insufficient health and education resources to support the Centre, there have been significant local voluntary efforts to welcome and integrate the Centre's residents. Nevertheless, Farid, who had been a resident of a large Syrian city, describes his lack of integration with the community and an unfamiliarity with its scale and socialisation culture.

> In Syria there's maybe like a hundred thousand or something in these areas, so there's always somebody also to go and get out with, like regarding for my age and there's always something to do. Whereas here I realise in [place name], they're not really mixing together.

Farid's story demonstrates multiple agents and acts of exclusion, ranging from the international conflict zone of his home country to Irish national place-based resettlement policy to the local institutional and rural cultural contexts. It also demonstrates that powerlessness can be reflected in the inability of communities to access appropriate supports. In this manner, Farid's story is likely to reflect the complex layering of agency that is experienced by other older populations displaced to rural communities in international jurisdictions (see Hoffman and Roos, Chapter 19).

Conclusion

Through the application of the four attributes of exclusion to the illustrative cases the analysis has illuminated some of the traditional concerns of critical gerontology within the lived experiences of rural ageing contexts (Barrs et al., 2006). This includes the role of structural mechanisms, separate and as a part of globalisation processes, within Tess's and Farid's story – amounting to *unequal spatial opportunity structures* as described by Bernard et al. (2019). It includes the role of normative and cultural forces in the social construction of personal identities and status – whether this relates to Tess's "blow-in" status, Imelda's perceived stigmatisation as a homeless older woman, Mary's expectations with respect to normative integration, or indeed Farid's position as a refugee in transition. It also includes, perhaps most critically for this chapter, the nature of inequalities, where the use of these conceptual attributes has provided a more comprehensive and textured view of disadvantage for rural-dwelling older people.

The scope of the chapter did not allow for a full analysis of individual and place diversity, or to fully disentangle the intersectional processes related to rurality and ageing. It did, however, capture aspects of the interplay between individual and social change and point to the need to attend to personal experiences and interpretations, as well as structural mechanisms, to more fully reflect the construction of and response to disadvantage (Barrs and Phillipson, 2013).

These attributes on their own are not proposed as an in-depth analysis framework. They do provide a helpful set of explanatory and conceptual unlocking mechanisms for critically reflecting on complex social phenomena that give rise to exclusions and inclusions in rural communities. But there is also the potential to further enhance the role of these conceptual attributes, and the construct of exclusion itself, in supporting the development of a critical rural gerontology. First, there must be more active empirical engagement with multidimensional, dynamic, relative and agentic attributes of disadvantage in the study of inequalities for rural-dwelling older people. Second, the lived experience of heterogenous ageing populations in diverse places must be central within empirical research, with a view to more accurately capturing the interpretation of meaning and local responses to social and cultural strictures and contexts. Third, and in tandem with these data-led programmes as an interactive process of development, these attributes must be embedded within a more active effort to theoretically elaborate on rural ageing experiences.

References

Ali, M.A., Alam, K. and Taylor, B. (2020). Measuring the concentration of information and communication technology infrastructure in Australia: Do affordability and remoteness matter? *Socio-Economic Planning Sciences*, 70, p. 100737.

Atkinson, A. B. (1998). *Social Exclusion, Poverty and Unemployment* (CASE paper). http://sticerd.lse.ac.uk/dps/case/cp/Paper4.pdf

Barrs, J. and Phillipson, C. (2013). Introduction. In: J. Barrs and J. Dohmen, eds., *Ageing, Meaning and Social Structure: Connecting Critical and Humanistic Perspectives*. Bristol: Policy Press, 1–10.

Barrs, J., Dannefer, D., Phillipson, C., Walker, A., (eds.). (2006). *Aging, Globalization and Inequality: The New Critical Gerontology*. New York: Routledge.

Bernard, J., Contzen, S., Decker, A. and Shucksmith, M. (2019). Poverty and social exclusion in diversified rural contexts. *Sociologia Ruralis*, 59, pp. 353–368.

Cholat, F. and Daconto, L. (2020). Reversed mobilities as a means to combat older people's exclusion from services: insights from two alpine territories in France and Italy. In: K. Walsh, T. Scharf, S. Van Regenmortel and A. Wanka, eds., *Social Exclusion in Later Life: Interdisciplinary and Policy Perspectives: International Perspectives on Aging*, vol. 28. Cham: Springer.

Curry, N., Burholt, V., Hagan Hennessy, C. (2014) Conceptualising rural connectivities in later life. In: C. Hagan Hennessy, R. Means and V. Burholt, eds., *Countryside Connections: Older People, Community and Place in Rural Britain*. Bristol: Policy Press, pp. 95–124.

Dwyer, P. and Hardill, I. (2011) Promoting social inclusion? The impact of village services on the lives of older people living in rural England. *Ageing & Society*, 31, pp. 243–264.

Giarchi, G. (2006) Older people 'on the edge' in the countrysides of Europe. *Social Policy and Administration*, 40(6), pp. 705–721.

Grant, R. and Walker, B. (2020) Older Lesbians' experiences of ageing in place in rural Tasmania, Australia: an exploratory qualitative investigation. *Health and Social Care Community*. Epub ahead of print 22 May. DOI: 10.1111/hsc.13032.

Grigoryeva, I., Parfenova, O. and Dmitrieva, A. (2020). Social policy for older people in the post-Soviet space: how do pension systems and social services influence social exclusion? In: K. Walsh, T. Scharf, S. Van Regenmortel and A. Wanka, eds., *Social Exclusion in Later Life: Interdisciplinary and Policy Perspectives: International Perspectives on Aging*, vol. 28. Cham: Springer.

Hagan Hennessy, and C., Means, R. (2018). Connectivity of older people in rural areas. In: A. Walker, ed., *The New Dynamics of Ageing*. Policy Press, pp. 147–166.

Neville, S., Napier, S., Adams, J., Shannon, K. and Clair, V. (2020). Older people's views about ageing well in a rural community. *Ageing & Society*. Epub ahead of print 22 April. DOI: 10.1017/S0144686X20000458.

Phillipson, C. (2007) The 'elected' and the 'excluded': sociological perspectives on the experience of place and community in old age. *Ageing & Society*, 27(3), pp. 321–342.

Scharf, T. and Bartlam, B. (2008). Ageing and social exclusion in rural communities. In: N. Keating, ed., *Rural Ageing: A Good Place to Grow Old?*. Bristol: Policy Press, pp. 97–108.

Scharf, T. and Keating, N. (2012). Social exclusion in later life: a global challenge. In: T. Scharf and N. Keating, eds., *From Exclusion to Inclusion in Old Age: A Global Challenge*. Bristol: Policy Press, pp. 1–16.

Scharf, T., Walsh, K. and O'Shea, E. (2016). Ageing in rural places. In: D. Shucksmith and D. L. Brown, eds., *Routledge International Handbook of Rural Studies*. London: Routledge, pp. 50–61.

Shucksmith, M. (2012), Class, power and inequality in rural areas: beyond social exclusion? *Sociologia Ruralis*, 52: pp. 377–397.

Shucksmith, M., (2018). Re-imagining the rural: from rural idyll to Good Countryside. *Journal of Rural Studies*, 59, pp. 163–172.

Silver, H. (2019). Social exclusion. In: A. Orum, ed., *The Wiley Blackwell Encyclopedia of Urban and Regional Studies*. New York: John Wiley & Sons, pp. 1–6.

Skinner, M. and Winterton, R. (2018). Interrogating the contested spaces of rural aging: Implications for research, policy, and practice. *The Gerontologist*, 58(1), pp. 15–25.

Stockdale, A. (2017) From 'trailing wives' to the emergence of a 'trailing husbands' phenomenon: retirement migration to rural areas. *Population, Space and Place*, 23, p. e2022.

UNHCR (2018). *Syria Emergency*. Available at: www.unhcr.org/en-ie/syria-emergency.html

Van Regenmortel, S., De Donder, L., Dury, S., Smetcoren, A. S., De Witte, N., and Verté, D. (2016). Social exclusion in later life: a systematic review of the literature. *Journal of Population Ageing*, 9, pp. 315–344.

Walsh, K., O'Shea, E. and Scharf, T. (2019). Rural old-age social exclusion: a conceptual framework on mediators of exclusion across the life course. *Ageing & Society*. *Ageing & Society*. Epub ahead of print 2 July. DOI: 10.1017/S0144686X19000606.

Walsh, K., Scharf, T. and Keating, N. (2017). Social exclusion of older persons: a scoping review and conceptual framework. *European Journal of Ageing*, 14, pp. 81–98.

Warburton, J, Cowan, S, Winterton, R and Hodgkins, S. (2014) Building social inclusion for rural older people using information and communication technologies: perspectives of rural practitioners. *Australian Social Work*, 67, pp. 479–494.

Winter, B. and Burholt, V. (2018). The welsh – Y Cymry cymreig: A study of cultural exclusion among rural dwelling older people using a critical human ecological framework. *International Journal of Ageing and Later Life*, 12(2), pp. 1–33.

23 Defining the relationship between active citizenship and rural healthy ageing

A critical perspective

Rachel Winterton and Jeni Warburton

Introduction

As Burholt and Scharf have noted earlier in this volume (Chapter 6), critical gerontological perspectives seek to articulate how older people are able to interact with their environments to influence outcomes that directly affect them. The present chapter applies this critical perspective to interrogate the relationship between active citizenship and rural healthy ageing. Ageing policy discourses across international contexts romanticize the role of active citizenship in keeping older people healthy and engaged (Walker, 2002). Active citizenship reflects the participation of individuals in civil society, community and political life (Hoskins and Mascherini, 2009) and encompasses activities such as volunteering, civic, social and political participation (Walker, 2002; Mendes, 2013). Within contemporary neoliberal government ideologies, active citizenship is implicitly depicted as a responsibility and obligation that older adults are required to assume, with the aim of reducing the anticipated pressure on social security systems posed by population ageing (Mendes, 2013). Across global rural settings, older adults' active citizenship practices play a critical role in keeping rural communities sustainable and in enabling key health and aged care services to be provided (Winterton and Warburton, 2014; Skinner et al., 2016). Neoliberal policies enacted across many countries have resulted in continued rural restructuring, service centralisation and service withdrawal (Argent, 2011; Ryser and Halseth, 2014), with the civic and voluntary contributions of rural residents critical in both filling service gaps and in advocating for the needs of rural populations (Cheshire and Woods, 2009). As a result, volunteering is much higher in rural than urban areas, and, as rural populations are proportionally older than their urban counterparts (United Nations, 2015), responsibility for active citizenship practices commonly falls to older people in rural areas (Skinner, 2014; Warburton and Winterton, 2017).

However, critical gerontology also demands that scholars interrogate widely held assumptions, norms and discourses relating to ageing (Skinner and Winterton, 2018), and this is particularly applicable to rural active citizenship. Recent studies have identified significant challenges associated with relying on the capacity and desire of rural older people to become, or remain active citizens (Jones and Heley, 2016; Skinner and Winterton, 2018). Simultaneously, the role of active citizenship in supporting the health and well-being of ageing populations, both in rural and urban settings, is also being questioned (Mendes, 2013; Stephens, Breheny and Mansvelt, 2015; Skinner and Winterton, 2018). Through reviewing the contemporary academic literature related to rural ageing and active citizenship, this chapter explores how active citizenship trends among rural older adults support or hinder the capacity of rural settings to support healthy ageing. In doing so, we propose a conceptual framework for understanding the mutually constitutive and complex relationship between the civic activities of older rural adults and their capacity to age well in these environments. This is in line with contemporary critical gerontological approaches, which seek to explore dynamics that facilitate positive or negative health or well-being outcomes for diverse older adults (see Burholt and Scharf, Chapter 6). While some of the findings reported here are consistent with findings reported in the broader literature on active citizenship in older age (see Warburton, 2015), examining the findings in isolation provides a more detailed view of the complexity of the interactions between rural people and rural places.

Active citizenship, rurality and healthy ageing: a conceptual framework

The intersection of rural older adults' active citizenship practices with healthy ageing is depicted in Figure 23.1. This framework is premised on three interrelated factors:

- Active citizenship practices among rural older adults are mediated by micro- and macro-level factors;
- Active citizenship practices among rural older adults influence the capacity of rural places to facilitate healthy ageing, which subsequently impacts on individual health and well-being outcomes and the capacity of other rural older adults to engage in active citizenship; and
- Active citizenship directly influences health and well-being outcomes for older adults, which impacts on both their subsequent active citizenship practices and the capacity of rural communities to support healthy ageing.

Active citizenship is mediated by micro and macro-level factors

In line with the emphasis on critical human ecology within the rural gerontological literature (see Keating et al., Chapter 5), factors associated with both the micro and macro environments of rural older adults can mediate involvement in active citizenship.

First, factors that are specific to the rural older person, mediate both level and type of involvement. Consistent with the broader volunteer literature which shows that older volunteers are more likely to have higher levels of individual resources than non-volunteers (e.g., Warburton, 2015), the rural literature also demonstrates the importance of personal resources in supporting active citizenship involvement among older people. Key resources include high levels of self-rated health,

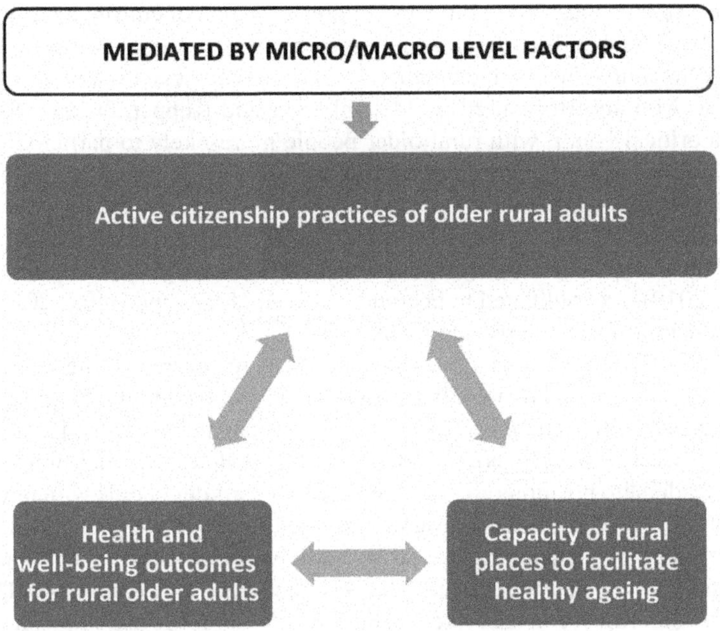

Figure 23.1 Conceptual framework of the intersection of active citizenship, rurality and healthy ageing

education, access to personal transport, higher incomes, greater social networks and proximal family networks (Farmer et al., 2011; Curry and Fisher, 2013; Curry et al., 2014; Munoz et al., 2014; Levasseur et al., 2015; He et al., 2017; Kafková, Vidovićová and Wija, 2018). Involvement is restricted where individuals have poorer health, limited financial resources, access to personal transport, or social networks (Davis et al., 2012; Curry and Fisher, 2013; Glasgow, Min and Brown, 2013; Mettenberger and Küpper, 2019). Time is also significant, with rural older people with existing employment or care responsibilities restricted in their engagement (Davis et al., 2012; Rozanova et al., 2012). Older people with higher levels of community trust are also more likely to engage in rural civic activity (Davis et al., 2012), where older people who report feeling unsafe in their rural communities are less likely to be involved (Curry and Fisher, 2013).

Individual demographic characteristics are also linked with active citizenship, with older people who are married and in the younger-old cohort more likely to be engaged (Farmer et al., 2011; Hodgkin, 2012; Curry and Fisher, 2013). Less likely to be involved are those in the older-old cohort (Glasgow, Min and Brown, 2013), are new to their rural communities (Farmer et al., 2011; Davis et al., 2012; Buys et al., 2015) or experience loss of a spouse (Jones and Heley, 2016). Lifecourse histories also play a role, with longer length of residence (Davis et al., 2012;

Munoz et al., 2014; Levasseur et al., 2015) and a history of voluntary involvement (Farmer, Muñoz et al., 2011) linked with greater levels of engagement. Individuals who have held high pressure employment roles are often less engaged, due to a lack of lifecourse involvement and a desire to have a less structured retirement (Glasgow, Min and Brown, 2013; Mettenberger and Küpper, 2019). Motivation is also a critical factor, with rural older people more likely to participate if they enjoy active citizenship, have a desire for socialisation, want to learn something new, contribute to community or keep busy (Davis et al., 2012; Heley and Jones, 2013; Jones and Heley, 2016; He et al., 2017; Mettenberger and Küpper, 2019). Conversely, they are less likely to engage if they lack interest (Mettenberger and Küpper, 2019), feel obligated to become involved (Jones and Heley, 2016) or are not confident to do so (Curry and Fisher, 2013).

Certain individual characteristics also mediate type of involvement. For example, older rural women are more likely to be involved in community, group-based, social forms of active citizenship and men more likely to be involved in individual forms of civic activity (Davis et al., 2012; Jones and Heley, 2016). Rural volunteer or civic leadership positions are often undertaken by older people with high levels of education, professional background, good health and in the younger-old cohort (Davis et al., 2012; Munoz et al., 2014; Kafková, Vidovićová and Wija, 2018). These individuals also tend to be what Mettenberger and Küpper (2019) have termed as "multiply engaged" key players, undertaking leadership roles across different organisations, and often not within traditional rural ageing groups and organisations. Conversely, Rozanova et al. (2012) have suggested that older people with lower levels of economic capital are more likely to engage in more informal, caregiving volunteer activity.

Community or place related factors can mediate both level and type of involvement of rural older people as active citizens. Socio-demographic and spatial community characteristics are associated with level of involvement, with higher engagement observed in rural communities with higher proportions of older adults, more highly educated residents and growing populations (Keating, Eales and Phillips 2013). Munoz et al. (2014) have identified that participation is higher in remoter rural communities, with Curry and Fisher (2013) noting that people in remote or deprived areas are more likely to be volunteering or actively assisting others. Some declining rural communities have higher levels of older people participating in group, civic or community activities, as they are more likely to see the benefits of participation (Davis et al., 2012) and local need associated with service withdrawal (Keating, Eales and Phillips, 2013). Associated cuts to health and social services within rural regions, while both increasing need for volunteers and opportunities for volunteering, can also lead to pressure to undertake active citizenship activity, and thus promote feelings of obligation (Keating, Swindle and Fletcher, 2011; Rozanova et al., 2012; Winterton and Warburton, 2014). Conversely, however, there may be also lower levels of participation in disadvantaged communities, particularly where they comprise high proportions of poorly resourced older people in relation to transport, education, health, age and education (Keating, Eales and Phillips, 2013).

In relation to the rural resource environment, research highlights factors associated with lower levels of engagement, such as limited physical and transport infrastructure (Majee et al., 2018; Neville et al., 2018). Non-participation was also linked with limited available options for active citizenship (Davis et al., 2012; Rozanova, Keating and Eales, 2012), particularly in more remote rural areas (Curry, Burholt et al., 2014). However, interrogation of this data suggests that this is highly subjective and mediated by the fit between older people and their rural places (Keating, Eales and Phillips, 2013). It may be that rural older people are not necessarily interested in activities aimed at sustaining essential rural health and social infrastructure, with rural service providers struggling to offer volunteer activity that both meets the preferences of older people and the needs of rural communities (Winterton and Warburton, 2014; Mettenberger and Küpper, 2019). Conversely, Keating, Eales and Phillips (2013) have noted that in retirement communities with high levels of resources, community-active rural older adults are generally satisfied with opportunities for engagement.

Macro-level trends and policies can also influence active citizenship trends among rural older adults, but they largely mediate level of involvement rather than type. These trends are generally observed to have a "negative" impact on active citizenship among older people in rural areas, in terms of both their capacity to engage and the voluntary nature of this engagement. While certain demographic trends increase potential opportunities for rural older adults to be engaged as active citizens, such as reduced numbers of younger people volunteering and the out-migration of younger-old cohorts from rural areas, this also places more pressure on older-old volunteer cohorts (Wiersma and Koster, 2013; Jones and Heley, 2016). This will be especially pressing where older rural in-migrants do not feel obligated or willing to volunteer (Farmer et al., 2011), or where older rural residents undertake seasonal travel and are not available as volunteers (Wiersma and Koster, 2013; Winterton and Warburton, 2014).

Macro-level economic trends and events, such as recession and drought, may also impact negatively on older adults' resource capacity to engage as active citizens (Davis et al., 2012; Walsh et al., 2014; Jones and Heley, 2016). These events, coupled with changing government policies around retirement age, result in older adults having to work longer rather than volunteer (Jones and Heley, 2016; Liu et al., 2019). Increasing engagement of rural older adults in providing care for grandchildren also limits their capacity to commit to active citizenship (Winterton and Warburton, 2014). These factors are exacerbated by a general lack of political recognition or resourcing for the financial costs involved with volunteering (Winterton and Warburton, 2014; Mettenberger and Küpper, 2019). As Jones and Heley (2016) also note, policy attempts to formalise active citizenship within rural policy agendas also shift the emphasis on active citizenship from one of choice to one of obligation. Research has noted that rural older people are actively involved in contesting discourses relating to co-production of services, in that they want to choose what they engage in (Mettenberger and Küpper 2019), or view service provision as a basic entitlement rather than a civic responsibility (Wiersma and Koster, 2013). Similarly, policy discourses aimed at formalising the volunteer sector, such as introduction of formal police checks, may dissuade older rural volunteers and undermine informal codes of conduct relating to active citizenship (Farmer et al., 2011; Winterton and Warburton, 2014; Jones and Heley, 2016).

Active citizenship influences health and well-being of rural older adults

Consistent with findings from the broader literature on active citizenship and ageing, where rural older people do engage in active citizenship it has differential effects on their health and well-being. This is associated with the macro and micro-level factors that mediate their involvement. Active citizenship among rural older adults is associated with higher quality of life (He et al., 2017), better self-rated health (Liu et al., 2019) and better mental health (Takagi et al., 2013; Yuasa et al., 2014; Guo, Bai and Feng, 2018). However, this positive affect occurs where rural older adults want to be productive and active within their communities (Davis et al., 2012; Keating, Eales and Phillips, 2013; Buys et al., 2015; Jones and Heley, 2016) and feel that their contributions are filling a need, thus providing a sense of achievement, self-worth and community respect (Davis et al., 2012; Glasgow, Min and Brown, 2013; Winterton and Warburton, 2014; Jones and Heley, 2016). Benefits also accrue where activities provide the desired level of social interaction and a sense of belonging to a community (Farmer et al., 2011; Davis et al., 2012; Winterton and Warburton, 2014; Buys et al., 2015). Importantly, benefits are observed where the level and type of active citizenship reflects individual preferences for involvement (Mettenberger and Küpper, 2019; Nonaka et al., 2019).

However, involvement in active citizenship can have negative, or negligible, outcomes for rural older adults where it impinges on ideals of what retirement or rural living should be about (Farmer et al., 2011) and is based on a sense of obligation or expectation rather than choice (Farmer et al., 2011; Keating, Swindle and Fletcher, 2011; Munoz et al., 2014; Jones and Heley, 2016; Nonaka et al., 2019). It also negatively impacts on well-being where time or health costs of the activity are perceived as too high, or incongruous with personal preferences for involvement (Keating, Swindle and Fletcher, 2011; Wiersma and Koster, 2013; Mettenberger and Küpper, 2019). As Figure 23.1 suggests, these health and well-being impacts will, in turn, influence their ongoing active citizenship practices, which will subsequently impact on capacity of rural places to support healthy ageing.

Active citizenship influences the capacity of rural places to support healthy ageing

Active citizenship practices of rural older people can impact the capacity of rural communities to support healthy ageing in three ways – through positive impacts associated with participation, through negative impacts associated with non-participation and through negative impacts associated with participation. First, participation in active citizenship by rural older adults may bolster the capacity of rural settings to better support their ageing populations. The voluntary and civic contributions of rural older adults contribute to filling gaps in health and aged care infrastructure, through providing support services to vulnerable older adults within their communities (Heley and Jones, 2013; Skinner et al., 2014, Winterton and Warburton, 2014; Mettenberger and Küpper, 2019). They advocate for the needs of older people within local planning and policy and in doing so increase the age-friendliness of their communities (Kuokkanen, 2018; Mettenberger and Küpper, 2019) and contribute to the development of new and existing rural community institutions that support older people (Glasgow, Min and Brown, 2013; Keating, Eales and Phillips, 2013). They also contribute to broader rural community sustainability so that older people can age in place, through participating in economic development and tourism initiatives (Heley and Jones, 2013; Skinner et al., 2014; Winterton and Warburton, 2014).

However, where older people choose not to, or are unable to engage as active citizens, this may have implications for local rural health and social infrastructure. Lack of active citizenship engagement among older adults may, when considered in concert with other institutional factors, impact on local health and aged care service availability and responsiveness (Skinner et al., 2014; Winterton and Warburton, 2014) and can contribute to an increased community reliance on externally provided services (Farmer et al., 2011). This may have implications for those older adults within rural communities who do not have the personal resources to travel or gain formal or informal support. Further, where certain cohorts of rural older adults do engage in active citizenship, it may have a negative impact on community capacity to support health and well-being for other older people within their communities. As this chapter has already established, certain cohorts of rural older people are over-represented within active citizenship, which primarily reflects those with high levels of personal resources. As a result, their advocacy efforts may not represent the needs of broader cohorts of older adults (Glasgow, Min and Brown, 2013; Curry, Burholt et al., 2014), or the best interests of rural communities, with rural older people often resistant to community change (Jones and Heley, 2016).

Consequently, engagement of rural older adults in active citizenship has the capacity to foster contested spaces of rural ageing (Skinner and Winterton, 2018), in that it can create community tensions and facilitate processes of exclusion for other older adults (Glasgow, Min and Brown, 2013; Jones and Heley, 2016). Mettenberger and Küpper (2019) note tensions where rural older adults who are heavily involved in civic engagement clash with those who prefer passive forms of active citizenship, and other studies note conflicts between groups of older adults in relation to how rural spaces should be occupied and used (Glasgow, Min and Brown, 2013; Skinner and Winterton, 2018). Further, where rural older adults engage in active citizenship, but prioritise activities that are not aligned with direct rural service provision or elect to participate in activities that require low levels of commitment, this will threaten the capacity of rural communities to provide health and aged care services (Glasgow, Min and Brown, 2013; Munoz et al., 2014; Mettenberger and Küpper, 2019). In some communities, the enthusiasm of older people for volunteering may also undermine the market for paid professionals, with communities not having a contingency plan to cater for declining numbers of volunteers (Winterton and Warburton, 2014; Jones and Heley, 2016).

Therefore, as Figure 23.1 suggests, the role of older adults' active citizenship practices in supporting rural places to promote healthy ageing is important in maintaining health outcomes for rural older adults, given the pivotal role rural community, health and aged care services play in supporting older adults' well-being (Winterton et al., 2016). The role active citizenship plays in maintaining rural services and infrastructure that support healthy ageing is also pivotal in addressing the micro and macro-level factors that influence rural older people's engagement in active citizenship, through providing opportunities and addressing constraints associated with involvement. Consequently, as depicted in Figure 23.1, the dynamic is reciprocal and circular in nature.

Towards critical perspectives on rural ageing and active citizenship

In line with its critical approach, this chapter has sought to articulate how dynamics associated with rural older people's active citizenship trends facilitate positive or negative outcomes, both at the individual and community level. The framework outlined in this chapter, which has highlighted the dynamic relationship between older people, active citizenship and healthy ageing, raises three critical questions for rural gerontology that must be considered:

1. Is active citizenship "good" for rural older people, and in what circumstances?
2. Is active citizenship a valid option for all older people in rural communities?
3. Is active citizenship among rural older people "good" for rural communities, and in what circumstances?

The literature indicates that the answers to these questions are highly dependent on both person and place-related characteristics, in line with Keating, Eales and Phillips' (2013) assertion that age-friendliness is mediated by the intersection between characteristics of rural people and rural places. Therefore, assumptions around the benefits of rural active citizenship in older age must be closely interrogated, in terms of considering whether these benefits extend to the diversity of rural ageing populations and consequently who is excluded and the resultant implications. For example, involvement in active citizenship may be good for rural older people where they have the individual resources to support choice and agency and where their rural places can support preferences and motivations for involvement. However, it will likely not be advantageous where place-related or macro-level characteristics, coupled with individual resources, may pressure rural older people into involvement, or facilitate a lack of fit between people's needs and preferences for involvement.

It must also be questioned whether active citizenship is a valid option for all rural older adults, given that they may have significant health constraints, or be undertaking paid employment or informal care roles. This chapter has highlighted that both individual and place-related factors negatively influence level and type involvement in active citizenship, and these must be interrogated. Specifically, there is a need to explore individual-level factors that preclude involvement, in order to determine whether these are actual constraints, or certain cohorts of rural older people are simply exercising "the right not to participate" (Curry and Fisher, 2013). Given that certain forms of active citizenship among rural older people (particularly those related to high-level decision making and control) appear to be undertaken by more socio-economically advantaged cohorts, it must be questioned whether certain cohorts of older adults are being excluded from active citizenship, and the implications of this for rural communities must be considered.

Consequently, a critical perspective would suggest that the notion that active citizenship among older people is universally "good" for rural communities must be carefully considered, particularly where rural older people are not acting in accordance with the needs and preferences of broader rural ageing cohorts, or in line with rural community sustainability objectives. At a base level, contributions of older adults are essential in providing community contexts that support healthy ageing. However, critical perspectives on rural ageing and active citizenship must consider how processes and practices associated with older adults' participation (or non-participation) in active citizenship facilitate uneven geographies of rural disadvantage, through contributing to contested spaces of rural ageing (Skinner and Winterton, 2018). While active citizenship plays a critical role in supporting the health and well-being of older adults, it also has the capacity to widen the chasm of rural disadvantage and must be carefully mediated within and across diverse rural contexts. This is particularly essential in the context of sustained rural and aged care policy discourses that promote the responsibility of older people in facilitating the services and support they need to age in their rural communities.

References

Argent, N. (2011). Trouble in paradise? Governing Australia's multifunctional rural landscapes. *Australian Geographer*, 42(2), pp. 183–205.

Buys, L., Burton, L., Cuthill, M., Hogan, A., Wilson, B. and Baker, D. (2015). Establishing and maintaining social connectivity: an understanding of the lived experiences of older adults residing in regional and rural communities. *Australian Journal of Rural Health*, 23(5), pp. 291–294.

Cheshire, L. and Woods, M. (2009). *Citizenship and Governmentality, Rural: International Encyclopedia of Human Geography*. Amsterdam: Elsevier.

Curry, N., Burholt, V., and Hennessy, C. (2014). Conceptualising rural connectivities in later life. In: C. H. Hennessy, R. Means and V. Burholt, eds., *Countryside Connections: Older People, Community and Place in Rural Britain*. Bristol: Policy Press, pp. 31–62.

Curry, N. and Fisher, R. (2013). Being, belonging and bestowing: differing degrees of community involvement amongst rural elders in England and Wales. *European journal of ageing*, 10(4), pp. 325–333.

Davis, S., Crothers, N., Grant, J., Young, S. and Smith, K. (2012). Being involved in the country: productive ageing in different types of rural communities. *Journal of Rural Studies*, 28(4), pp. 338–346.

Farmer, J., Muñoz, S-A., Steinerowski, A. and Bradley, S. (2011). Health, wellbeing and community involvement of older people in rural Scotland. In: Q. Le, ed., *Health and Wellbeing: A Social and Cultural Perspective*. New York: Nova Science Publishers, pp. 127–142.

Glasgow, N., Min, H. and Brown, D. L. (2013). Volunteerism and social entrepreneurship among older in-migrants to rural areas. In: N. Glasgow and E. H. Berry, eds., *Rural Aging in 21st Century America*. New York: Springer, pp. 231–250.

Guo, Q., Bai, X. and Feng, N. (2018). Social participation and depressive symptoms among Chinese older adults: a study on rural – urban differences. *Journal of affective disorders*, 239, pp. 124–130.

He, Q., Cui, Y., Liang, L., Zhong, Q., Li, J., Li, Y., Lv, X. and Huang, F. (2017). Social participation, willingness and quality of life: a population-based study among older adults in rural areas of China. *Geriatrics & Gerontology International*, 17(10), pp. 1593–1602.

Heley, J. and Jones, L. (2013). Growing older and social sustainability: considering the 'serious leisure' practices of the over 60s in rural communities. *Social & Cultural Geography*, 14(3), pp. 276–299.

Hodgkin, S. (2012). 'I'm older and more interested in my community': older people's contributions to social capital. *Australasian Journal on Ageing*, 31(1), pp. 34–39.

Hoskins, B. L. and Mascherini, M. (2009). Measuring active citizenship through the development of a composite indicator. *Social indicators research*, 90(3), pp. 459–488.

Jones, L. and Heley, J. (2016). Practices of participation and voluntarism among older people in rural Wales: choice, obligation and constraints to active ageing. *Sociologia Ruralis*, 56(2), pp. 176–196.

Kafková, M. P., Vidovićová, L. and Wija, P. (2018). Older Adults and civic engagement in rural areas of the Czech Republic. *European Countryside*, 10(2), pp. 247–262.

Keating, N., Eales, J. and Phillips, J. E. (2013). Age-friendly rural communities: conceptualizing 'best-fit'. *Canadian Journal on Aging*, 32(4), pp. 319–332.

Keating, N., Swindle, J. and Fletcher, S. (2011). Aging in rural Canada: a retrospective and review. *Canadian Journal on Aging*, 30(3), pp. 323–338.

Kuokkanen, R. (2018). The role of the local community in promoting discursive participation: a reflection on elderly people's meetings in a small rural community in Finland. *Journal of Public Deliberation*, 14(1), p. 9.

Levasseur, M., Cohen, A. A., Dubois, M-F., Généreux, M., Richard, L., Therrien, F-H. and Payette, H. (2015). Environmental factors associated with social participation of older adults living in metropolitan, urban, and rural areas: the NuAge study. *American Journal of Public Health*, 105(8), pp. 1718–1725.

Liu, J., Rozelle, S., Xu, Q., Yu, N. and Zhou, T. (2019). Social engagement and elderly health in China: Evidence from the China Health and Retirement Longitudinal Survey (CHARLS). *International Journal of Environmental Research and Public Health*, 16(2), p. 278.

Majee, W., Aziato, L., Jooste, K. and Anakwe, A. (2018). The graying of rural America: community engagement and health promotion challenges. *Health Promotion Practice*, 19(2), pp. 267–276.

Mendes, F. (2013). Active ageing: a right or a duty? *Health Sociology Review*, 22(2), pp. 174–185.

Mettenberger, T. and Küpper, P. (2019). Potential and impediments to senior citizens' volunteering to maintain basic services in shrinking regions. *Sociologia Ruralis*, 59(4), pp. 739–762.

Munoz, S-A., Farmer, J., Warburton, J. and Hall, J. (2014). Involving rural older people in service co-production: is there an untapped pool of potential participants? *Journal of Rural Studies*, 34, pp. 212–222.

Neville, S., Adams, J., Napier, S., Shannon, K., and Jackson, D. (2018). Engaging in my rural community: perceptions of people aged 85 years and over. *International journal of Qualitative Studies on Health and Well-Being*, 13(1), p. 1503908.

Nonaka, K., Fujiwara, Y., Watanabe, S., Ishizaki, T., Iwasa, H., Amano, H., Yoshida, Y., Kobayashi, E., Sakurai, R. and Suzuki, H. (2019). Is unwilling volunteering protective for functional decline? The interactive effects of volunteer willingness and engagement on health in a 3-year longitudinal study of Japanese older adults. *Geriatrics & Gerontology International*, 19(7), pp. 673–678.

Rozanova, J., Keating, N. and Eales, J. (2012). Unequal social engagement for older adults: constraints on choice. *Canadian Journal on Aging*, 31(1), pp. 25–36.

Ryser, L. and Halseth, G. (2014). On the edge in rural Canada: the changing capacity and role of the voluntary sector. *Canadian Journal of Nonprofit and Social Economy Research*, 5(1), pp. 41–56.

Skinner, M. W. (2014). Ageing, place and voluntarism: towards a geographical perspective on third sector organisations and volunteers in ageing communities. *Voluntary Sector Review*, 5(2), pp. 161–179.

Skinner, M., Joseph, A., Hanlon, N., Halseth, G. and Ryser, L. (2016). Voluntarism, older people, and ageing places: pathways of integration and marginalization. In: M. Skinner and N. Hanlon, eds., *Ageing Resource Communities: New Frontiers of Rural Population Change, Community Development and Voluntarism*. London: Routledge, pp. 60–76.

Skinner, M. W., Joseph, A. E., Hanlon, N., Halseth. G. and Ryser, L. (2014). Growing old in resource communities: exploring the links among voluntarism, aging, and community development. *The Canadian Geographer*, 58(4), pp. 418–428.

Skinner, M. W. and Winterton, R. (2018). Interrogating the contested spaces of rural aging: implications for research, policy, and practice. *The Gerontologist*, 58(1), pp. 15–25.

Stephens, C., Breheny, M. and Mansvelt, J. (2015). Volunteering as reciprocity: beneficial and harmful effects of social policies to encourage contribution in older age. *Journal of Aging Studies*, 33, pp. 22–27.

Takagi, D., Kondo, K., and Kawachi, I. (2013). Social participation and mental health: moderating effects of gender, social role and rurality. *BMC public health*, 13(1), p. 701.

United Nations (2015). *World Population Ageing*. New York, Department of Economic and Social Affairs, Population Division.

Walker, A. (2002). A strategy for active ageing. *International Social Security Review*, 55(1), pp. 121–139.

Walsh, K., O'Shea, E., Scharf, T. and Shucksmith, M. (2014). Exploring the impact of informal practices on social exclusion and age-friendliness for older people in rural communities. *Journal of Community Applied Social Psychology*, 24(1), pp. 37–49.

Warburton, J. (2015). Volunteering and ageing. In: J. Twigg and W. Martin, eds., *Handbook of Cultural Gerontology*. London: Routledge, pp. 345–352.

Warburton, J. and Winterton, R. (2017). A far greater sense of community: the impact of volunteer behaviour on the wellness of rural older Australians. *Health & Place*, 48, pp. 132–138.

Wiersma, E. C. and Koster, R. (2013). Vulnerability, volunteerism, and age-friendly communities: placing rural northern communities into context. *Journal of Rural and Community Development*, 8(1). Available at: https://journals.brandonu.ca/jrcd/article/view/683

Winterton, R. and Warburton, J. (2014). Healthy ageing in Australia's rural places: the contribution of older volunteers. *Voluntary Sector Review*, 5(2), pp. 181–201.

Winterton, R., Warburton, J., Keating, N., Petersen, M., Berg, T. and Wilson, J. (2016). Understanding the influence of community characteristics on wellness for rural older adults: a meta-synthesis. *Journal of Rural Studies* 45, pp. 320–327.

Yuasa, M., Ukawa, S., Ikeno, T. and Kawabata, T. (2014). Multilevel, cross-sectional study on social capital with psychogeriatric health among older Japanese people dwelling in rural areas. *Australasian Journal on Ageing*, 33(3), pp. E13-E19.

24 A critical view of older voluntarism in ageing rural communities

Prospect, precarity and global pandemics

Amber Colibaba, Mark Skinner and Elizabeth McCrillis

Introduction

In an era of global population ageing, communities are facing the challenge of supporting their older residents, often through an increased reliance on volunteers and volunteer-based programs. Nowhere is the support of voluntarism more pivotal, yet precarious, then in rural and small town environments. As illustrated earlier in this book (see Ryser et al., Chapter 13; Menec and Novek, Chapter 14) the past two decades of research have highlighted how rural communities are addressing the challenges associated with rural population ageing. In a key feature of this scholarship, researchers from around the world have demonstrated how volunteers and volunteer-based programs are essential components of the fabric of rural communities. Not only does such voluntarism support older residents ageing in place, it can act as a catalyst for positive community development, as the interactions with individuals and communities alike help to sustain services within ageing rural communities (Davies, Lockstone-Binney and Holmes, 2018; Joseph and Skinner, 2012; Winterton and Warburton, 2014). Emerging critical approaches to understanding rural ageing and voluntarism are becoming sensitive to how individual experiences and interactions influence the complexity of ageing rural communities. This has highlighted how both older peoples' experiences and the dynamics of ageing communities are mediated through voluntarism as a means of both integration and marginalisation (Joseph and Skinner, 2012; Skinner et al., 2016). In particular, during times of global crises such as economic recessions, natural disasters and health care emergencies, such as the worldwide COVID-19 pandemic, the emergent element of precarity among voluntarism in rural communities becomes particularly exacerbated. Normally, volunteers supporting older residents are older themselves and are at risk of burnout given the downloading of responsibility onto volunteer-based programs (Skinner, 2014). Colibaba and Skinner (2019) have described the phenomena "older voluntarism", in which individual, older volunteers' activities and voluntary organisations featuring an older volunteer base provide essential services and supports to ageing communities, which has created fears of uncertainty regarding rural service sustainability and for the precarity of older volunteers themselves. Extending research on this phenomena, this chapter reviews the scholarship on voluntarism in ageing rural communities and recent developments towards critical approaches. It builds upon this concept to understand how acute crises such as pandemics may transform the everyday challenges of older voluntarism into graver situations, as older volunteers may begin to require additional community supports themselves and, further, become less able or unable to support vulnerable older adults. In this manner, acute crises magnify the importance of further exploring and understanding a critical view of rural ageing that seeks to elicit integrated, inclusive understandings of the connection between older people and ageing communities vis-a-vis voluntarism. The chapter concludes with a commentary on the prospect and precarity of older voluntarism in post-pandemic rural communities and what this means for rural ageing in the future.

Rural ageing and voluntarism

For the past two decades, the burgeoning field of rural ageing studies has made persistent calls for greater attention to the complexities of ageing in rural communities as these settings cope with unprecedented demographic change (Keating, 2008; Milbourne, 2012; Scharf, Walsh and O'Shea, 2016). Examining the rural ageing demographic brings to light critical questions about both the suitability and sustainability of ageing in place in rural communities (Keating, 2008). These questions speak to a situation poignantly referred to by Joseph and Cloutier-Fisher (2005, p. 137) as the "double jeopardy of vulnerable persons in a vulnerable community". Here, the experiences of older residents are seen as linked to the trajectories of the rural communities in which they age in place. A critical outcome of this double jeopardy has been the increasing responsibility placed upon rural service care providers, volunteers and both formal and informal caregivers for supporting ageing rural residents (Russell et al., 2019; Skinner, 2014).

Important for developing a critical understanding of ageing in rural communities, another outcome of the growing interest in rural ageing has been a call for a greater understanding of the diverse and complex experiences of older people as they relate to the dynamics of place in the communities where they live (Keating, Swindle and Fletcher, 2011; Keating et al., Chapter 5). Examining the integration between older people and ageing places provides a means of understanding both the experiences of individual volunteers and the development trajectories of ageing rural communities (Ryser et al. Chapter 13; Skinner and Hanlon, 2016). Indeed, rural ageing studies are already revealing how this integrative approach can act as a window into the diversity of older people's experiences, such as new retirees versus long-term residents, for example (Winterton and Warburton, 2012).

Parallel to the emergence of rural ageing studies has been an increasing interest within the social sciences to understand how the activities of volunteers and voluntary organisations relate to individual experiences of and community responses to ageing (Warburton, 2015). In the rural context, voluntarism has been seen as mediating the implications of population ageing for individuals' well-being, exemplified through increased participation in older age (see Winterton and Warburton, Chapter 23). This may be facilitated through community vitality, for instance in expanding services and supports (Lovell, 2009), which, in turn, may facilitate ageing in place and positive community development (Skinner and Hanlon, 2016; Winterton and Warburton, 2014).

Recent efforts to better understand this fundamental role of voluntarism have highlighted ageing rural volunteers, and challenges associated with voluntarism in ageing rural communities, as a key issue both for rural residents ageing in place and for the sustainability of ageing rural communities (Colibaba and Skinner, 2019; Davies, Lockstone-Binney and Holmes, 2018; Gieling and Haartsen, 2016; Jones and Heley, 2014; Walsh and O'Shea, 2008). The widespread prevalence of older voluntarism in rural communities – with organisations comprised of a pool of primarily older volunteers – creates growing uncertainty among many rural leaders. This is due to the inherent challenges of progressively burdening the voluntary sector given complications to service delivery sustainability, wherein volunteer populations are ageing at a similar rate to their patrons (Ryser and Halseth, 2014).

A critical view of rural older voluntarism

Out of the growing interest in rural ageing has emerged a concerted effort among key scholars to advocate for careful attention to developing critical perspectives that are sensitive to the diversity of older people's experiences – culturally, geographically, socio-economically – and their interactions with and influence on the complexity of ageing societies (e.g., Burholt and Dobbs, 2012; Scharf, Walsh and O'Shea, 2016; Skinner and Winterton, 2018). Indeed, recent calls for critical perspectives in rural gerontology emphasize fundamental gaps in our understanding of the macro processes that impact ageing communities and older rural residents. This includes the influence of rurality within structures that empower and disenfranchise older people; the differential roles and capacity of rural older people in creating and contesting their environments; and, perhaps most reflexively, the absence of marginalised older people, such as Indigenous, older lesbian, gay, bisexual, intersex or queer (LGBTIQ) and racialised older persons, within rural ageing scholarship, policy debates and public discourse (see Winterton et al., Chapter 29).

What this has meant for rural ageing studies of voluntarism is the adoption and development of new approaches. These emerging critical approaches seek to understand voluntary sector activities in support of older people, and volunteering by older residents, as intimately connected to their culturally diverse experiences, within particular settings and environments and among various societal processes and structures. Examples include Warburton and Winterton's (2010) examination of rural volunteering, role identity and cultural transition; Skinner and Hanlon's (2016) interrogation of the transformative role of voluntarism in resource hinterland communities; and Walsh, O'Shea and Scharf's (2019) multidimensional conceptualisation of rural social exclusion in older age (see also Burholt et al., 2013).

Of particular significance to building a critical view of older voluntarism in ageing rural communities has been an emerging emphasis on understanding of how the interactions between the diverse experiences of older people and complex dynamics of their ageing communities are mediated through voluntarism (Joseph and Skinner, 2012). This understanding is facilitated both positively, as a means of *integration*, and negatively, as a means of *marginalisation* (Skinner et al., 2016). From this perspective, voluntarism is understood as mutually transformative for older people and their communities, whereby volunteer-based programs and volunteering are seen to connect the interests of older residents (e.g., healthy ageing in place) within their community's development trajectories (e.g., retirement economy diversification) (Hanlon et al., 2014; Skinner, 2014). Also from this perspective, however, voluntarism has been observed and problematised as a very real means by which some older people are excluded from social participation as volunteers and community development priorities (Colibaba and Skinner, 2019; Skinner et al., 2016). In this way, voluntarism can be viewed, in some cases, as a means of othering particular groups of rural older people (Chalmers and Joseph, 2006). Through this body of work, Skinner, Joseph and their colleagues (Skinner et al., 2014; Skinner et al., 2016) have provided a new lens towards a more nuanced, critical understanding. Led by the innovative thinking of Alun Joseph, their perspective is sensitive to how integration and marginalisation are often explicit within volunteer-based initiatives, and how voluntarism is often complicit in integrating some older people and groups within ageing rural communities, and in marginalising others.

These critical views raise important questions about the prospects of voluntarism in ageing rural communities. In these spaces, the demographic reality of ageing rural populations means that most rural volunteers are older residents themselves, often coping with the precarity of age-related health and mobility issues combined with the burden of care and burnout associated with rural volunteering. Concern for the latter is informed by recent conceptualisations of, and investigations into, the multiple dimensions of "precarity in later life" (Grenier et al., 2020), attention to which has the potential to further advance our critical understanding of older voluntarism as a means of integration and marginalisation.

It is particularly important to pre-emptively identify factors that may magnify challenges associated with the precarious nature of older voluntarism in rural communities, as unexpected local, national or world events may be acutely experienced by older adults, as well as older volunteers and the rural communities dependent upon them. For example, the fallout of natural disasters, economic recession or a health crisis may be negatively experienced by all people in an affected area (see Carroll and Walker, Chapter 28); however, the dependency of rural areas on the day to day work of older volunteers may be amplified during these times of crises. Volunteers may face limitations to working in their communities or may be entirely unable to do so, especially given the pervasive nature of older voluntarism in conjunction with underlying medical and chronic conditions experienced by many older volunteers. Further, older adults who rely on volunteers may become progressively isolated or may be especially impacted by challenging events. In this way, it is critical to understand the precarity associated with older rural voluntarism as well as the prospects.

Prospect and precarity of rural older voluntarism in an era of pandemic

With the emergence of the COVID-19 pandemic, it is increasingly important to understand its impacts on the older adult population and their communities. For older adults, and those with underlying medical and chronic conditions who are at a higher risk of serious illness and death from contracting the COVID-19 virus, everyday implications of the pandemic may be more serious (WHO, 2020) (see also Carroll and Walker, Chapter 28). For example, Reynolds (2020) has referred to the pandemic as a "focusing event" that has highlighted the specific challenges associated with the pandemic for older adults, primarily stemming from the "extreme social consequences of ageism" (p. 1), which have clearly hampered efforts to support older adults who are generally more vulnerable to the virus (Ehni and Wahl, 2020). While recent literature points to the importance of understanding the impacts of COVID-19 on long-term care settings and policies (Behrens and Naylor, 2020; Béland and Marier, 2020) and assisted living communities (Dobbs, Petersen and Hyer, 2020), nowhere is the need to understand the impacts of COVID-19 on older voluntarism more pressing than in underserviced small towns and rural communities.

Due to the vulnerability of older adults, government enforced social and physical distancing measures are critical; however, recent international literature and media reports have shown that distancing measures are creating new, and often insurmountable challenges in supporting older people and for volunteers themselves, including a risk of increased feelings of isolation (Armitage and Nellums, 2020; Henning-Smith, 2020; Hill, 2020; Ireland, 2020; Muscedere, 2020). In response to this risk, the UN (2020) stated,

> although physical distancing is necessary to reduce the spread of the disease, if not implemented with supports in place, it can also lead to increased social isolation of older persons at a time when they may be at most need of support.

Importantly, this emphasises the need to ensure local and community responses are in place to help counter these feelings of isolation. For older adults, isolation can be particularly prominent and impactful, as many older adults live alone, where especially in rural communities, technology access may be limited (see Kosurko et al., Chapter 27 in this volume for more on rural ageing and technology). This may increase the risk of depression and anxiety among isolated older adults, especially among those who lack family or friends nearby and/or who may rely on the support of voluntary services (Armitage and Nellums, 2020). Community-based organisations have demonstrated their ability to quickly adapt to providing modified services to older adults (Wilson et al., 2020); however, it is unclear how this success has translated in rural settings, given additional social and demographic challenges. Municipalities have also demonstrated quick and effective coordination with their local voluntary sector relevant to older adults (Angel and Mudrazija, 2020) but limitations associated with small, rural tax bases (often comprised of older adults who are offered reduced taxation rates) and larger geographic settlement areas may challenge rural municipalities to apply this flexibility. With voluntary-based programs and activities often relied upon to counter feelings of isolation (Colibaba and Skinner, 2019), physical distancing measures provide an added challenge – as many are temporarily (or permanently) shut down.

Understanding the impacts that COVID-19 and physical distancing measures pose not only on rural older adults more generally, but older volunteers and volunteer-based programs, is of importance to researchers, local governments and volunteer-based programs and voluntary organisations alike. To push these developing themes forward, we suggest some emergent questions to help frame future thinking about how we can explore older voluntarism in an era of global pandemic and post-pandemic. Table 24.1 outlines these questions and how they may guide inquiry in research, policy and practice. Emergent study of the experiences of older adults, older volunteers and ageing rural communities can support rural ageing studies in uncovering the personal impacts of the pandemic. Further, they may clarify how ageing rural communities continue to support their ageing populations in a time where support is of utmost importance and may help identify what supports look like post-pandemic. Technology has been relied upon as a critical tool for people and programs to adapt to life in a pandemic and post-pandemic environment; however, there are additional challenges associated with translating

Table 24.1 Emergent questions about rural older voluntarism in an era of global pandemic

Emergent questions for research	*Emergent questions for policy*	*Emergent questions for practice*
How are older rural volunteers coping with the physical distancing measures implemented to help stop the spread of COVID-19? How are rural communities working to help support their older residents during the pandemic? What is older voluntarism in a post-pandemic era?	How can local health policy focus directly on reducing the risk of social isolation caused by physical distancing measures? What policies can volunteer-based programs and organisations implement to ensure that older volunteers are able to continue with voluntary activities, while adhering to physical distancing measures?	What are the challenges facing volunteer-based programs supporting older rural residents during COVID-19? How can local governments help to reduce social isolation of older rural adults and volunteers during a global pandemic? What organisation, community and individual solutions are in place in rural communities to mediate the impacts of COVID-19 on older voluntarism (during and post)?

this to an older adult population (Xie et al., 2020). Furthermore, shifting to complete technological reliance in rural environments where internet services may not be dependable or even available may compound challenges associated with the evolving logistics of voluntarism during a pandemic. Questions relating to supports and programs can aid in developing policies and practical programs to help older volunteers and volunteer-based programs become co-dependent, in ways where older adults may continue to volunteer and thus sustain the volunteer-program, all while following physical distancing measures.

To tackle the questions posed by the 2020 global pandemic, critical approaches to understanding the implications of COVID-19 on older voluntarism and ageing rural communities are being established in rural gerontology, advancing avenues for future research on the subject. With a call by the UN (2020) that urges governments to consult with older adults in policy decisions that affect their lives and to put in place supportive measures that guarantee their inclusion in society, researchers too can align future endeavours to this call. By aligning policy and research in this way, it highlights the importance of research conducted both *with* and *for* older adults, ensuring that their voices help inform policies that will directly affect their lives (for example, see Farmer, Hill and Muñoz, 2012).

A Canadian example

One example that helps illustrates the importance of understanding the implications of COVID-19 on older voluntarism is from Trent University (Canada), led by this chapter's co-authors and working in partnership with the administration of a local rural township. The project outlined here aimed to understand the impacts of and responses to the challenges of physical and social distancing on rural older volunteers and the volunteer-based programs they support and rely on. Using a case study of three rural volunteer-based programs in the township, namely a library, a fire department and a housing initiative, the research examined the experiences and perspectives of older volunteers and program administrators within these voluntary sectors on the COVID-19 pandemic. Preliminary findings from the research on the impacts of COVID-19 on older voluntarism begin to answer some of the emergent questions outlined in Table 24.1.

Based in the Township of Selwyn (pop. 17,060) in Peterborough County, Ontario, the three programs are ideal for researching rural older voluntarism as they serve a typical Canadian rural municipality, comprised of three communities that each showcase different rural North American typologies; Bridgenorth (rural-recreational), Ennismore (rural-agriculture) and Lakefield (small town) (Statistics Canada, 2017). Selwyn Township is also ideal as the three programs are comprised mostly of volunteers, many of them older (over the age of 65). With ethics approval from the Trent University Research Ethics Board and informed consent from participants, the project involved interviews with administrators and volunteers from each program, who are quoted anonymously in the sections that follow to illustrate their personal experience and perspective of the implications of the COVID-19 pandemic.[1]

Older volunteer experiences

Interviews with older volunteers reveal their experiences during this time of uncertainty and how they are coping. For some, voluntary activities have changed little since COVID-19. The Abbeyfield House Society of Lakefield, a volunteer-based housing initiative, has continued their monthly steering committee meetings, switching from face-to-face to online, virtual meetings. In reflecting on the switch, and the lack of stability of rural technology, one Abbeyfield volunteer stated, "It's been great. Rural internet is sketchy but when there's a will, there's a way" (Female, early 70s).

Volunteers at the Selwyn Fire Department have continued responding to emergency calls throughout the pandemic, although they described new policies around Personal Protective Equipment (PPE) and the ability for the older volunteers to self-identify (meaning, deciding whether or not they would like to continue to respond to calls). One firefighter describes his experience with deciding whether to continue volunteering during this time:

> I must be a true volunteer because if someone is in need, I want to be there, so yes I go to all the calls. The department tried to remind me how old I was, but I know that we're protected so I don't fret on that. I just love to volunteer because you never know, it could be your neighbour or your family who needs help, but if they need help, they need help and I'll be there for them.
>
> (Male, late 70s)

Another participant reflected upon whether she could volunteer in a post-pandemic era. As a volunteer at the Selwyn Public Library, she has been unable to fulfil her weekly shift at the library. Although she missed the library and her fellow volunteers, she was apprehensive about returning and how her new role would look:

> I do miss it and will be glad when I can volunteer at the library again, but I might be a little nervous. . . . I'm careful when I go to the grocery store, but I'm not sure how it's going to work in the library with the patrons and my volunteering partner.
>
> (Female, early 80s)

Although she is excited to return to the library, she wonders how she will be supported when she returns.

Volunteer-based programs perspectives

Interviews with program administrators reveal how, amid the COVID-19 pandemic, their programs are ensuring older rural residents and older volunteers remain supported. Preliminary findings demonstrate that even in a time of uncertainty, there is a "show must go on" attitude to ensure residents continue to receive the services they know and rely upon. For example, although the library has suspended all volunteer activity in their branches, staff are tirelessly working to restart library services, as stated by the library administrator, "Older community members rely on our services, so [the library] is working hard to provide it to them in some shape or form" (Female, mid 30s).

Our findings also shed light on promising practices that volunteer-based programs are implementing to ensure that their older volunteers are able to participate, while still adhering to physical and social distancing measures. Whether it is new policies surrounding the use of PPE, self-isolation or the use of online meeting platforms, volunteer-based programs are ensuring their older volunteers are at a low-risk of contracting the virus. For example, a fire department administrator emphasised how important these policies are to his volunteers, as they continue to respond to emergency calls around the township. Regarding continually evolving policies, he stated, "You gotta keep looking ahead and keep your nose in the news so you know how to react and adapt" (Male, late 50s).

As the preliminary data shows, volunteers and volunteer-based program administrators are beginning to feel the implications of the COVID-19 pandemic and these implications are adding to the both the integration and marginalisation aspects of older voluntarism. Interviews reveal how volunteers are remaining active, despite their age; however, they may later become disenfranchised by fears of post-pandemic volunteering. Future research looking at the impacts of the COVID-19 pandemic must take into account both the prospects and precarity of older voluntarism and whether a global pandemic enhances the integration of older adults and rural communities, or increases the marginalisation already felt by many. In particular, questions of diminished sense of community or community integration, or how fear of contracting or spreading the virus may impact the ability to volunteer, may lead to marginalisation of currently integrated volunteers, further enhancing the precarity and dynamics surrounding rural ageing.

Concluding comments

Rural communities worldwide rely on volunteers and volunteer-based organisations. While an essential component to the fabric of ageing rural and small town communities, older voluntarism is faced with key challenges, namely in sustaining a volunteer base comprised of older adults, who are supporting their peers as they age in place. To further the line of inquiry in understanding older voluntarism in ageing rural communities, this chapter shared critical approaches, emergent questions and promising practices on how these connections and supports can be understood in an era of global pandemic, where the challenges of older voluntarism are heightened due to government-implemented physical and social distancing measures.

The chapter contributes to our understanding of rural ageing and ageing rural communities through the construct of voluntarism. The literature on voluntarism reveals how volunteering in older age is linked to older peoples' increased health and participation in their communities, as volunteering provides a means of building and maintaining social networks and supports (Winterton and Warburton, 2014; Warburton, 2015). Additionally, rural ageing literature points to the critical role voluntarism plays in ageing rural communities as a means of providing care and support to older rural residents and positively contributing to rural community development (Skinner, 2014; Skinner and Hanlon, 2016). Critically examining these themes, the chapter demonstrates that in an era of global pandemic, the dynamics of rural older voluntarism can change, as physical distancing measures may inhibit voluntary support provision, especially face-to-face interactions. Additionally, as older rural residents commonly act as volunteers, their increased vulnerability to COVID-19 puts older voluntarism and volunteer-based programs at risk of potential closure. For the older volunteer, a global pandemic can increase individual precarity through disenfranchisement and diminished sense of community that may occur alongside health-related decisions not to volunteer.

The chapter also provides contributions to the burgeoning field of critical rural gerontology and the calls for understanding the macro-level processes that impact ageing communities and older rural residents. Research underway in Canada, and elsewhere, addresses these gaps by examining the emergent questions surrounding older voluntarism in ageing rural communities in an era of global pandemic. These critical approaches and future research will produce new knowledge on how older volunteers are coping with the changes brought on by COVID-19 and the ways in which both policy and volunteer-based programs help to support older volunteers and ageing rural population. Outcomes from the Canadian rural township study examining three examples of volunteer-based initiatives not only provides new knowledge on the impacts of COVID-19 on older voluntarism in ageing rural communities, insights relevant to other research studying similar challenges and issues. These rural perspectives may also be applicable to their urban counterparts, potentially revealing how COVID-19 may differentially impact older voluntarism across various geographical locales. Last, the methodological approach may be applied to different realms of the voluntary sector; for example, in community care, recreation and heritage.

Such research endeavours into understanding the challenges and opportunities of older voluntarism and ageing rural communities adapts and expands a conceptualisation of "precarity in later life" (Grenier et al., 2020), illuminating the implications of older voluntarism as a pivotal, multi-facetted and complex dimension of rural ageing and rural community development. It allows research to uncover and understand both the prospect (rewards) and precarity (risk) of older voluntarism, and the role of older volunteers in particular, as a means of creating positive outcomes for ageing in place and community development in ageing rural communities.

Acknowledgements

Funding for the research presented in this chapter was provided, in part, by the Trent University Office of Research and Innovation's Rapid Response to COVID-19 Pandemic Grant program and the Canada Research Chairs program (Mark Skinner).

Note

1 For more information about this and related studies in Selwyn Township, see Colibaba and Skinner (2019) and Rutherford et al. (2018), and www.trentu.ca/ruralaging

References

Angel, J. L. and Mudrazija, S. (2020). Local government efforts to mitigate the novel coronavirus pandemic among older adults. *Journal of Aging & Social Policy*, 32(4–5), pp. 439–449.

Armitage, R. and Nellums, L. B. (2020). COVID-19 and the consequences of isolating the elderly. *The Lancet*, 5, p. e256.

Behrens, L. L. and Naylor, M. D. (2020). "We are alone in this battle": a framework for a coordinated response to COVID-19 in nursing homes. *Journal of Aging & Social Policy*, 32(4–5), pp. 316–322.

Béland, D., and Marier, P. (2020). COVOID-19 and long-term care policy for older people in Canada. *Journal of Aging and Social Policy*, 32, pp. 358–364.

Burholt, V. and Dobbs, C. (2012). Research on rural ageing: where have we got to and where are we going in Europe? *Journal of Rural Studies*, 28(4), pp. 432–446.

Burholt, V., Scharf, T. and Walsh, K. (2013). Imagery and imaginary of islander identity: older people and migration in Irish small-island communities. *Journal of Rural Studies*, 31, pp. 1–12.

Chalmers, A. I. and Joseph, A. E. (2006). Rural change and the production of otherness: the elderly in New Zealand. In: P. Cloke, T. Marsden and P. Mooney, eds., *The Handbook of Rural Studies*. London: SAGE Publications, pp. 388–400.

Colibaba, A. and Skinner, M. W. (2019). Rural public libraries as contested spaces of older voluntarism in ageing communities. *Journal of Rural Studies*, 70, pp. 117–124.

Davies, A., Lockstone-Binney, L. and Holmes, K. (2018). Who are the future volunteers in rural places? Understanding the demographic background characteristics of non-retired rural volunteers, why they volunteer and their future migration intentions. *Journal of Rural Studies*, 60, pp. 167–175.

Dobbs, D., Petersen, L. and Hyer, K. (2020). The unique challenges faced by assisted living communities to meet federal guidelines for COVID-19. *Journal of Aging & Social Policy*, 32(4–5), pp. 334–342.

Ehni, H-J. and Wahl, H-W. (2020). Six propositions against ageism in the COVID-19 pandemic. *Journal of Aging & Social Policy*, 23(4–5), pp. 515–525.

Farmer, J., Hill, C. and Muñoz, S-A., (eds). (2012). *Community Co-Production: Social Enterprise in Remote and Rural Communities*. Cheltenham: Edward Elgar.

Gieling, J. and Haartsen, T. (2016). Liveable villages: the relationship between volunteering, and liveability in the perceptions of rural residents. *Sociologia Ruralis*, 57(S1), pp. 576–597.

Grenier, A., Phillipson, C., and Settersten, R. A., (eds). (2020). *Precarity and Ageing: Understanding Insecurity and Risk in Later Life*. Bristol: Policy Press.

Hanlon, N., Skinner, M. W., Joseph, A. E., Ryser, L. and Halseth, G. (2014). Place integration through efforts to support healthy aging in resource frontier communities: the role of voluntary sector leadership. *Health and Place*, 29, pp. 132–139.

Henning-Smith, C. (2020). The unique impact of COVID-19 on older adults in rural areas. *Journal of Aging and Social Policy*, 32, pp. 396–402.

Hill, A. (2020). Elderly people set up support networks to cope with UK lockdown. *The Guardian*, 4 April. Available at: www.theguardian.com/society/2020/apr/04/elderly-people-in-uk-set-up-support-networks-to-tackle-coronavirus.

Ireland, N. (2020). How to help seniors get through the COVID-19 pandemic. *CBC News*, 4 April. Available at: www.cbc.ca/radio/whitecoat/the-dose-how-we-can-help-seniors-get-through-covid-19–1.5519909

Jones, L. and Heley, J. (2014). Practices of participation and voluntarism among older people in rural Wales: choice, obligation and constraints to active ageing. *Sociologia Ruralis*, 56(2), pp. 176–196.

Joseph, A. E. and Cloutier-Fisher, D. (2005). Ageing in rural communities: vulnerable people in vulnerable places. In: G. J Andrews and D. R Phillips, eds., *Ageing and Place: Perspectives, Policy and Practice*. London: Routledge, pp. 133–149.

Joseph, A. E., and Skinner, M. W. (2012). Voluntarism as a mediator of the experience of growing old in evolving rural spaces and changing rural places. *Journal of Rural Studies*, 28, 380–388.

Keating, N. (2008). *Rural Ageing: A Good Place to Grow Old?* Bristol: Policy Press.

Keating, N., Swindle, J. and Fletcher, S. (2011). Ageing in rural Canada: a retrospective and review. *Canadian Journal on Aging*, 30(3), pp. 323–338.

Lovell, S. A. (2009). Social capital: the panacea for community? *Geography Compass*, 3(2), pp. 781–796.

Milbourne, P. (2012). Growing old in rural places. *Journal of Rural Studies*, 28, pp. 315–317.

Muscedere, J. (2020). Seniors with frailty at greatest risk from virus. *Peterborough Examiner*, 23 March. Available at: www.thepeterboroughexaminer.com/opinion-story/9913972-seniors-with-frailty-at-greatest-risk-from-virus/

Reynolds, L. (2020). The COVID-19 pandemic exposes limited understanding of ageism. *Journal of Aging & Social Policy*, 32, pp. 499–505.

Russell, E., Skinner, M. W. and Fowler, K. (2019). Emergent challenges and opportunities to sustaining age-friendly initiatives: qualitative findings from a Canadian age-friendly funding program. *Journal of Aging and Social Policy*. Epub ahead of print 6 July. DOI: 10.1080/08959420.2019.1636595.

Rutherford, K., Pirrie, L., Smith, A., Jennings, N., Russell, E. and Marris, J. (2018). *A Community-Based Approach to Retirement Living Development Projects*. Lakefield, ON: Trent University, in partnership with AHSL.

Ryser, L. and Halseth, G. (2014). On the edge in rural Canada: the changing capacity and role of the voluntary sector. *Canadian Journal of Nonprofit and Social Economy Research*, 5(1), pp. 41–56.

Scharf, T., Walsh, K. and O'Shea, E. (2016). Ageing in rural places. In: M. Shucksmith and D. L Brown, eds., *Routledge International Handbook of Rural Studies*. London: Routledge, pp. 50–61.

Skinner, M. W. (2014). Ageing, place and voluntarism: towards a geographical perspective on third sector organizations and volunteers in ageing communities. *Voluntary Sector Review*, 5(2), pp. 161–179.

Skinner, M., and Hanlon, N., (eds). (2016). *Ageing Resource Communities: New Frontiers of Rural Population Change, Community Development and Voluntarism*. London: Routledge.

Skinner, M. W. Joseph, A. E., Hanlon, N., Halseth, G., and Ryser, L. (2014). Growing old in resource communities: exploring the links among voluntarism, aging, and community development. *The Canadian Geographer*, 58(4), pp. 418–428.

Skinner, M., Joseph, A., Hanlon, N., Halseth, G., and Ryser, L. (2016). Voluntarism, older people and ageing places. In: M. Skinner and N. Hanlon, eds., *Ageing Resource Communities: New Frontiers of Rural Population Change, Community Development and Voluntarism*. London: Routledge, pp. 38–54.

Skinner, M. W. and Winterton, R. (2018). Interrogating the contested spaces of rural ageing: implications for research, policy, and practice. *The Gerontologist*, 58(1), pp. 15–25.

Statistics Canada. (2017). *Selwyn, TP [Census subdivision], Census Profile: 2016 Census* (Statistics Canada Catalogue no. 98–316-X2016001). Ottawa, ON: Statistics Canada [Accessed 29 November 2017].

UN (2020). *Policy Brief: The Impact of COVID-19 on Older Persons*. New York: United Nations.

Walsh, K. and O'Shea, E. (2008). Responding to rural social care needs: older people empowering themselves, others and their community. *Health and Place*, 14(4), pp. 795–805.

Walsh, K., O'Shea, E. and Scharf, T. (2019). Rural old-age social exclusion: a conceptual framework on mediators of exclusion across the lifecourse. *Ageing & Society*. *Ageing & Society*. Epub ahead of print 2 July. DOI: 10.1017/S0144686X19000606.

Warburton, J. (2015). Volunteering in older age. In: J. Twigg and W. Martin, eds., *Handbook of Cultural Gerontology*. London: Routledge, pp. 345–352.

Warburton, J., and Winterton, R. (2010). The role of volunteering in an era of cultural transition: can it provide a rode identity for older people from Asian cultures? *Diversity*, 2, pp. 1048–1058.

WHO (2020). *Coronavirus Disease 2019 (COVID-19): Situation Report – 51*. Available at: www.who.int/docs/default-source/coronaviruse/situation-reports/20200311-sitrep-51-covid-19.pdf?sfvrsn=1ba62e57_8

Wilson, T. L., Scala-Foley, M., Kunkel, S. R. and Brewster, A. L. (2020). Fast-track innovation: area agencies on aging respond to the COVID-19 pandemic. *Journal of Aging & Social Policy*, 32, pp. 432–438.

Winterton, R. and Warburton, J. (2012). Ageing in the bush: the role of rural places in maintaining identity for long term rural residents and retirement migrants in north-east Victoria, Australia. *Journal of Rural Studies*, 28(4), pp. 329–337.

Winterton, R. and Warburton, J. (2014). Health ageing in Australia's rural places: the contribution of older volunteers. *Voluntary Sector Review*, 5(2), pp. 181–201.

Xie, B., Charness, N., Fingerman, K., Kaye, J., Kim, M. T. and Khurshid, A. (2020). When going digital becomes a necessity: ensuring older adults' needs for information, services, and social inclusion during COVID-19. *Journal of Aging & Social Policy*, 32, pp. 460–470.

25 Older people and poverty

Making critical connections in rural places

Paul Milbourne

Introduction

In a recent review of the changing state of research on rural ageing in Canada over the previous couple of decades, Keating, Swindle and Fletcher (2011) identify two key trends: first, the increased volume of rural gerontological scholarship and its engagement with a broader range of themes; and second, the development of more critical and sophisticated approaches to making sense of these themes. In relation to the focus of this chapter, their review indicates that the marginalisation of rural elders has emerged as a more significant research area during recent years, with attention given to older people at risk of poor health and those unable to access key services. However, they suggest that this increased academic interest in marginalisation has not really extended to issues of poverty, with the authors commenting that "the socio-economic status of older adults warrants updating. Lack of attention to employment of older workers, to processes of retirement, or to income security [of older people] means that little is known about the extent of rural privation" (334).

This situation is not unique to Canada nor to the time period covered by Keating, Swindle and Fletcher's (2011) review. More recent searches for published research on poverty among rural elders across a number of global North countries reveal that older people remain a relatively marginalised group within academic work on rural poverty and that poverty represents a rather peripheral aspect of research within rural gerontology. In addition, almost all of the published research on poverty among rural elders has been focused on the UK and the US. That this situation still exists is somewhat surprising for a couple of reasons. First, there have been important previous studies of poverty among older people in rural places that should have acted as springboards for further scholarship. Second, programmes of welfare reform and the implementation of austerity policies in many global North countries during the last couple of decades have shrunk the welfare state, cut public sector budgets and reduced levels of social support, which have created additional difficulties for impoverished and disadvantaged groups within society, including the elderly.

My intention with this chapter is to raise the profile of poverty among older people in rural places by providing a critical review of published research in this area across the last four decades. The chapter is structured around five sections. The first provides an overview of the changing scale, profile and geographies of rural elders living in poverty, based on analyses of official statistics and survey data from several countries. Second, consideration is given to processes of marginalisation and the lived experiences of older people in poverty in rural areas. Attention in the third section shifts to the complex relations between poverty and social exclusion among rural elders. Fourth, the impacts of welfare reform and austerity on rural elders' poverty are discussed. The chapter concludes by setting out an agenda for future research on older people poverty in rural areas, highlighting gaps in the extant research evidence base and new opportunities for future scholarship.

Statistical scales, geographies and profiles

Poverty among rural elders first began to be recognised by researchers in the 1980s. One of the first published accounts was provided by Bradley (1986), based on a survey of rural deprivation in five rural localities in England (see McLaughlin, 1986). The research revealed that elderly households were "far and away the most vulnerable to poverty" (Bradley, 1986: 164), with between 40 and 67% of single elderly men and 45–85% of single elderly women living in or on the margins of poverty across the study areas. A decade later, a major household survey in 16 rural areas of England and Wales confirmed the numerical significance of the elderly within the rural poor population, with elderly single households comprising 42% and elderly couple households 28% of all households living in or on the margins of poverty. Furthermore, the survey showed that 86% of retired households in poverty were solely reliant on the state pension as their income source (see Cloke, Goodwin and Milbourne, 1997a; Cloke, Milbourne and Thomas, 1997b; Milbourne, 2004).

The early 1990s also witnessed the publication of the first evidence on the scale and profile of poverty among elders in rural America (Rural Sociological Society Task Force on Persistent Rural Poverty, 1993; Dudenhefer, 1994; Glasgow, 1993; McLaughlin and Jensen, 1995; Rowles and Johansson, 1993). Not only did this work reveal that a significant minority of older households were living below the poverty line but that poverty levels among older people were higher in rural areas than in cities. For example, Glasgow (1993) states that while the poverty rate among older people in rural areas had fallen during recent decades, it had remained consistently above that recorded for urban areas, and in 1990, 16.1% of older people were in poverty in non-metropolitan areas compared with 10.8% in metropolitan areas and 14.6% in central cities. Other statistical analyses provided detail on the characteristics of poverty among older households in rural America. An analysis by McLaughlin and Jensen (1995) highlighted that rural elders spent more time in poverty in their retirement years than their counterparts in metropolitan areas. In addition, it was found that there was a "lower likelihood of poverty exit among nonmetro than metro elders" (Jensen and McLaughlin, 1997: 467). Further work explored the geographies of elderly poverty within rural America, with Rogers (1999) commenting that "among non metro counties, the poverty rate for older persons increases with greater rurality – ranging from 12.8 percent for counties of 20,000 population, adjacent to a metro area, to 20.6 percent for non adjacent, completely rural counties" (11).

For Glasgow (1993), higher levels of poverty in rural America were seen to result from the structure of local labour markets and particularly the dominance of smaller employers offering lower wage jobs and fewer occupational pension opportunities. It is also the case that rural elders are less likely to have health insurance than those in metropolitan areas as a result of the structure of local economies (Butler, 2006). Other analyses of the geographies of elderly poverty in the US highlight additional distinctive features of this poverty in rural places. For example, Smith and Trevelyn (2019) reveal that "the rural older population was less racially and ethnically diverse, less likely to live in nursing homes, and less likely to have educational attainment beyond a high school degree than their urban counterparts" (20).

Further spatial analyses of official income data have provided more recent comparisons of pensioner poverty rates in rural and urban areas, although these have been largely focused on the UK and the US. Beginning with the UK, Palmer (2009) calculated that older households constitute 21% of all low-income households in rural districts in England compared with 14% in urban districts. However, a more recent analysis of poverty rates in England (Department for Food, Environment and Rural Affairs, 2019) reveals a different geographical picture, with 17% of pensioner households in rural areas living in relative poverty (before housing costs) in 2017–18 compared with 19% in urban areas. As with the situation in the US, the highest rates of poverty among older people would appear to exist in remoter rural areas of the UK (see also Gilbert, Philip and Shucksmith, 2006). In addition to these analyses of income data, work undertaken on costs of living reveals that elderly households are generally faced with higher levels of expenditure in rural places, with these resulting largely from the increased reliance on private transport and higher domestic fuel costs (Smith, Davis and Hirsch, 2010).

In the US, recent analyses of official poverty data provide a similarly confusing account of the significance of older person poverty in rural and urban areas. A report by the US Department of Agricultural Economic Research Service (2019) finds that 10.1% of older people in rural areas are living in poverty, which is higher than the rate in metropolitan areas (9.1%). By contrast, an analysis of official income data for the period 2012–16, by Smith and Trevelyn (2019), reveals that 8.4% of the older rural population are living below the poverty line compared with 9.6% in urban areas. That said, their more detailed spatial account of rural – urban variations indicates a higher poverty rate among elders in "completely rural counties" (11.4%) than in "mostly urban counties" (9.2%).

Beyond the US and the UK, there exists little published data on levels of poverty among rural elders. In fact, the only other country where such evidence exist is Australia. Again, the situation regarding the scale of elder poverty in rural places is rather complex. Using a 50% of average income indicator, 12% of people aged 65 years or older in rural areas are shown to be living in poverty compared with 13% for the whole of the population, but adopting a 60% of mean income threshold reveals a higher rural poverty rate, with 28% of older people in rural areas living in poverty compared with 21% for the overall population. It is clear then that further work is needed to develop spatial analyses of poverty data in a larger number of countries in order to generate a more robust account of the scale of poverty among older people in rural areas. What is also evident is that more attention needs to be given to the ways in which the use of different thresholds of poverty produces particular rural–urban distributions of poverty among older populations in different countries.

Marginalisation processes and the lived experiences of low income

Beyond these statistical assessments, attention has been given to the lived experiences of poverty among older people in rural areas as well as processes of marginalisation bound up with this poverty. It is clear that rural elders in poverty remain a marginalised group in several senses (see Milbourne and Doheny, 2012). They are denied any real visibility in physical terms, living in isolated dwellings and in small towns and villages scattered across the countryside. They remain largely hidden within official mappings of poverty that tend to highlight proportions (and thus spatial concentrations) rather than absolute numbers of people living in poverty. The older rural poor are also largely absent within dominant political, policy and media discourses of both poverty and rurality, where there remains a fixation on the more visible forms of poverty in the inner cities. In addition, narratives of idyll continue to dominate cultural constructions of rurality in certain countries, within which elderly people are seen to be well-cared for within close-knit communities (Scharf and Bartlam, 2008). Indeed, it would seem that Bradley's (1986) assessment of the marginalised position of elderly poverty in the English countryside remains largely true today:

> The old and white in village England cannot even claim political recognition. Fragmented, weak and deprived of social networks, their fate is to serve-out their final years in loneliness and isolation, their silent poverty disrupted only by the occasional disappearance of further services and the social landmarks of their lives.
>
> (171)

In a similar vein, writing more recently about single elderly women living on low income in rural America, Reig (2002) refers to them as the "unspoken poor" (284), commenting that "in the final stages of their lives, they live alone and must make do on inadequate fixed-incomes. They are women who have found themselves unaccompanied and impoverished by the cycle of life, and must now struggle to survive" (258).

Turning to the lived experiences of poverty among rural elders, research indicates that the "normalisation" of low income during working-age years extends into retirement, with older households generally accepting the careful management of household finances, low standards of living and a strong sense of self-sufficiency as key components of living on low incomes in their retirement years (Scharf and Bartlam, 2006, 2008). In terms of the last of these features, Scharf and Bartlam (2006) report that poverty tends also to be "internalised" so that "even where finances were stretched, participants sought to manage without seeking additional support from the state, relying instead on the help of informal sources" (3). Similarly, a recent study of older persons living on low income in rural Wales (Milbourne and Doheny, 2012; Doheny and Milbourne, 2014) reveals that material hardships were largely constructed as being an essential component of rural living, with struggles associated with low income frequently restricted to the private sphere, perpetuating the "silence of poverty" among older people in rural places. Moreover, Milbourne and Doheny (2012) comment that many of the older people living below the officially recognised poverty threshold in their study were often unwilling to recognise the presence of poverty within their household and community.

This reliance on informal support and denials of poverty relate to particular cultures and moral discourses of rurality (see Milbourne, 2014; Sherman, 2006). As Bradley (1986) argues in the context of rural England, attitudes towards (elderly) poverty in many rural places have been shaped by historical ideologies of paternalism and self-sufficiency, which have led to the limited acceptance of formal welfare services in many rural places and an unhelpful reliance on informal, and often inadequate, forms of social support. These attitudes have been reinforced by political structures and processes in rural places, with the (historical) control of councils by conservative politicians leading to a less developed local welfare support systems (Milbourne, 2004) and producing what Bradley (1986:169) refers to as the "self-fulfilling prophesy" that "rural society dispenses its own welfare". A similar situation has been reported in rural America, with Rowles and Johansson (1993) suggesting that older people's "resistance to assistance from the outside . . . is part of the historically ingrained and socially reinforced rural culture" (364).

It is claimed that place plays a particularly important role in shaping experiences of poverty in later life in rural areas. Glasgow (1993) argues that elderly poverty is largely linked to the specific attributes of rural places, including small and dispersed settlement structures, spatial isolation, the limited capacity of local government and the local low-wage economy. Such attributes act to "constrain lifelong opportunities for individual economic accumulation; limit effective public sector responses to the elderly's social and economic needs; and, in certain instances, reduce elders' access to informal helper networks that can provide social, economic, and psychological support" (Glasgow, 1993: 310). Other research points to a more complex relationship between older people, poverty and place, involving strong and supportive social networks that provide older people with high levels of social participation but also feelings of isolation emanating from the withdrawal of services and retail facilities (see Scharf and Bartlam, 2006; Milbourne and Doheny, 2012).

It is also the case that the older poor in rural areas tend to report high levels of satisfaction with the social contexts of place, with the close-knit nature of rural living, the density of informal local support networks and inclusion within local society being frequently referenced. As Milbourne and Doheny (2012) comment, "in these senses, it is probably more appropriate to point to the connections between material poverty and social inclusion rather than social exclusion within the lives of the older poor in rural places". In addition, landscape and proximity to nature and open spaces are often cited by rural elders as positive aspects of rural living, offering a form of "psychic income" as well as spaces of escape from everyday material hardships (Doheny and Milbourne, 2014).

It is widely acknowledged that ageing is accompanied by processes of "spatial constriction" (Rowles, 1978), with the life worlds of older people becoming ever more focused on the local spaces of the community and the home. Consequently, it is claimed that changes to the physical, social and cultural composition of older people's communities can have significant impacts on their everyday lives and senses of security and well-being. For example, the in-movement of more affluent "outsider" retirees can accentuate feelings of relative poverty as well as develop new forms of social isolation, while the reduction in local service and retail provision may accentuate older people's senses of physical and social isolation. In these terms, it is claimed that "aspects of place may in effect serve as [both] a domain and a mediator of exclusionary experiences" (Walsh et al., 2019: 9) as the "relative weighting given to problems and challenges of rural living [is dependent] on people's connection and relationship to their place" (Walsh et al., 2019: 355).

From poverty to social exclusion

It has long been recognised that the quality of life of rural elders is not just impacted by low income but also by "social isolation, lack of access to essential services, and lack of geographic mobility" (Glasgow, 1993: 316). More recently, attempts have been made to engage with the concept of social exclusion to explore these broader forms of disadvantage and marginalisation experienced by older people in rural areas. While there is not room here to discuss the various understandings of social exclusion, it is generally agreed that it adopts a more multi-dimensional, processual and dynamic approach to making sense of disadvantage than is poverty (Byrne, 1999; Walker and Walker, 1997), and that it approaches exclusion in terms of a series of inter-connected domains related to material resources, labour markets, social relations, essential services and civil participation.

It is claimed that social exclusion represents a particularly useful concept for making sense of the multiple forms of disadvantage experienced by older people (Scharf et al., 2001). However, the emphasis on working-age groups within mainstream research on social exclusion raises questions about the relevance of particular domains of exclusion in the context of older people. Clearly, participation in the labour market is a less significant indicator of social inclusion for older people, many of whom are retired, than working-age groups. In addition, the emphasis that social exclusion researchers place on the dynamics of low income – movements into and out of poverty – is less relevant to elderly households, where the persistence of low income in retirement is a more significant issue (Phillipson and Scharf, 2004; see also Scharf and Keating, 2012). It is also suggested that other dimensions of social exclusion need to be considered in relation to older people. Scharf et al. (2001), for example, propose that increased attention be given to their participation in social networks; structures and processes of institutional disengagement and, in particular, the withdrawal of everyday services; and older people's social constructions of local space.

More recently, Walsh et al. (2017, 2019) have sought to provide more sophisticated accounts of the relations between social exclusion and older people in rural places. Following a comprehensive review of published work on social exclusion and ageing, they propose a framework of older age exclusion composed of the following interconnected domains: material and financial resources; social relations; socio-cultural aspects; neighbourhood and community; civic participation; and services, amenities and mobility. Walsh et al. (2019) then proceed to examine this framework through a study of the experiences of rural elders in rural Ireland and Northern Ireland, producing a conceptual framework on rural old age social exclusion. This proposes five domains of exclusion related to financial and material resources, social relations, service infrastructure, transport and mobility and safety, security and crime. These domains are linked to a set of mediating factors – macro-economic forces, place and community, life course trajectories and individual capacities – that, Walsh et al. (2019) argue, "operate singularly or in combination to intensify or protect against domain-specific experiences of disadvantage" (19).

Welfare reform and austerity

The last two or three decades have witnessed the implementation of welfare reform programmes and, more recently, austerity policies in many global North countries. Important efforts have been made by national governments to cut public spending and to move to workfare policies, whereby welfare support is viewed mainly/solely in terms of bringing the unemployed into paid work. Although there has been some research on the impacts of welfare reform in rural areas (see Milbourne, 2010; Pickering et al., 2006), this has largely ignored older households. In addition, there has been virtually no work undertaken on rural impacts of austerity, let alone its impacts on the lives of poorer older people in rural places (although see Milbourne, 2016 and Walsh et al., 2019 for notable exceptions).

Perhaps it is understandable why research on welfare reform has largely excluded the elderly given the focus of such reform on working-age groups. This is not the case with austerity, which has impacted detrimentally on a broader range of low incomed and disadvantaged groups, including older people. Smaller government – a key component of the austerity agenda – has long been a feature of rural areas, with many of the welfare services taken for granted in the city largely absent in rural places and increased reliance placed on community and other informal systems of support in order to fill gaps in the local welfare system. In one sense, this makes rural places more resilient to the austerity agenda. In another, austerity may impact harder on rural places, with funding cuts leading not only to reductions in, but the closure of, already under-resourced welfare and other support services for older people.

Only one study has explicitly explored austerity in the context of impoverished older people in rural areas. Milbourne's (2016) research in four rural places in Wales indicates that the national austerity agenda has acted to reinforce longer-standing local policies and cultures of austerity in rural places, perpetuating the invisibility of rural poverty and making older groups on low incomes more reluctant to claim the state benefits to which they are entitled. However, this research also reveals some more positive local consequences of austerity, as public sector funding cuts have encouraged (or forced) local welfare agencies to rebalance responsibilities between the state, communities and individuals, which has led, in some cases, to the development of more effective policy interventions for both local government and older people. In particular, important efforts have been made to include older people and rural communities in policy deliberations over welfare support and to build on community assets to enable (vulnerable) older people to stay longer and lead more productive lives in their own homes and local communities.

Additional work needs to be undertaken in order to provide broader and deeper understandings of the impacts of austerity on the lives of older people in rural areas. This involves exploring austerity impacts in other countries, where different national austerity policy and rural contexts may have resulted in different outcomes to those mentioned earlier. It would also be worthwhile to consider how austerity has impacted on different types of rural places, for example, in (larger) accessible and (smaller) remote rural communities and in places characterised by different levels of general service provision and types of (formal and community-based) of welfare support. In addition, attention could be directed towards the impacts of austerity on particular vulnerable groups of the older poor population in rural areas, including single households, the "older old", those with health problems and older people with mobility difficulties. Following on from Milbourne's (2016) study, it would be useful to examine the ways in which the local state, voluntary groups, local communities and individuals have responded to the national austerity policy agenda in rural places. This would enable researchers to consider the extent to which national austerity has been implemented wholesale, moderated or resisted in different local contexts and in what ways local responses have produced more effective solutions to poverty among older people or reinforced local denials of poverty and cultures of self-sufficiency among the older poor in rural areas.

Conclusion

Returning to Keating, Swindle and Fletcher's (2011) review of rural gerontological research in Canada that began this chapter, it is clear that increased academic attention has been given to poverty among rural elders during the last decade. A larger body of statistical evidence now exists on the changing scale, profile and geography of low income among older people in rural areas. It is also apparent that we have a better understanding of the lived experiences of older life poverty, the broader forms of marginalisation, disadvantage and exclusion encountered by older people in rural places and some of the impacts of welfare reform and austerity on the provision of welfare support for rural elders. Another reading of the current situation, though, is less positive. Beyond the UK and the US, there has been little published work on the poverty of rural elders, and even in these two countries, such poverty still remains a rather marginal feature of recent scholarship in rural studies, rural gerontology and poverty studies. Indeed, previous references to the poverty of rural elders being "silent" (Bradley, 1986) or "unspoken" (Reig, 2002) within political and media discourse would appear equally relevant today in relation to academic scholarship.

It is clear that further research is required in order to develop more comprehensive, critical and sophisticated understandings of the scale, nature and experiences of poverty among older people in rural places, as well as the impacts of austerity and welfare policy interventions on such poverty. The geographical focus of research also needs to be broadened to provide a picture of older age rural poverty in a wider range of global North countries. For example, rural depopulation and higher pensioner poverty rates are evident in many central and eastern European countries (as well as in the global South), with the older poor very much a group that has been left behind in rural communities. In extending the evidence base on later life poverty in rural areas, it would be useful to develop more multi-scalar and multi-method research approaches that analyse official national datasets (where available), generate detailed survey-based materials on the (changing) incidence and profile of older age poverty in different spatial contexts, and provide in-depth, qualitative and place-based accounts of the lived experiences of poverty. It is also the case that further work is required on the rather complex relations between poverty, marginalisation and (social) exclusion in order to understand how the changing economic, technological[1], political, socio-cultural and natural contexts of particular rural places perpetuate and/or relieve situations of low income among older residents.

Note

1 The term technological is used here to cover such things as the digitisation of services, internet-based shopping and advances in mobile technologies.

References

Bradley, T. (1986). Poverty and dependency in village England. In: P. Lowe, T. Bradley and S. Wright, eds., *Deprivation and Welfare in Rural Areas*. Norwich: Geo Books, pp. 151–178.

Butler, S. S. (2006). Low income, rural elders' perceptions of financial security and health care costs. *Journal of Poverty*, 10(1), pp. 25–43.

Byrne, D., (ed.). (1999). *Social Exclusion*. Buckingham: Open University Press.

Cloke, P., Goodwin, M. and Milbourne, P., (eds). (1997a). *Rural Wales: Community and Marginalization*. Cardiff: University of Wales Press.

Cloke, P., Milbourne, P. and Thomas, C. (1997b). Living lives in different ways? Deprivation, marginalisation and changing lifestyles in rural England. *Transactions of the Institute of British Geographers*, 22(3), pp. 210–230.

Department of Agricultural Economic Research Service. (2019). *Rural poverty and well-being*. https://www.ers.usda.gov/topics/rural-economy-population/rural-poverty-well-being/ [Accessed 21 January 2020].

Department for Food, Environment and Rural Affairs. (2019). *Rural poverty – to 2017/18*. https://www.gov.uk/government/statistics/rural-poverty [Accessed 20 January 2020].

Doheny, S. and Milbourne, P. (2014). Older people, low income and place: making connections in rural Britain. In: V. Burholt, C. Hennessy and R. Means, eds., *Countryside Connections: Older People, Community and Place in Rural Britain*. Bristol: Policy Press, pp. 193–220.

Dudenhefer, P. (1994). Poverty in the United States. *Rural Sociologist*, 14(1), pp. 4–25.

Gilbert, A., Philip, L. J. and Shucksmith, M. (2006). Rich and poor in the countryside. In: P. Lowe and L. Speakman, eds., *The Ageing Countryside: The Growing Older Population of Rural England*. London: Age Concern Books, pp. 63–93.

Glasgow, N. (1993). Poverty among rural elders: trends, context, and directions for policy. *Journal of Applied Gerontology*, 12(3), pp. 302–319.

Jensen, L. and McLaughlin, D. K. (1997). The escape from poverty among rural and urban elders. *The Gerontological Society of America*. 37(4), pp. 462–468.

Keating, N., Swindle, J. and Fletcher, S. (2011). Aging in rural Canada: a retrospective and review. *Canadian Journal on Aging*. 30(3), pp. 323–338.

McLaughlin, B. (1986). The rhetoric and reality of rural deprivation. *Journal of Rural Studies*, 2(4), pp. 291–307.

McLaughlin, D. K. and Jensen, L. (1995). Becoming poor: the experiences of elders. *Rural Sociology*, 60(2), pp. 202–223.

Milbourne, P., (ed.). (2004). *Rural Poverty: Marginalisation and Exclusion in Britain and the United States*. London: Routledge.

Milbourne, P., (ed.). (2010). *Welfare Reform in Rural Places: Comparative Perspectives*. Bingley: Emerald Press.

Milbourne, P. (2014). Poverty, place and rurality: material and socio-cultural disconnections. *Environment and Planning A*, 46(3), pp. 566–80.

Milbourne, P. (2016). Austerity, welfare and older people in rural places: competing discourses of voluntarism and community? In: M. Skinner and N. Hannon, eds, *Ageing Resource Communities: New Frontiers of Rural Population Change, Community Development and Voluntarism*. London: Routledge.

Milbourne, P. and Doheny, S. (2012). Older people and poverty in rural Britain: material hardships, cultural denials and social inclusions. *Journal of Rural Studies*, 28(4), pp. 389–397.

Palmer, G. (2009). *Indicators of Poverty and Social Exclusion in Rural England: 2009*. London: Commission for Rural Communities.

Pickering, K., Harvey, M. H., Summers, G.F and Mushinksi, D., (eds). (2006). *Welfare Reform in Persistent Rural Poverty: Dreams, Disenchantments, and Diversity*. State College, PA: Penn State University Press.

Phillipson, C. and Scharf, T. (2004). *The Impact of Government Policy on Social Exclusion among Older People*. London: Office of the Deputy Prime Minister.

Reig, M. L. (2002). The unspoken poor: single elderly women surviving in rural America. *The Elder Law Journal*, 9, pp. 257–284.

Rogers, C. C. (1999). *Changes in the Older Population and Implications for Rural Areas* (Rural Development Research report number 90). Washington, DC: Food and Rural Economics Division, Economic Research service, U.S. Department of Agriculture.

Rowles, G. (1978). *Prisoners of Space? Exploring the Geographical Experience of Older People*. Boulder, CO: Westview Press.

Rowles, G. and Johansson, H. K. (1993). Persistent elderly poverty in rural Appalachia. *Journal of Applied Gerontology*, 12(3), pp. 349–367.

Rural Sociological Society Task Force on Persistent Rural Poverty, (ed.). (1993). *Persistent Poverty in Rural America*. Boulder, CO: Westview Press.

Scharf, T. and Bartlam, B. (2006). *Rural Disadvantage: Quality of Life and Disadvantage amongst Older People: A Pilot Study*. London: Commission for Rural Communities.

Scharf, T. and Bartlam, B. (2008). Ageing and social exclusion in rural communities. In: N. Keating, ed., *Rural Ageing: A Good Place to Grow Old?* Bristol: Policy Press, pp. 1–11.

Scharf, T., Bartlam, B., Hislop, J., Bernard, M., Dunning, A. amd Sim, J., (eds). (2006). *Necessities of Life: Older People's Experiences of Poverty*. London: Help the Aged.

Scharf, T. and Keating, N., (eds). (2012). *From Exclusion to Inclusion in Old Age: A Global Challenge*. Bristol: Policy Press.

Scharf, T., Phillipson, C., Kingston, P. and Smith, A. E. (2001) Social exclusion and older people: exploring the connections. *Education and Ageing*, 16(3), pp. 303–320.

Sherman, J. (2006). Coping with rural poverty: economic survival and moral capital in rural America. *Social Forces*, 85, pp. 891–913.

Smith, N., Davis, A. and Hirsch, D. (2010). *A Minimum Income Standard for Rural Households*. London: Commission for Rural Communities; Joseph Rowntree Foundation.

Smith, A. S. and Trevelyn, E. (2019). *Older Population in Rural America*. United States Census Bureau. Available at: www.census.gov/library/stories/2019/10/older-population-in-rural-america.html [Accessed 20 January 2020].

Walker, A. and Walker, C. (1997). *Britain Divided: The Growth of Social Exclusion in the 1980s and 1990s*. London: CPAG.

Walsh, K., O'Shea, E. and Scharf, T. (2019). Rural old-age social exclusion: a conceptual framework on mediators of exclusion across the lifecourse. *Ageing & Society*. *Ageing & Society*. Epub ahead of print 2 July. DOI: 10.1017/S0144686X19000606.

Walsh, K., Scharf, T. and Keating, N. (2017). Social exclusion of older people: a scoping review and conceptual framework. *European Journal of Ageing*, 14, pp. 81–98.

26 Critical perspectives on mental health, dementia and rural ageing

Rachel Herron and Eamon O'Shea

Introduction

Research on mental health, dementia and rural ageing remains dominated by biomedical approaches, in spite of the growth and diversity of critical perspectives in gerontology more broadly. By locating the problem of dementia within the brain and the body, biomedical approaches play a powerful role in directing science and society to focus on prevention, diagnosis and treatment, sometimes at the expense of understanding social causes and responses to dementia (Keohane and Grace, 2019). We begin this chapter from a different starting point by acknowledging that dementia is not just a growing health condition; it is also a disability. Although some people with a dementia diagnosis may not see themselves as disabled, they become disabled by the social attitudes, policies and practices in their surrounding environments (Shakespeare et al., 2019). Disability is the overlapping territory of mental health, dementia and rural ageing, but studies of rural dementia care have been slow to embrace the insights of disability scholarship. Over the last three decades, critical gerontologists have critiqued the disabling effects of biomedical discourses and called for more attention to be paid to the world outside the brain and the body (Gaines and Whitehouse, 2006). By choosing disability as the starting point for this chapter on critical rural gerontology and dementia care, we advocate for a deeper understanding of how places, policies and programs affect the opportunities of diverse people living with dementia (see Hennessy and Innes, Chapter 17 in this volume).

Rural places can present distinct challenges for different people living with dementia. Research on rural dementia care has identified challenges related to rural service use including distance to services, weak or non-existent public transportation systems, weather conditions, shrinking social support networks and socio-cultural beliefs (Morgan, Innes and Kosteniuk, 2011). In addition, people living in rural places often face difficulties getting a diagnosis, wait times to see a specialist, limited and rigid service options and lack of educational resources (Dal Bello-Haas et al., 2014). Some Canadian studies indicate that people living in rural areas with dementia typically use fewer community-based services because of competing priorities, inappropriate services, not knowing how to access services, stigma and privacy issues (Bayly et al., 2020; Forbes et al., 2006). Although the availability of formal health and social supports is important, there is a need to more carefully consider opportunities for personhood and social inclusion beyond formal service use in rural places.

Place can be a nebulous concept, but for many older people it is linked to the space, real or imagined, that people call home. Places are dynamic, changing in meaning across the lifecourse as opportunities, circumstances, distances and relationships evolve over time (Wiles et al., 2012). Ageing *in place* is often associated with living at home in the community, with some level of independence, rather than in residential care (Davey et al., 2004; Davey, 2006). Ageing in place is often preferred because of the costs of institutional care. Thus, it is often favoured by policy makers, health providers and by older people themselves (Spasova et al., 2018). Two of the United Nations General Assembly (1991) principles for older people specifically relate to this concept: older persons should be able to live in environments that are safe and adaptable to personal preferences and changing capacities; and older persons should be able to reside at home for as long as possible.

Remaining in rural places is often difficult to achieve, mainly due to the absence of home and community supports (Herron and Rosenberg, 2019). Economies of scale arguments are frequently invoked to explain spatial differences in social service provision for people with dementia living in rural areas compared to those living in urban areas (O'Shea, 2009). It is often considered too costly to provide the same level of service provision to people living with dementia in rural areas because they are too scattered and sometimes live in remote places (O'Shea and Monaghan, 2017). This cost argument is not just applied to health and social care provision for people with dementia. It is part of a wider efficiency rationale, which is commonly used to justify the rationalisation and curtailment of more general social provision in rural areas (Wenger, 2001; Walsh et al., 2012). The primacy of the scale argument is rarely contested in Ireland and Canada because the public expenditure debate has never moved beyond an economic efficiency framework (O'Shea and Keane, 2002; O'Shea, 2009; O'Shea et al., 2012). While equity and citizenship may still matter for government, they do not matter at any cost.

In this chapter, we focus on how critical gerontological perspectives on place and relationships can expand the study of dementia and rural ageing. Specifically, we discuss person-centred care in research and policy and the need to understand the diversity of rural people and contexts. Next, we outline several under-theorised areas of rural dementia care research: specifically studies of social location and diversity among people living with dementia and social inclusion and meaningful engagement beyond familial relations of care. We conclude with a careful consideration of supply side and demand-side measures for mobilising rural communities to support innovative approaches to the social production of dementia care. In doing so, we attempt to draw together the logic of economics and questions of equity and citizenship.

Critical perspectives on person-centred care

Over the last 30 years there has been a growing awareness that effective medical and social responses to dementia must be person-centred; that means, recognising the history and identity of the individual rather than seeing actions and behaviours solely as a product of disease (Kitwood, 1997). In gerontology, person-centred care represents a fundamental paradigm shift towards understanding the person and recognising personhood. The latter of which refers to the enduring selfhood of the person living with dementia, including the expression of emotions, character and preferences in the most advanced stages of dementia (Kontos, 2004). Although personhood is at the centre of person-centred philosophy, it is not always central in policy and practice (Hennelly and O'Shea, 2019). For example, a recent UK policy review questioned reductionist applications of person-centred care in the form of personalisation and individualisation, which emphasize choice and control over social care through personal budgets (Manthorpe and Samsi, 2016). Importantly, individual choice, control and identity are not the only values of person-centred care; other values include love, attachment and inclusion, but there has been a tendency to focus on the individual rather than more relational values (Fazio et al., 2018). As a consequence, person-centred care has been criticised for failing to consider the importance of relationships and context to the provision of quality care (Kontos et al., 2017).

In response to the limitations of person-centred care, there has been a growing emphasis on understanding the relationality of care; "good care" takes place within a range of relationships to family, friends, cultural groups, institutions and the state. There are many different forms of engagement that people living with dementia can exercise in these different relationships. Individual rights, expression and retention of purpose and belonging can only take place within the context of relationships. In addition, such relational qualities are highly place-specific; for example, sense of belonging may look very different throughout one's lifecourse. Furthermore, sense of belonging may be complicated by the multiple social locations that people identify with. As such, critical rural gerontology must draw on person-centred, relational and place-based approaches to critically examine the experiences and outcomes of policies and programs for people living with dementia and carers in rural contexts.

Personhood and social location

The intersections of multiple social positions and their influence on experiences of dementia in rural places are undertheorised (Hulko, 2009). The term "people with dementia" has dominated dementia care research for several decades, receiving little criticism until recently. For example, Bartlett and colleagues (2016) have called for critical feminist research on dementia to shed light on how disability is influenced by gendered locations and relations. Worldwide women make up a larger proportion of people living with dementia (approximately two-thirds of those with a diagnosis in Canada and Ireland are women) as well as a larger proportion of people caring for someone with dementia (Alzheimer Society of Canada, 2019; Pierce, Cahill and O'Shea, 2013). Gender is a very important social location and consideration for dementia care planning, programming and policy – it influences how diverse people living with dementia experience ageing in place – yet it is seldom given consideration in national dementia care strategies and much of the research on rural dementia care.

Studies exploring gender and caregiving in rural Canada and Ireland have shown that female carers for a person living with dementia typically report a greater burden and severity of distress than male carers (McDonnell and Ryan, 2014; Stewart et al., 2016). In contrast, male carers typically struggle to facilitate the autonomy of a female spouse living with dementia and are more likely to place their spouse in a long-term residential care facility earlier in the disease process (Boyle, 2013). Gender also influences the day to day activities of people living with dementia, their participation in the community and their service use and acceptance of community services (Herron and Rosenberg, 2017). For example, some men find day program activities emasculating or inconsistent with rural life ways. In contrast, gendered responsibilities (e.g., cleaning, cooking and other chores) can prevent women from accessing support. Gender is just one social location that influences experiences of ageing in place and social inclusion. Understanding the social location and relationships of people living with dementia is important in advancing person-centred care and addressing health inequities and the differential power relations that shape access to support as well as inclusion in rural communities.

The diversity of rural people and places and the implications for effective services, service use and broader participation in the community are not well understood. Rural people living with dementia are not only genderless in vast majority of research; it is also generally assumed that they share a rural cultural identity with less consideration of other forms of cultural diversity such as sexuality, "race" and ethnicity. For example, in rural Canada, there is still a relative lack of research with Indigenous communities about their experiences and understandings of dementia, in spite of the fact that studies have shown a higher prevalence of dementia in Indigenous communities compared to non-Indigenous communities (Jacklin et al., 2013). Continuing colonial policies and discrimination challenge the personhood and quality of care for Indigenous people living with dementia, particularly the centralisation of supports outside Indigenous communities (Jacklin et al., 2015). Place has an intrinsic link with the preservation of cultural lifeways, including eating traditional foods and engaging with cultural support networks (Pace and Grenier, 2017). Addressing the health inequities and exclusion experienced by older Indigenous peoples in rural areas involves attentiveness to place, power relations and culture. Moreover, supporting personhood for diverse people living with dementia requires an understanding of how people became the people they are through their bonds with cultural groups and places.

Improving social inclusion and meaningful engagement

Most research and support services focus on familial relationships of care; however, this may inadvertently isolate both people living with dementia and carers in their homes (Herron et al., 2019). Furthermore, people living with dementia and family carers often need different forms of support and do not necessarily agree on what is needed, which places strain on informal care relationships. To develop more sustainable care in rural contexts, researchers and policy makers must support the maintenance of other relationships and spaces of care outside the home. Social inclusion provides a way of conceptualising this need.

At the individual level, social inclusion refers to an individuals' ability to participate in particular places and relationships with others (Wilton et al., 2018). In this way, it links the individual experience to the local environment and relationships beyond informal and formal care. Many people living with dementia want to continue to, not only participate in their communities but also contribute through paid, voluntary and community work; yet formal support services are often geared towards helping rather than enabling more reciprocal relationships of care. For diverse people living with dementia to feel like services are appropriate (particularly in earlier stages), their capacity to contribute as well as removing barriers to contributing must be an integral part of planning and programming (Herron and Rosenberg, 2017).

Many people living with dementia actively resist exclusion; however, they face prescribed disengagement where they are told to stop working, stop driving and plan for future decline (Swaffer, 2015). Unlike other disabilities, there is much less emphasis on accommodating people living with dementia at a structural level. Resisting the isolation and exclusion associated with the loss of a driver's license in rural places can be particularly difficult because of the distances associated with rural living. Walking can enable people living with dementia to continue to get out in the neighbourhood and enjoy leisure activities as well as socialize. Although several studies have identified some tensions between safety and independence as people living with dementia continue walking in isolated areas and small towns, close social ties in the community can enable people living with dementia to continue activities as well as maintain their safety (Clarke and Bailey, 2016; Herron and Rosenberg, 2017).

Rural places are often associated with stronger social ties, familiarity with place and familiarity among people, which can be supportive of social inclusion, but this same familiarity can also make going out in the community more difficult (Clarke and Bailey, 2016). People living with dementia in rural communities are often treated differently by others in the community after disclosing their diagnosis. They experience pity and loss of autonomy when others take over activities because they assume people with a dementia diagnosis are unable. These social responses can lead to internalised stigma and self-withdrawal as a means of protecting ones-self. Studies have also identified difficulties continuing day-to-day activities such as shopping and banking, particularly because of expectations about the pace at which these activities should be done and challenges in relation to the ever-changing retail environment (Clarke and Bailey, 2016; Herron and Rosenberg, 2017). People living with dementia in rural (and urban communities) have felt stigmatised within these everyday environments, generally because of communication challenges. Addressing these challenges requires not only continued dementia education but also restructuring services in ways that accommodate people living with dementia and recognize dementia as a disability.

Dementia care should be sensitive and responsive to the needs, expectations and human rights of people with dementia. People with dementia need integrated, person-centred, accessible, affordable health and social care to maintain a level of functional ability consistent with their basic rights, fundamental freedoms and human dignity (WHO, 2017: 23). Prince et al. (2013) agree, characterising quality care as "the maintenance of personhood and well-being through a conducive physical and social care environment". Care should be comprehensive in the sense of supporting the person with dementia across their care journey. Importantly, Van der Roest et al. (2007) argue that health care systems should be reformed from service-based to "demand-directed care" and "experience-orientated care". An important first step, therefore, is to ascertain the preferences and expressed needs of people with dementia living in rural areas. In short, what do they want, when and where and delivered by whom? For most people, to feel included, connected and meaningfully engaged are central to their quality of life.

Mobilising rural communities and innovative approaches

Dementia care involves more than just health and social services, it involves mobilising rural communities and supporting innovative approaches that enable people living with dementia in rural areas. Programs must involve multi-sectoral and community-based collaborations that enhance every-day spaces and experiences for diverse people living with dementia. For example, Skinner and colleagues (2018) have recently examined the inclusive potential of video-streaming a dance program from urban to rural communities for people living with dementia and carers (see Kosurko et al., Chapter 27 for more about the technological implementation of this dance program). There are several innovative elements of this project that could be applied elsewhere, including engaging non-traditional sectors in social inclusion (e.g., looking outside health and social care to the arts sector), combining local and remote expertise to produce solutions and supporting forms of engagement that do not rely on verbal communication. Ultimately, mobilising rural communities to enable people living with dementia involves recognising the many different ways that people living with dementia can relate and connect with others as well as the different setting in which these relationships can take place. In Ireland, recent evaluations of Genio-funded demonstrator programmes have highlighted the advantages of a more community-oriented, personalised approach to care in rural areas (O'Shea and Monaghan, 2017), but it will take time for the overall culture to change in relation to home care provision for people with dementia.

When markets fail, or are thinly provided, and when the resulting public sector response is weak, or inadequate, social innovation is necessary to deal with the resulting impasse. We have seen in the previous section the importance of local responses to social needs as a means of addressing, albeit imperfectly, the needs of vulnerable people. Social entrepreneurs are an important source of social innovation. They identify social needs, but more importantly, they identify new ways of addressing these needs. Social entrepreneurs are likely to be complex, multi-dimensional people, who are driven by a need to do something about the social problems in their immediate areas. They find ways and means to address these problems through using under-utilised resources – people, buildings, equipment – and putting these to good effect to solve problems, particularly a lack of opportunities for people living with dementia in rural areas. They identify sources of funding for particular projects and find new ways of delivering services to people in need. Leadbeater (1997) describes the core assets of social entrepreneurs in terms of social capital, meaning relationships, networks, trust and co-operation, which gives them access to physical and financial capital. Social capital is associated with shared values, trust, networking and co-operation (Putnam, 1993), all of which are necessary to transform blighted social landscapes into dynamic living social organisms. The more extensive the social contacts, the more complex the networking, the greater the co-operation, then the more likely it is that social needs can be addressed in an effective way. Social entrepreneurs harness and develop social capital for social productive purposes.

Unfortunately, social entrepreneurs are in scarce supply, especially those interested in working with people living with dementia in rural places. This is not to deny that many communities have people who care and try to do something to respond to social needs related to dementia, as evident from the Genio social economy model in Ireland (O'Shea and Monagahn, 2015; Keogh et al., 2018). Voluntary effort is still highly valued in rural places (Skinner and Joseph, 2011), so a general framework does exist for the emergence of social entrepreneurship. Sometimes, however, communities find it difficult to get beyond the documentation of problems, and, even when they do, their response is often too isolated, too narrowly focused and too poorly funded to impact on the populations they wish to serve. The desire to effect change is there, but the skills, expertise and resources necessary to transform social life are underdeveloped. In Ireland, Genio overcame these problems through comprehensive and consistent action designed to support the creation of local social entrepreneurs and consortia, through the development of an institutional framework to support their activities and the provision of significant financial resources (O'Shea and Murphy, 2014). In the Genio project, training in organisation, management and business oriented skills was also used to encourage potential entrepreneurs and embryonic community groups to think in social enterprise terms in relation to dementia care, and, ultimately, to develop sustainable socially oriented projects (O'Shea and Monaghan, 2015).

The lesson from the Genio programme on dementia care in rural areas in Ireland is that social entrepreneurship can be mobilised to effect real change for people with dementia, especially those on the boundary between community-living and long-stay facilities (Cullen and Keogh, 2018). Extending this model for people with dementia living in rural areas will require the provision of seed capital and start-up grants for social production in dementia care, using similar schemes to those currently available to economic entrepreneurs in almost all countries. Interested parties should be given support in identifying latent commercial opportunities within the social care field and generating realistic business plans that match economic imperatives with the realities of social economy provision for people with dementia. Some failures will undoubtedly occur, but that is only to be expected given the nature of the production and the risks associated with new initiatives in this area. Very often, part of the concern of health professionals with the development of the social economy in social care provision is the potential for inferior quality of care to be delivered to prospective clients. But education and training programmes, coupled with the development of an appropriate regulatory structure can help to realise the quality of care that is fundamental to any new initiatives in the social economy sector. If such programmes can work for people with disabilities, there is no reason why they cannot be used to address the needs of people with dementia.

One approach to expanding the potential of the social economy in dementia care might be to introduce a designated social care voucher scheme whereby publicly financed vouchers could be cashed in return for care from dedicated social economy providers (O'Shea et al., 2017). As well as developing the potential of the social economy, designated voucher schemes would encourage autonomy and consumer choice in dementia, something that is absent under current provider-driven arrangements in many countries. We might find that older people with dementia and their families prefer to spend their vouchers on satisfying social and psychological needs rather than on medical or residential forms of care. This, in turn, might stimulate an appropriate supply side response within the social economy for psychosocial care, such as cognitive stimulation therapy, counselling, reminiscence, creativity programmes and so on (Cahill et al., 2012).

Calling for critical studies of rural dementia care

A more critical approach to rural dementia care is needed to address the continuing dominance of biomedical approaches in research, policy and programming. This approach would address the misapplication of person-centred care; that is programs and policies that focus on the individual without consideration of how the individual is positioned in the world in terms of gender, class, "race" and ethnicity. Research must move beyond the homogenising terminology of people with dementia. While some scholars may see rural places as particularly homogenous, this fails to appreciate the diversity of people and culture. If we do not consider these differences, we risk overlooking and reinforcing disparities among people living with dementia in rural places.

The sustainability of care is often framed in economic and medical terms, referring to the costs of formal care and the burden of informal care. What is missing is a more careful consideration of personhood and citizenship for people with dementia (Bartlett and O'Connor, 2010), particularly for those living in rural areas. Sustainable care requires a multidimensional approach to care beyond the family and the home. Social inclusion may provide a lens for evaluating and addressing sustainable care with attentiveness to individual experiences as well as structural barriers to living well with dementia. Social inclusion extends the focus of care beyond the interpersonal lens in which it is so often trapped, by looking at everyday environments and relationships with multiple people and groups. Finally, sometimes inclusion needs to be stimulated on the supply side through incentives for social entrepreneurship and on the demand side with real purchasing power. Both approaches may be necessary within rural communities to ensure that people with dementia get to exercise their personhood through a sustained attachment to place.

References

Alzheimer Society of Canada. (2019) *Dementia Numbers in Canada*. Available at: https://alzheimer.ca/en/Home/About-dementia/What-is-dementia/Dementia-numbers

Bartlett, R. and O'Connor, D., (eds). (2010). *Broadening the Dementia Debate: Towards Social Citizenship*. London: Policy Press.

Bartlett, R. L., Gjernes, T. K. O, Lotherington, A. T. and Obstfelder, A. (2016). Gender, citizenship and dementia care: a scoping review of studies to inform policy and future research. *Health and Social Care in the Community*, 26(1), pp. 14–26.

Bayly, M., Morgan, D., Froehlich Chow, A., Kosteniuk, J. and Elliot, V. (2020). Dementia-related education and support service availability, accessibility, and use in rural areas: barriers and solutions. *Canadian Journal on Aging*. Epub ahead of print 24 January. DOI: 10.1017/S0714980819000564.

Boyle G. (2013). Facilitating decision-making by people with dementia: is spousal support gendered? *Journal of Social Welfare and Family Law*, 35(2), pp. 227–243.

Cahill, S., O'Shea, E. and Pierce, M. (2012). *Creating Excellence in Dementia Care*. Dublin. Trinity College Dublin.

Clarke, C. L. and Bailey, C. (2016). Narrative citizenship, resilience and inclusion with dementia: on the inside or on the outside of physical and social places. *Dementia*, 15(3), pp. 434–452.

Cullen, K and Keogh, F. (2018). *Personalised Supports and Care for People with Dementia*. Dublin: Genio Ltd.

Dal Bello-Haas, V., Cammer, A., Morgan, D., Stewart, N. and Kosteniuk, J. (2014). Rural and remote dementia care challenges and needs: perspectives of formal and informal care providers residing in Saskatchewan, Canada. *Rural Remote Health*, 14(3), pp. 27–47.

Davey, J. (2006). 'Ageing in place': the views of older homeowners on maintenance, renovation and adaptation. *Social Policy Journal of New Zealand*, 27, p. 128.

Davey, J., de Joux, V., Nana, G. and Aracus, M. (2004). *Accomodation Options for Older People in Aoetearoa/New Zealand*. Wellington, New Zealand: Centre for Housing Research Aotearoa.

Fazio, S., Pace, D., Flinner, J. and Kallmyer, B. (2018). The fundamentals of person-centered care for individuals with dementia. *The Gerontologist*, 58(Suppl. 1), pp. S10–S19.

Forbes, D. A., Morgan, D. and Janzen, B. L. (2006). Rural and urban Canadians with dementia: use of health care services. *Canadian Journal on Aging*, 25(3), pp. 321–330.

Gaines, A. and Whitehouse, P.J. (2006) Building a mystery: Alzheimer's disease, mild cognitive impairment, and beyond. *Philosophy, Psychiatry and Psychology*, 13(1), pp. 61–74.

Hennelly, N. and O'Shea, E. (2019). Personhood, dementia policy and the Irish National Dementia Strategy. *Dementia: The International Journal for Research and Practice*, 18(5), pp. 1810–1825.

Herron, R., Funk, L. and Spencer, D. (2019). Responding the "wrong way": The emotion work of caring for a family member with dementia. *The Gerontologist*, 59(5), e470 – e478.

Herron, R. and Rosenberg, M. (2017). "Not there yet": examining community support from the perspective of people with dementia and their partners in care. *Social Science & Medicine*, 173, pp. 81–87.

Herron, R. and Rosenberg, M. (2019). Dementia in rural settings: examining the experiences of former partners in care. *Ageing & Society*, 39(2), pp. 340–357.

Hulko, W. (2009). From 'not a big deal' to 'hellish': experiences of older people with dementia. *Journal of Aging Studies*, 23(3), pp. 131–144.

Jacklin, K., Pace, J. and Warry, W. (2015). Informal dementia caregiving among indigenous communities in Ontario, Canada. *Care Management Journals*, 16(2), pp. 106–120.

Jacklin, K., Walker, M. and Shawande, J. (2013). The emergence of dementia as a health concern among first nations populations in Alberta, Canada. *Canadian Journal of Public Health*, 104(1), pp. E39–E44.

Keogh, F., Pierce, M., Neylon., Fleming, P., Carter, L., O'Neill, S. and O'Shea, E. (2018). *Supporting Older People with Complex Needs at Home: Evaluation of the Intensive Home Care Package Initiative*. Dublin: Genio.

Keohane, K., and Grace, V. (2019). What is 'Alzheimer's disease'? The 'Auguste d' case re-opened. *Culture, Medicine, and Psychiatry*, 43(2), pp. 336–359.

Kitwood, T., (ed.). (1997). *Dementia Reconsidered: The Person Comes First*. Buckingham: Open University Press.

Kontos, P. (2004). Ethnographic reflections on selfhood, embodiment and Alzheimer's disease, *Ageing & Society*, 24, pp. 829–849.

Kontos, P., Miller, K.L. and Kontos, A. P. (2017). Relational citizenship: supporting embodied selfhood and relationality in dementia care. *Sociology of Health and Illness*, 39(2), pp. 182–198.

Leadbeater, C., (1997). *The Rise of the Social Entrepreneur*. London: Damos.

Manthorpe, J. and Samsi, K. (2016). Person-centered dementia care: current practices. *Clinical interventions in aging*, 11, pp. 1733–1740.

McDonnell, E. and Ryan, A. A. (2014). The experience of sons caring for a parent with dementia. *Dementia*, 13(6), pp. 788–802.

Morgan, D., Innes, A. and Kosteniuk, J. (2011). Dementia care in rural and remote settings: A systematic review of formal or paid care. *Maturitas*, 68(1), 17–33.

O'Shea, E. and Keane, M. (2002). Social entrepreneurship and social services provision in Gaeltacht regions in Ireland. *European Research In Regional Science*, 12, pp. 58–75.

O'Shea, E. (2009) Rural ageing and public policy in Ireland. In: J. McDonagh, T. Varley and S. Shortall, eds., *A living countryside: The politics of sustainable development in rural Ireland*. Surrey: Ashgate, pp. 269–285.

O'Shea, E. and Murphy, E. (2014). *Evaluaton of the Genio Dementia Programme: Year 1*. Galway: Irish Centre for Social Gerontology.

O'Shea, E. and Monaghan, C. (2015). *Genio Dementia Programme Evaluation of Year 2*. Galway: Irish Centre for Social Gerontology.

O'Shea, E. and Monaghan, C. (2017). An economic analysis of a community-based model for dementia care in Ireland: a balance of care approach. *International psychogeriatrics*, 29, pp. 1175–1184.

O'Shea, E., Wash, K. and Scharf, T. (2012) Exploring community perceptions of the relationship between age and social exclusion in rural areas. *Quality In Ageing*, 13(1), pp. 16–26.

Pace, J. and Grenier, A. (2017). Expanding the circle of knowledge: Reconceptualising successful aging among north american older Indigenous peoples. *Journals of Gerontology Series B: Psychological Sciences and Social Sciences*, 72(2), 248–258.

Pierce, M., Cahill, S. and O'Shea, E. (2013). Planning dementia services: New estimates of current and future prevalence rates of dementia for Ireland. *Irish Journal of Psychological Medicine*, 30(1), pp. 13–20.

Prince, M., Bryce, R., Albanese, E., Wimo, A., Ribeiro, W. and Ferri, C. (2013). The global prevalence of dementia: a systematic review and meta analysis. *Alzheimer's and Dementia*, 9, pp. 63–75.

Putnam, R., (ed.). (1993). *Making Democracy Work: Civic Traditions in Modern Italy*. Princeton, NJ: Princeton University Press.

Shakespeare, T., Zeilig, H. and Mittler, P. (2019). Rights in mind: thinking differently about dementia and disability. *Dementia*, 18(3), pp. 1075–1088.

Skinner, M., Herron, R., Bar, R., Kontos, P. and Menec, V. (2018). Improving social inclusion for people with dementia and carers through sharing dance: a qualitative sequential continuum of care pilot study protocol. *BMJ Open*, 8(11), p. e026912.

Skinner, M. W. Joseph, A. E. (2011). Voluntarism within evolving spaces of care in ageing rural communities. *GeoJournal*, 76, pp. 151–162.

Spasova, S., Baeten, R., Coster, S., Ghailani, D., Peña-Casas, R. and Vanhercke, B. (2018). *Challenges in long-Term Care in Europe. A Study of National Policies, European Social Policy Network (ESPN)*. Brussels: European Commission.

Stewart, N., Morgan, D., Karunanayake, C., Wickenhauser, J., Cammer, A., Minish, D., . . . Hayduk, L. (2016). Rural caregivers for a family member with dementia: models of burden and distress differ for women and men. *Journal of Applied Gerontology*, 35(2), pp. 150–178.

Swaffer, K. (2015). Dementia and Prescribed Disengagement™, *Dementia*, 14(1), pp. 3–6.

United Nations General Assembly (1991). *United Nations principles for older persons*. Adopted by General Assembly resolution 46/91 of 16 December 1991. Available at https://www.ohchr.org/en/professionalinterest/pages/olderpersons.aspx.

Van der Roest, H. G., Meiland, F. J., Maroccini, R., Comijs, H. C. Jonker, C. and Droes, R-M. (2007). Subjective needs of people with dementia: a review of the literature. *International Psychogeriatrics*, 19(3), pp. 559–592.

Walsh, K., O'Shea, E. and Scharf, T. (2012) Ageing in changing community contexts: cross-border perspectives from rural Ireland and Northern Ireland. *Journal of Rural Studies*, 28(4), pp. 347–357.

Wenger, C (2001). Myths and realities og ageing in rural Britain. *Ageing & Society*, 21, p. 1170136.

Wiles, J., Leibing, A., Guberman, N., Reeve, J. and Allen, R. (2012). The meaning of "aging in place" to older people. *The Gerontologist*, 52(3), pp. 357–366.

Wilton, R., Fudge Schormans, A. and Marquis, N. (2018). Shopping, social inclusion and the urban geographies of people with intellectual disability. *Social and Cultural Geography*, 19(2), pp. 230–252.

WHO (2017). *Global Action Plan on the Public Health Response to Dementia 2017–2025*. Geneva: World Health Organisation.

27 Rural gerontechnology

Arts-based insights into rural ageing and the use of technology

An Kosurko, Mark Skinner, Rachel Herron, Rachel J. Bar, Alisa Grigorovich, Pia Kontos and Verena Menec

Introduction

Technology is traditionally positioned as a solution to challenges of care and support associated with ageing, and in the rural context there may lie many opportunities for this prevailing approach, given geographical and resource challenges. Yet ageing and technology scholars are invited to move beyond this interventionist logic that creates a negative view of ageing; to consider how the ageing experience is co-constituted by gerontechnology, as older people use, change and produce gerontechnology (Peine and Neven, 2019). An interdisciplinary field, gerontechnology aims to bring together developing technologies with needs and aspirations of older people in alignment with public health goals (Bronswijk et al., 2009). Some researchers have pointed to the importance of supporting the social nature of the human being through the use of gerontechnology (Pekkarinen et al., 2013). There has been a call for greater debate regarding the role of social scientists in the interdisciplinary development of the field and in interrogating images of older people that are embedded in gerontechnology design and use (Gallistl and Wanka, 2019). As rural populations are ageing more quickly around the world, researchers in the field of gerontechnology look to the rural context in anticipation of older-peoples' needs (O'Connell et al., 2018). Social isolation and social inclusion can be challenges related to rural ageing, and evidence suggests that information and communication technology (ICT) specifically (e.g., computers, smart phones, the internet) may, in some circumstances, be effective in addressing these challenges (Khosravi et al., 2016).

In rural ageing studies, researchers increasingly recognize the potential of technology to enable greater choices and preferences in relation to ageing in place (O'Shea, 2009). As outlined elsewhere in this volume (see, Burholt and Scharf, Chapter 6 in this volume) critical rural scholars emphasize the importance of deeper insights into the diverse perspectives of older people, with sensitivity to how different rural spaces, through their facilitation of technological infrastructure can influence the ageing experience and empower or disempower older people (Skinner and Winterton, 2018). Gerontechnology researchers have identified that older people in rural areas can experience a double digital divide due first to their lack of access to broadband or high-speed internet infrastructure and second, due to lower levels of digital literacy and familiarity that result from the lack of exposure to technology (O'Connell et al., 2018). For rural scholars, this exemplifies the relational and multi-level nature of the social exclusion of rural older people that manifests in rural areas – in obstructed access to services, resources, social relations and socio-cultural aspects of society (Walsh et al., 2019). ICT may have proven potential to connect older rural people to their communities, but older populations face many challenges in capitalising on those opportunities such as infrastructure connectivity, lack of training and support for practitioners at the organisational level; negative attitudes and anxiety about ICT and low digital literacy (Warburton et al., 2013). More research is needed in rural contexts that provides evidence as to the meaningful effectiveness of ICT in addressing challenges, specifically those related to social inclusion for older people in rural places.

With these critical challenges in mind, this chapter marks an initial effort to address this gap in the rural gerontology literature with respect to gerontechnology, by exploring how ICT has been used to address social connectivity and thereby enhance social inclusion for older people, including people living with dementia. We outline the conceptual scope of rural ageing and technology by identifying rural threads in gerontechnology and technology threads in rural gerontology, precipitating a discussion of thematic intersections in bodies of literature for each field. As an example of a critical, research informed, approach, we turn to an arts-based empirical project, *Improving social inclusion for Canadians with dementia and their carers through Sharing Dance*, to demonstrate how the expansion of an ICT-delivered dance program provides insights into the use of technology to support social connectivity in ageing rural communities.

Parallel lines in the fields of gerontechnology and rural gerontology

Interestingly, literature in the fields of gerontechnology and rural gerontology rarely converge. O'Shea (2009) in his book chapter on rural ageing and public policy in Ireland noted that "We do not know what older people want with respect to technology because we rarely ask them directly" (O'Shea, 2009: 279). Meanwhile, gerontechnology scholars had two years previously asserted that the central accomplishment of their field was to prioritize the older person's perspective and their needs as the driving force of the technology research and development agenda (Bouma et al., 2007). This illuminates a gap in the lack of engagement of older people in gerontechnology design in specific rural contexts at the time.

The field of gerontechnology emerged in the early 1990s, in order to promote the role of technology in improving day-to-day functions of older people (Bouma and Graafmans, 1992). Academics, along with professionals and engineers, have come to collaborate in this field to study older people in technological society, while supporting the development of technology to support older people's everyday lives and participation as full citizens (Bouma et al., 2007). This collaboration has developed in its focus and purpose to minimize impacts of ageing populations on health care systems (Bjering, Curry and Maeder, 2014) and to facilitate ageing in place (Gallistl and Wanka, 2019). Deeper insight into the nature of interventionist digital health collaboration illuminates some of the challenges of this interdisciplinarity; where differing approaches in research design necessitate the exploration of common ground for future growth (Blandford et al., 2018). Insights from science and technology studies in collaboration with social gerontology suggest that using a co-constitution logic involves study in new contexts to reveal richer images of older people as technology users (Peine and Neven, 2019). Waycott and colleagues (2019), have challenged the interventionist approach in their framework for evaluating gerontechnology by emphasising social connectedness rather than focusing on isolation and loneliness in compensating for deficiencies. They identified personal relationships, community connections and societal engagement as three dimensions of social connectedness according to older people (Waycott et al., 2019).

Where gerontechnologists conduct studies of ageing and technology, rural gerontologists situate studies in space and place. Place is important in interpreting the experiences of rural ageing, particularly in relationships between what older people need and the capacity of their communities to provide for those needs (Herron and Skinner, 2018). Older people in rural areas have been identified as quintessential "end-users" for technology development (O'Connell et al., 2018), but a lack of exposure to technology has limited digital social development in a sector historically vulnerable to disadvantages both caused by and as a result of limited access to technology (Joseph and Cloutier-Fisher, 2005). ICT has been identified as having the potential to compensate for challenges related to rural ageing such as social isolation, through building micro- and macro-level capacities for individuals to connect to services, socialisation and support in the context of health and wellness (Berg et al., 2017). But the reality of uneven distribution and quality of digital infrastructure in rural areas creates barriers to equitable connectivity and inclusion, as lower literacy and skills discourage its uptake and use (Salemink et al., 2017). While rural older people may be particularly in need of technology, they are least likely to have access to it (O'Connell et al., 2018), a situation made increasingly clear during the self-isolation and physical distancing required by the COVID-19 pandemic (see Colibaba et al., Chapter 24).

Common themes in gerontechnology and rural gerontology

Recognition of older peoples' perspectives is a thematic intersection for the two fields. Scholars in the field of gerontechnology have asserted the prioritisation of the "end-user" perspective of the older person (Bouma and Graafmans, 1992). However, this has been criticised as tokenism and noted as one of the challenges of the end-user driven research agenda (Sixsmith, 2017). Perhaps use of the term "end-user" in and of itself denotes a lack of inclusion in the conception and processes of technology development; as gerontechnology from some perspectives is intended to provide for older persons' needs and deficiencies rather than to engage them as co-creators in life-enhancing technology. In a rural context, participating in the development of technology is one way that older people can contribute to shaping their communities in what Skinner and Winterton (2018) refer to as the mutually co-constitutive relationship between ageing and rural sustainability. For the gerontechnologist, understanding these perspectives may be a top priority; however, this could also be complicated by processes of technology development as older people interact with multiple systems and services, necessitating an understanding of other stakeholders' needs and agendas such as carers, service providers and policy makers (Sixsmith, 2017).

Social inclusion and exclusion are also common themes in both fields. Given that ICT has the potential to both include and exclude older people and others in rural areas (O'Connell et al., 2018; Pekkarinen et al., 2013), greater attention to connectivity and inclusion research in rural studies is important to help set the research agenda for better understanding the impacts of rapid technological developments (Salemink et al., 2017). Many studies in both fields acknowledge the potential for ICT to connect older people in their communities (Kilpeläinen and Seppänen, 2014; O'Connell et al., 2018; Warburton et al., 2013); and the importance of ICT for accessing services and socialisation to age well in place (Berg et al., 2017), as calls have been made for better internet infrastructure to enable virtual connections (O'Shea, 2009; Salemink et al., 2017). While there has been some research on rural ageing and technology, including rural older peoples' experiences with technology, more studies are needed to examine the outcomes and the effectiveness of ICT to address social connectivity (Khosravi et al., 2016) particularly with respect to older people living in rural areas and those with cognitive impairments (Stojanovic et al., 2017). There is more work to be done to understand the increasingly important role ICT can play in providing social and physical activities and supports (Berg et al., 2017), especially those that older people living with dementia and their carers want and need (Herron and Rosenberg, 2017).

ICT-delivered dance to enhance social inclusion

To address the gaps introduced earlier, we turn now to an example of critical, arts-based research into the experiences of older people, including persons living with dementia, in a dance program that was delivered to rural areas via ICT. Findings from our four-year Canadian Institutes of Health Research and Alzheimer Society of Canada CIHR/ASC funded study, *Improving social inclusion for older Canadians with dementia and carers through Sharing Dance*, provide insights from older adult perspectives about rural ageing and technology. The project was a qualitative sequential continuum of care pilot study focused on evaluating the potential of an innovative program developed by Baycrest and Canada's National Ballet School (NBS) called Sharing Dance Seniors (Skinner et al., 2018). Our multi-method, qualitative approach focused on older adult experiences of Sharing Dance by collecting in-depth information through diaries, observations, video-recording, interviews and focus groups from older participants and their carers; facilitators; staff; and volunteers of host organisations. Between 2017 and 2019, we collected data from 12 rural small town sites, (including rural regional service centres) in non-metropolitan (rural) regions of two Canadian provinces: Peterborough, Ontario and Brandon, Manitoba. Distances of these sites from larger or urban population centres ranged from 16.8 km to 99.9 km. Three pilot studies took place in sequential phases in each of the two regions (referred to as P1, P2 and P3 in Peterborough, and as B1, B2 and B3 in Brandon). We analysed the data to be attentive to different dimensions of social inclusion using Walsh and colleagues' (2019) conceptualisation of social exclusion to help identify themes of potential exclusion or inclusion. Full details of the research methods including the theoretical framework and data analysis are published in the study protocol (see Skinner et al., 2018) and can be found on the project website, along with project reports on early findings from each of the project phases at www.sdseniorsresearch.ca.

About Baycrest NBS Sharing Dance Seniors

Sharing Dance Seniors aims to make dance accessible to older people with a range of physical and cognitive abilities, including people living with dementia. Developed by Canada's National Ballet School (NBS) and Baycrest Health Sciences, the program is offered in terms (e.g., Fall, Winter, Spring). Terms are designed to build weekly from class to class. Each dance within a class includes physical and artistic goals such as physical awareness and mobility; coordination; strength; confidence; eye focus; storytelling through movement and gesture; joy; and engagement with music. The program was developed and is delivered in-person but also has a suite of remotely led (through video streaming) dance sessions available for participants in institutional and community settings with on-site facilitators supporting participants. The program was piloted for nationwide expansion under the Public Health Agency of Canada's Multi-Sectoral Partnerships to Promote Healthy Living and Prevent Chronic Disease approach, in collaboration with long-term (residential) care homes, regional home-care providers and community support agencies. During the study, the Sharing Dance Seniors program was delivered digitally (via live-stream video, pre-recorded videos for download, or pre-recorded video stream) and in-person during special events. Sharing Dance Seniors is produced as a weekly video series for streaming to multiple remote settings from a studio at NBS in Toronto, Ontario. In-person facilitators are identified at each remote site (in both community and institutional settings) and are integral to the digital delivery of the program. Program orientation and training for on-site facilitators, via online course modules are offered through NBS's online learning platform.

Arts-based insights about rural ageing and the use of technology

When the Baycrest NBS Sharing Dance Seniors program was initiated, there was excitement about having such a program available in rural areas.

> I appreciate that we can get the expertise of the ballet group through live streaming – [It's] awesome and amazing – particularly for elderly in the rural. Being able to access this without having to drive [to the city] is incredible.
>
> (Participant, Focus Group Transcript P3)

Figure 27.1 Baycrest NBS Sharing Dance Seniors in session in a Manitoba personal care home

The opportunity for NBS to bring dance to older people in rural areas was enabled by ICT-delivery using on-screen instructors (see Figure 27.1), demonstrating how ICT can foster social inclusion in rural areas by providing a program that otherwise would not have been available. Personalised aspects of the program helped to connect participants in a shared artistic experience. On-screen instructors acknowledged the place names of participating sites and referred to individual participants at intervals as they led dance sequences for participants to follow.

> During the program, when the welcome introduction was initiated by an on-screen instructor via video stream, participants enjoyed clapping and waving as they heard [their towns and locations] being introduced, responding with "That's us!"
>
> (Field note, B2)

> They did personal little things. . . . It made the whole group feel like they knew us individually. There was just two or three times [that they did that], but that cemented the group as a special group. It made us feel like we were important.
>
> (Participant, Interview Transcript, P3)

> It was just a real sense of belonging, of . . . being in a place where you can be content. . . . I think small towns are having a real struggle to keep going. Small towns need something like this to sort of put us back on the map . . . I really feel a part of this place now and this program helped.
>
> (Older Adult Participant, Interview Transcript, B3)

This data shows how the NBS Baycrest Sharing Dance program, enabled by ICT, was effective in helping participants feel connected to each other, within their communities and to the larger world from the places they participated from. This aligns with three key dimensions of social connectivity identified by older people in Waycott and colleagues' (2019) framework that uses a non-interventionist approach by looking not at problems to be solved, but at how connectivity is experienced by older people. In a rural context, this application of a gerontechnology evaluation framework illuminates the unique aspects of the challenges faced by ageing people at local, community and macro-levels, reinforcing the relevance of geographical scale and the complex relational aspects of social inclusion that intersect through people, spaces and places.

The double digital divide and diminished potential of rural ICT-delivered programs

When the program was introduced in a small town in Ontario in P2, the number of older adult participants started with 27 in attendance. Over a period of several weeks, the number of participants decreased (to 15 in Week Two; to 12 in Week Three; to eight in Week Four), partly as a result of frustration due to technical difficulties in live-streaming the program.

1: We had a bit of problem with the technology too am I right?
2: Yes
3: Yeah, quite a bit
1: Lost the sound one day, and
3: That was frustrating to a lot of people
1: I'm sure it was, yeah
2: I think that's why some people got. . .

Do you think the technical issues – you said it frustrated people – but do you think folks may have dropped out because of them?

2: I do, I certainly do.

(Focus Group Transcript, P2)

In response to technical difficulties during the P2 phase, NBS adjusted the live-stream delivery of the program by providing a back-up of pre-recorded, downloadable sessions that would not be dependent on internet connectivity. One facilitator described how the pre-recorded version was a preferred format because it eliminated set-up time in trying to get the computer connected, which allowed her to socially connect in-person with her participants.

> *Would you say that being in charge of the computer-related tasks negatively affected your involvement as a facilitator?*
>
> Yes, because I didn't feel like I could give as much as I would have liked to the participants because I spent so much time getting it set up . . . the TV set up, the chair set up, . . . and I didn't really get an opportunity to personally welcome people and connect with them because my focus was just making sure that we have a class going [i.e., internet connectivity]. I found that quite stressful at times . . . that's what I liked with the pre-recorded [version]. I didn't have the worry, and it enabled me to connect a bit better with the group from a social point of view.
>
> (Facilitator, Interview Transcript, P2)

The pre-recorded version was intended as a back-up, but as the previous statement shows, a facilitator preferred it to the live-streamed version because it was more user-friendly, even though it significantly changed the experience for participants. One participant described this difference as, "When you knew they were live it had a different feeling, very positive – like when you're by somebody rather than an empty chair . . . you felt a greater connection". While the live-stream version may have been preferred in the experience of a program participant, it was not preferred from the perspective of who was operating it.

These findings demonstrate, consistently with other rural studies (Warburton et al., 2013), how rural technological infrastructure and attitudes towards technology can create barriers to the operation of ICT-delivered programs resulting in loss of participants; frustration and anxiety on the part of people running the programs; and program providers having to adjust their programs and sacrifice types of experiences for participants. While ICT has been identified as effective in addressing challenges of social isolation and connectivity (Khosravi et al., 2016), these examples demonstrate how its potential is diminished in the rural context due to the double digital divide (O'Connell et al., 2018). ICT in this context can be seen as a mediating force that both enhances and creates barriers to social inclusion for older people in rural areas (Walsh et al., 2019).

Key role of facilitators and volunteers to support ICT-delivery

From on-screen instructors' perspectives, they perform for the cameras from the studio and cannot see how participants are following movements and interacting with each other, as can be seen in Figure 27.2. They rely on facilitators at each

Figure 27.2 Baycrest NBS Sharing Dance Seniors instructors, Canada's National Ballet School

site to complement the program's delivery and to follow up with feedback via the online module. Participants expressed that the in-person facilitator augmented the digital delivery of the program by "making it good"; by reinforcing the right actions; and by connecting everyone to each other.

> It was nice to have to an actual person in the room – the video was great, the music was great, but it was nice to have an actual person there. I guess I'm not as much into technology a lot yet – I'm 70 – I want the person.
>
> (Community Participant, Interview Transcript, B3)

> It took a while for them to just mimic the actions from the TV. It did help that [the facilitator] was doing it at the front too, some of them would look at her and then do it, just the reinforcement of the right action, and they're used to her encouraging them and being open to new stuff. Just by calling their name and saying "yes I know you can do this" that little extra one-on-one, face-to-face.
>
> (Volunteer, Interview Transcript, B3)

In-person interaction made a difference to improve the on-screen instruction from an older participant perspective. This suggests that older people see the value of technology in assisting, but not replacing, in-person social connections, consistent with studies that found that technology, or technology-mediated social connectivity, is not considered a substitute for face-to-face interaction. In the context of care, this reinforces findings that activities should be "facilitated but not dominated by technical means. The driving force should be the creation of an opportunity to improve the effectiveness and quality of the care provided to clients" (Van der Heide et al., 2012: 290). On a practical level, facilitators helped to reinforce instructions for participants, while on a personal level, they helped to encourage connection and interaction. The challenge of remote delivery for on-screen instructors, along with feedback from older adult participants points to the critical role of the in-person facilitator.

Staff in rural long-term care facilities navigated the demands of running the Sharing Dance Seniors program by also engaging local volunteers, most of whom identified as older people, to demonstrate by participating in the program and to encourage participants to follow along.

> We needed many hands on. The first [session], I looked at the layout and how many volunteers we were looking for and I thought, "Good luck with this." I understand why now because we [used] everybody. We did.
>
> (Facilitator, Interview Transcript, B3)

> As volunteers, we were the ones they could see what we were doing and if they couldn't get all the instructions – we could help.
>
> (Volunteer, Interview Transcript, B2)

Our data also shows how facilitators and volunteers at local levels, in responding to challenges to support such initiatives, contribute to the delivery and development of arts-based and technology innovation for older people in rural areas. As they enabled the ICT-program to run effectively, older-adult volunteers were identified as key stakeholders in the social connectivity of their communities and innovation in the rural context. This finding is supported in rural gerontology more broadly where voluntarism is acknowledged as a form of agency (Joseph and Skinner, 2012) and reinforces the idea that older people are active contributors in the co-constitution of program development in their communities and therefore key players in rural community sustainability (Skinner and Winterton, 2018). In discussions in gerontechnology, how technology influences these social relationships is an important consideration when developing ICT solutions for isolation (Sixsmith, 2017). In addition to facilitators, volunteers in rural communities are part of a network of support in caring for older people. Their needs and interests in the opportunities and challenges of ICT-delivery are important considerations in the development of gerontechnology and arts-based programs to enhance social inclusion.

Conclusion

In this chapter we have sought to outline the scope of rural ageing and technology as a field of study, drawing together two bodies of literature in rural gerontology and gerontechnology, then illustrating how these can be put into practice through the empirical example of a Canadian arts-based study of Sharing Dance. This work contributes to critical rural gerontology through deeper consideration of the diverse perspectives of older people and the network of carers who support them in unique rural contexts. We have demonstrated how ICT may be effective in delivering social connectivity opportunities for older people in rural areas, but that it is not a panacea. Careful consideration of the interconnectivity of complex processes affecting ageing people in rural places should be made in implementing further technological connectivity in an attempt to solve problems. Implementing services and programs like Sharing Dance Seniors via ICT will be dependent on the capacity and willingness of local facilitation by staff and volunteers, who will determine their acceptance based on their applicability in unique rural contexts according to the people who will deliver and use them. One of the limitations of this chapter in introducing our social inclusion study into the conversations around rural gerontechnology is its relevance to more general forms of ICT intervention. We did not touch upon other assistive technologies such as robotics, telemedicine, sensor technology, video games, etc. (Khosravi et al., 2016). In terms of social inclusion, our scope was limited to social connectivity in place. More research is needed in understanding ICT effectiveness for social connectivity, particularly in rural contexts and for people living with cognitive challenges (Stojanovic et al., 2017) and with the prevalence of population ageing taking place in rural contexts, the opportunity for study and development is a critical one (Skinner and Winterton, 2018). The geographical approach that considers the interrelated effects that take place at individual, local and macro levels, influenced by social gerontology to consider such processes and outcomes of social inclusion have moved towards a critical rural geographical gerontology (Skinner and Winterton, 2018). Researchers in the field of gerontechnology have much to contribute to this interdisciplinary field as they consider how challenges that require technological solutions may be well conceptualised directly by the older people in the unique rural places where they live, in the co-constitution of gerontechnology (Peine and Neven, 2019) for better health and quality of life.

Acknowledgements

Funding for the empirical research presented in this chapter was provided by the Canadian Institutes of Health Research and Alzheimer Society of Canada CIHR/ASC operating grant, *Improving Social Inclusion for Older Canadians with Dementia and Carers through Sharing Dance* (M. Skinner and R. Herron, co-principal investigators).

References

Berg, T., Winterton, R., Petersen, M. and Warburton, J. (2017). 'Although we're isolated, we're not really isolated': the value of information and communication technology for older people in rural Australia. *Australasian Journal on Ageing*, 36(4), pp. 313–317.

Bjering, H., Curry, J. and Maeder, A. J. (2014). Gerontechnology: the importance of user participation in ICT development for older adults. *Studies in Health Technology and Informatics*, 204, pp. 7–12.

Blandford, A., Gibbs, J., Newhouse, N., Perski, O., Singh, A. and Murray, E. (2018). Seven lessons for interdisciplinary research on interactive digital health interventions. *Digital Health*, 4, pp. 1–13.

Bouma, H. and Graafmans, J. A., (eds). (1992). *Gerontechnology*, vol. 3. Amsterdam: IOS Press.

Bouma, H., Fozard, J. L., Bouwhuis, D. G. and Taipale, V. (2007). Gerontechnology in perspective. *Gerontechnology*, 6(4), pp. 190–216.

Bronswijk, J. E., Bouma, H., Fozard, J. L., Kearns, W. D., Davison, G. C. and Tuan, P. C. (2009). Defining gerontechnology for R&D purposes. *Gerontechnology*, 8(1), p. 3.

Gallistl, V. and Wanka, A. (2019). Representing the' older end user'? Challenging the role of social scientists in the field of'active and assisted living'. *International Journal of Care and Caring*, 3(1), pp. 123–128.

Herron, R. V. and Rosenberg, M. W. (2017) "Not there yet": examining community support from the perspective of people with dementia and their partners in care. *Social Science & Medicine*, 173, pp. 81–87.

Herron, R. and Skinner, M. (2018). Rural places and spaces of health and health care. In: V. Crooks, J. Pearce and G. Andrews, eds., *Handbook of Health Geography*. New York: Routledge, pp. 267–272.

Joseph, A. E. and Cloutier-Fisher, D. (2005). Ageing in rural communities: vulnerable people in vulnerable places. In: G. J. Andrews and D. R. Phillips, eds, *Ageing and Place*. London: Routledge, pp. 149–162.

Joseph, A. E. and Skinner, M. W. (2012). Voluntarism as a mediator of the experience of growing old in evolving rural spaces and changing rural places. *Journal of Rural Studies*, 28(4), pp. 380–388.

Khosravi, P. and Ghapanchi, A. H. (2016). Investigating the effectiveness of technologies applied to assist seniors: a systematic literature review. *International journal of medical informatics*, 85(1), pp. 17–26.

Kilpeläinen, A. and Seppänen, M., (2014). Information technology and everyday life in ageing rural villages. *Journal of Rural Studies*, 33, pp. 1–8.

O'Connell, M. E., Scerbe, A., Wiley, K., Gould, B., Carter, J., Bourassa, C. and Warry, W. (2018). Anticipated needs and worries about maintaining independence of rural/remote older adults: opportunities for technology development in the context of the double digital divide. *Gerontechnology*, 17(3), pp. 126–138.

O'Shea, E. (2009). Rural ageing and public policy in Ireland. In: J. McDonagh, T. Varley and S. Shortall, eds, *A Living Countryside? The Politics of Sustainable Development in Rural Ireland*. Farnham: Ashgate, pp. 269–285.

Peine, A. and Neven, L. (2019). From intervention to co-constitution: new directions in theorizing about aging and technology. *The Gerontologist*, 59(1), pp. 15–21.

Pekkarinen, S., Melkas, H., Kuosmanen, P., Karisto, A. and Valve, R. (2013). Towards a more social orientation in gerontechnology: case Study of the "reminiscence Stick". *Journal of Technology in Human Services*, 31(4), pp. 337–354.

Salemink, K., Strijker, D. and Bosworth, G. (2017). Rural development in the digital age: a systematic literature review on unequal ICT availability, adoption, and use in rural areas. *Journal of Rural Studies*, 54, pp. 360–371.

Sixsmith, A., Mihailidis, A. and Simeonov, D. (2017). Aging and technology: taking the research into the real world. *Public Policy & Aging Report*, 27(2), pp. 74–78.

Skinner, M. W., Herron, R. V., Bar, R. J., Kontos, P. and Menec, V. (2018). Improving social inclusion for people with dementia and carers through sharing dance: a qualitative sequential continuum of care pilot study protocol. *BMJ Open*, 8(11), p. e026912.

Skinner, M. W. and Winterton, R. (2018). Rural ageing: Contested spaces, dynamic places. In: M. W. Skinner, G. J. Andrews and M. P. Cutchin, eds., *Geographical Gerontology: Perspectives, Concepts, Approaches*. New York: Routledge, pp. 136–148.

Stojanovic, J., Collamati, A., Mariusz, D., Onder, G., La Milia, D. I., Ricciardi, W . . . and Poscia, A. (2017). Decreasing loneliness and social isolation among the older people: Systematic search and narrative review. *Epidemiology, Biostatistics and Public Health*, 14(2). Available at: https://ebph.it/article/view/12408

Van der Heide, L. A., Willems, C. G., Spreeuwenberg, M. D., Rietman, J. and de Witte, L. P., 2012. Implementation of CareTV in care for the elderly: the effects on feelings of loneliness and safety and future challenges. *Technology and Disability*, 24(4), pp. 283–291.

Walsh, K., O'Shea, E. and Scharf, T. (2019). Rural old-age social exclusion: A conceptual framework on mediators of exclusion across the lifecourse. *Ageing & Society*. *Ageing & Society*. Epub ahead of print 2 July. DOI: 10.1017/S0144686X19000606.

Warburton, J., Cowan, S. and Bathgate, T. (2013). Building social capital among rural, older Australians through information and communication technologies: a review article. *Australasian journal on ageing*, 32(1), pp. 8–14.

Waycott, J., Vetere, F. and Ozanne, E. (2019). Building social connections: a framework for enriching older adults' social connectedness through information and communication technologies. In: B. B Neves and F. Vetere, eds., *Ageing and Digital Technology: Designing and Evaluating Emerging Technologies for Older Adults*. Singapore: Springer, pp. 65–82.

28 Rural older people, climate change and disasters

Matthew Carroll and Judi Walker

Introduction

In this chapter, the intersection between ageing and disasters is explored, including predicted increases in climate-related disasters and associated health impacts.[1] Attention is drawn to the disproportionate impact of disasters on older people in the last 25 years. The interplay between rurality, chronic health, social support and other factors is explored, giving a more nuanced understanding of older people's vulnerability to disasters and climate events, particularly in more exposed rural communities. The necessity of communicating with and supporting older people during disaster events is emphasised and steps to improve communications are flagged. This analysis promotes a move beyond stereotypes to recognise the resilience and adaptive capacity of older people. Two case studies provide examples of the need to better understand the experiences and capacity of older people and ways to engage them in preparing for and responding to climate change and disasters.

The intersection between ageing, rurality and disasters

The frequency of disaster events is increasing, with Figure 28.1 depicting data from the EM-DAT International Disaster Database showing a clear increase in climate-related disasters over the past 60 years (Leaning and Guha-Sapir, 2013). There is growing recognition of the health implications of climate change, with the recent Intergovernmental Panel on Climate Change (IPCC) special report on the impacts of global warming of 1.5^{0}C above pre-industrial levels reporting high confidence in health impacts including ozone-related mortality, increases in heat-related deaths and vector-borne diseases (IPCC, 2018). Such extreme weather events are likely to impact rural areas disproportionately and exacerbate existing disadvantages (Hughes et al., 2016).

In parallel with this increasing number of disaster events is a major demographic shift taking place globally as a result of population ageing, driven by decreasing fertility rates and increasing life expectancy (Oven, Wistow and Curtis, 2019). According to the World Health Organization (WHO), the number of people aged 60 years or older is expected to double to two Billion by 2050, with 80% living in

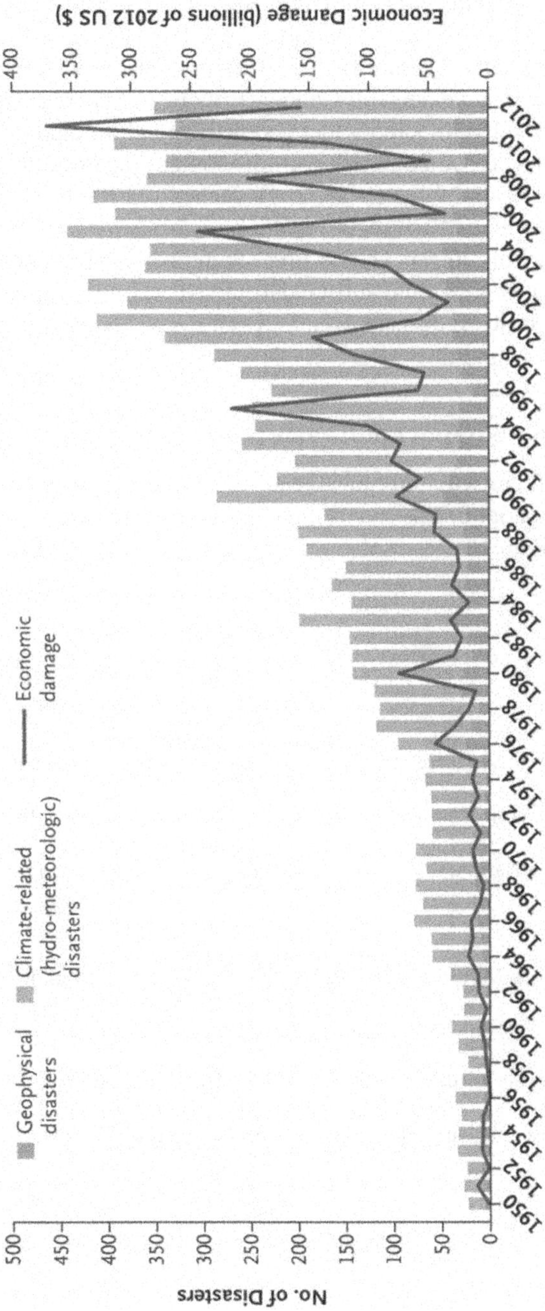

Figure 28.1 Numbers and types of natural disasters, 1950–2012

low- and middle-income countries (WHO, 2018). As outlined in Chapter 2 of this volume, the ageing of the population is even greater in rural settings, due to factors such as out-migration of younger people. The WHO also reports that 22% of older people worldwide face malnutrition, 180 million have visual or hearing impairments, 46.8 million have dementia, and nearly one in ten are at risk of depression (WHO, 2017). The likelihood of older people living alone also increases with age, particularly in developed countries and for women, for example peaking in Australia around 80 years of age (ABS, 2011; Reher and Requena, 2018). This puts older people at greater risk of social isolation, particularly after the loss of a spouse (Warburton and Lui, 2007). This is often combined with lower socio-economic status and living in small and substandard housing, both of which are known to increase vulnerability to disaster events (Kinney et al., 2008) and are more characteristic of rural communities (Benzie, 2014; Krawchenko et al., 2016).

Disproportionate impact of disasters on older people

Until recently, there was little awareness of the disproportionate impact of disasters on older people. This has changed as a result of major events in the last 25 years (see Figure 28.2).

- ☐ 1995 Kobe earthquake where 50% of casualties were older, subsequently accounting for 90% of deaths;
- ☐ 2003 Paris heatwave where 70% of the 14,800 deaths were people aged over 70 years;
- ☐ 2004 Aceh tsunami where those over 60 years had the highest mortality rates;
- ☐ 2005 Hurricane Katrina where 71% of deaths in Louisiana were people aged over 60 years;
- ☐ 2009 heatwave in Victoria, Australia, where the greatest number of deaths and emergency presentations were amongst people aged 75 years and older, with similar increases in deaths arising from the associated Black Saturday fires;
- ☐ 2011 Japanese earthquake where 60% of deaths in Rikuzentakata were people aged over 60 years, and;
- ☐ 2013 Typhoon Haiyan in the Philippines where 40% of deaths were people aged 60 years and over.

Figure 28.2 Major events in the last 25 years

There are multiple and overlapping reasons for older people being more prone to the impacts of different types of disasters, most of which are likely to be exacerbated in rural settings. Smith et al. (2009) highlighted the following characteristics of older people that can increase their susceptibility to disasters:

1 declining health and increased chronic diseases
2 limitations in vision, hearing and mobility
3 limited access to health care resources
4 low economic status due to fixed incomes, and
5 limited social networks.

The same factors make older people more likely to be impacted by the effects of climate change, including susceptibility to heat waves, particulate matter arising from bushfire smoke, exposure to toxic exposures following floods and other events. As outlined elsewhere in this volume, many of these are likely to be a greater concern for older people in rural locations (for example, see Glasgow and Doebler, Chapter 10; Ryser et al., Chapter 13; Menec and Novek, Chapter 14; Hennessy and Innes, Chapter 17).

In addition to immediate impacts, delayed outcomes are likely to present in older age with exposure to particulate matter, toxins and carcinogens increasing cancer risk (Prohaska and Peters, 2019). There is evidence of other longer-term outcomes from disaster events from analysis of diabetes service use data following Hurricane Katrina, where older people were less likely to access screening and support services for diabetes in the years following the disaster event (Quast and Feng, 2019). This reduction in the take up of services is in line with other research on older people suggesting that people may lower their health expectations as they age and can become reluctant to seek assistance, even when needed, and placing them more at risk (Canvin et al., 2018). Other research suggests that people from lower socio-economic backgrounds have reduced expectations for health and support as they age, increasing their marginalisation (Barrett, 2003).

The need for future planning to consider the needs of older people in disasters

Better understanding of the factors which make older people more susceptible to the impacts of climate change and disasters provides the opportunity to develop responses to improve future outcomes for older people and the wider community. From a public health perspective, this includes better management of chronic health conditions, better surveillance to detect and rapidly respond to disease outbreaks such as dengue fever, including improved control of disease pathways such as mosquito-control and improved food handling (Bambrick et al., 2011; Prohaska and Peters, 2019). While the focus of planning to date has largely been on urban areas, there is a pressing need to put strategies in place for climate-exposed rural communities (Krawchenko et al., 2016).

In addition, there is a requirement for enhanced health promotion, to improve uptake of warning messages (Hansen et al., 2011). Opportunities also exist for utilising and expanding existing health, social and community services pathways to monitor the health of frail older people and other at-risk groups, such as the establishment of phone registers for vulnerable people and promoting neighbour checks during heat events (Hansen et al., 2011). In rural Australian settings, volunteer-run services like Meals on Wheels have been found to provide an effective opportunity to check on older people during heat and disaster events (Loughnan and Carroll, 2015; Walker et al., 2017).

More broadly, consideration needs to be given to improved design to maximise the health and well-being of the community. Much of the focus to date has been on urban design, such as responding to the "urban heat island" effect, where urban areas experience hotter temperatures (Dialesandro et al., 2019). However, proposed solutions such as ensuring access to green space, water bodies, public transport and active transport (such as walking and cycling), to improve chronic health and social connection and reduce disaster vulnerability, (Rosenzweig et al., 2018) are just as important in rural and regional areas. Within this context, considerable effort will need to be put into planning for extensive in-migration in response to sea-level rises impacting on heavily populated coastal areas. It is projected that many millions of people will be displaced in coming years, with considerable potential for conflict and inequity (Burzynski et al., 2018). Ryser et al. (Chapter 13) in this book provides insight into the need for targeted community development approaches meeting the needs of older people, and it is vital that these activities consider future climate risk and the potential for greater impacts on older rural residents. While the Age Friendly Communities program (discussed in detail in Chapter 14), has been implemented in many locations internationally, a Canadian study of rural communities prone to storm surges has recommended that the framework be expanded to include consideration climate and disaster risk (Krawchenko et al., 2016).

While the challenges associated with climate change are vast, the range of policy responses are similarly vast, as outlined later in a table of health-related responses developed by Bambrick et al. (2011).

Communicating with and supporting rural older people during disaster events

While older people may be more susceptible to the impacts of climate change and disasters, there may be a number of factors that reduce their take up of emergency warnings and willingness to access support services. Sakauye et al., (2009) identified the following barriers to older people accessing disaster support:

1 stigma associated with receiving aid
2 concern about loss of other entitlements
3 self-reliance associated with the great depression causing some older adults to feel that others need the assistance more than they do

Table 28.1 Examples of urban policy responses to selected health impacts of climate change

Health impacts of climate change	*Urban Policy Responses*						
	Water	*Housing, buildings, and surrounds*	*Land use*	*Transport*	*Energy*	*Health systems*	*Other*
Thermal stress	Sufficient supply available for direct physical cooling (swimming pools, cool showers)	Building codes, including orientation, thermal properties Retrofitting, including insulation	Greening the city: public shade and "coolspots"	Active travel policies (reduced chronic disease burden and vulnerability)	Affordable, reliable energy for artificial cooling	Heat wave early warning systems Heat wave respite centres Service planning (seasonal/in response to a forecast event)	Reduce social disadvantage Neighbourliness and neighbourhood safety Social capital
Air quality			Antisprawl policies and mixed use urban design	Active travel policy (reduced local population, reduced chronic disease burden and vulnerability)	Renewable "clean" energy generation	Air pollution warning systems Service planning (seasonal/in response to forecast event)	
Urban vectors	Screen household water tanks Capture urban runoff	Flyscreens Improve drainage				Arboviral surveillance and control	
Enteric pathogens	Monitor supply quality	Design of food preparation areas in institutions	Produce food locally (shorter, simpler route to the table)			Improved traceability and recall	Food safety regulations, audits, education
Food security	Appropriate irrigation	Growing food on buildings	Provision for urban agriculture	Fewer "food miles"			

From Bambrick, H.J., Capon, A.G., Barnett, G.B., Beaty, R.M. & Burton, A.J., Asia Pacific Journal of Public Health, 23, 67S-79S., Copyright © 2020 by of SAGE Publications, Inc. Reprinted by Permission of SAGE Publications, Inc.

4 difficulty navigating complicated bureaucratic systems
5 being accustomed to having a spouse who took care of these things for them
6 lack of familiarity with online applications or personal computers, and
7 low reading skills or language barriers.

While social networks are generally held to contribute to positive health outcomes and climate change adaptation (Howard, Blakemore and Bevis, 2017), there is potential for increased risk in instances where the networks perpetuate poor awareness of health hazards. Wolf et al. (2010) conducted interviews regarding heat-health risk in the UK with older people living in the community and their primary social contacts. They found that the older residents valued their independence and were reticent to ask for help, and that they and their social contacts didn't perceive heat as a health risk, making the older respondents less likely to respond to heat-health alerts. This is likely to be more of an issue in rural settings where older people are even less likely to seek help from others (Loughnan and Carroll, 2015; Warburton, Scharf and Walsh, 2017).

Other factors which can impact on the capacity of older people to be aware of and respond to disaster-related information include lower levels of health literacy (Brown et al., 2014) and reduced access to transport to leave an impacted area (Duggan et al., 2010), both of which are lower in rural communities. Socio-economic status can play a critical role in determining how and when people respond to disaster events, with one assessment of the impacts of Hurricane Katrina noting that some older people had been reticent to evacuate because they were waiting on their next scheduled social security and other payments (Jenkins, Laska and Williamson, 2007). Finally, mistrust of government including emergency advice has come up as an issue in a number of studies (Duggan et al., 2010), with the disconnection between metropolitan-based decision makers and the impacted rural/regional community members being seen as a key factor in one study (Walker et al., 2017)

These factors can reduce the capacity of older people to access, understand and respond to disaster messaging. Recommendations made on how to improve communications include tailoring the messages for lower literacy levels and directly involving older people in the development and delivery of the messaging as well as piloting the communications and evaluating the outcomes (Boon, 2014; Brown et al., 2014). While there is a need to consider the interlocking factors that impact on older people's capacity to respond to a disaster event, emergency communications should not characterise older people as an homogenous vulnerable group as this may disenfranchise them and reduce their likelihood of attending to important messages (Duggan et al., 2010). Research on heat alerts issued for older people has found that older people often ignore messages as not being relevant to them because they do not consider themselves as "old" (Mayhorn, 2005; Nitschke et al., 2013). Walker et al. (2017) argue that public health messaging should instead focus on people with health and mobility issues *at all ages* rather than specifically targeting older people.

Moving beyond stereotypes

While there is clear evidence that there are physiological, societal and other factors that place older people at heightened risk during disaster events, older people should not be stereotyped as "vulnerable". Instead, the resilience and adaptive capacity of older people, needs to be recognised (Howard, Blakemore and Bevis, 2017). The fact that older people have had a longer life and are able to draw upon their previous experiences, including previous disaster events, supports the argument for involving them in disaster preparedness and response (Deeny et al., 2010; Hutton, 2008). As mentioned previously, Meals on Wheels volunteers in two rural Australian settings have already proven invaluable in providing heat-health checks (Loughnan and Carroll, 2015) and in checking on the well-being of older residents during a major pollution event (Walker et al., 2017). In both examples, many of the volunteers were older than the "older people" that they were supporting.

Research suggests that older people may be better able to cope mentally with disasters as a result of their previous experiences. The "Inoculation Hypothesis" (Adams et al., 2011) suggests that older people may draw upon emotional reserves as well as knowledge of strategies used successfully in the past. In a multi-year study of the impacts of Hurricane Katrina, Adams et al. (2011) found that older people were more positive in how they described themselves and were able to take a long-range view, drawing on their experiences of 'tougher times'. In contrast, middle aged participants in the study reported high levels of stress and the feeling that they might "break" under the pressure. This finding aligns with other research showing that older people are less likely to be psychologically impacted after natural disasters (Rafiey et al., 2016; Broder et al., 2020), highlighting the need for a more nuanced understanding of ageing and vulnerability.

The potential and moral imperative for older people to be agents of change with regard to climate change has been made most stridently by Rick Moody, who argued that the older generation bear more responsibility for the decisions and actions leading to climate change than any other current generation (Moody, 2017). He also argued that older people, especially those in the Baby Boomer generation, have the resources, capacity, purchasing and voting power, to make a real difference. While Moody may argue that older people have the responsibility and capacity to respond to climate change, research has suggested that older people may be less likely to see climate change as a threat and to support mitigation strategies (Moser, 2017).

The following case studies provide examples of the need to better understand the experience and capacity of older people and ways to engage them in responding to climate change and disasters.

Case study 1 – Engaging older people in climate change

Pillemer and Filiberto (2017), noting the evidence of reduced participation of older people in environmental organisations, reviewed the "age-friendliness" of relevant organisations in New York State and found most failed to make any accommodations for older people (such as accessible venues, offering daytime activities and providing opportunities for people with different interests or physical abilities). In response to these issues, the researchers worked with environmental organisations and ageing service providers to develop a model program for engaging older volunteers in environmental activities. The *Retirees in Service to the Environment* (RISE) program focuses on the provision of age-relevant information on environmental matters, leadership training to assist older people to optimise their contribution, development of a supportive social environment and opportunities to put new skills into practice.

The RISE website (http://citra-rise.human.cornell.edu/) provides free access to the program training manual for other groups to adopt and modify. The website also provides access to the core research publications arising from the group in the last decade. As befits a program bedded in research, the outcomes of the RISE program have been extensively evaluated (Pillemer et al., 2016). The program has now been implemented in 11 locations, including rural sites, with 149 participants, 84% of whom completed the extensive 30-hour program. For over two thirds, this was their first engagement in environmental volunteering, with 94% finding it extremely positive and 97% indicating that they would recommend the program to others. Pre-post analysis suggested that participants reported higher levels of generativity, attachment to others and social support two months after the program. It is estimated that the program has contributed up to 2,500 hours of volunteer activity.

Case study 2 – Learnings from the Hazelwood mine fire

The Hazelwood open-cut brown coal mine fire in February–March 2014 resulted in the nearby town of Morwell in rural/regional Victoria, Australia, being shrouded in smoke and ash for an extended six-week period. The event generated considerable community concern regarding immediate and long-term health impacts (Teague, Catford and Petering, 2014). The state government commissioned a wide-ranging health study which included an assessment of the impacts on older people, including consideration of policy and service responses at the time (Walker et al., 2017). This work, which involved interviews and focus groups with 91 older people and 17 service providers and decision makers, identified clear issues to do with the way that communications were conducted during the event, with older residents unhappy with the way that the response was managed by decision makers from the state capital rather than trusted local experts.

The messages targeting vulnerable older people was also found to risk disenfranchising the very group they were targeted at, echoing the concerns of Duggan et al. (2010). While the need for older people to be provided with additional supports was recognised, this support was targeted largely at those already receiving health and social services (and so already well looked after) and failed to support those "robust" older people living in the community (McCann, 2011). Walker et al. (2017) provided a number of other recommendations including using locally relevant spokespeople and age-relevant peers in communications, providing information as part of a more interactive dialogue rather than one-way updates. The research also highlighted the potential for older people to contribute to the development of community-led responses to future emergency events, drawing on their experiences of previous events.

Conclusion

In this chapter evidence has been cited about the association between older age and increased susceptibility to the impacts of climate change and disasters. This risk is often increased in rural settings, due to issues such as access to services, lower socio-economic status, as well as increased vulnerability to climate events through fire and flooding. Older people are more at risk during current and future disaster events and considerable effort is required to target the health, social and other risk factors and messaging to improve outcomes. However, the focus must not be solely on the risks associated with ageing and instead must focus on people of all ages with chronic health conditions and other risk factors. The potential for older people to participate in responding to disasters and climate change should be realised. While this potential has largely been missed in mainstream planning, it is inspiring to see that elders' groups are increasingly taking up the challenge. While there has been some research to date exploring the intersection between ageing, rurality and disasters, more is needed to better understand the implications and opportunities for rural older people, to better plan for and respond to future events.

At the time of finalising this chapter, COVID-19 was sweeping the world, with severity and mortality increasing with older age. Deaths of older residents in nursing homes account for almost half of all deaths in some regions, and yet they are frequently omitted, or only belatedly included, in mortality statistics. While the focus has been on metropolitan areas, rural settings are at risk because of the older age, poorer health and reduced access to services of these regions. Worryingly, in major developing countries lockdowns have seen millions of workers returning to their rural homes, increasing the risk to rural communities. Health care restrictions have led some regions to deny access to older people, due in part to a judgement that older people have less value, with less life ahead of them. While there has been much discussion about the vulnerability and value of older people, there has been very little heard from older people themselves. A notable exception was Texas Lieutenant Governor Dan Patrick, who asserted that older people would rather die than put the economy at risk. Similar to the disasters and climate impacts covered in this chapter, there is a pressing need to engage more appropriately with older people in the health, social and economic response to global pandemics (see Colibaba et al., Chapter 23).

Note

1 This chapter builds on and extends an earlier review of the literature on older people and disasters completed as part of a policy review of the impact of the Hazelwood mine fire on older people (Walker et al., 2017).

References

Abs. (2011). *Census of Population and Housing, Focusing on Age by Housing and Marital Status: Findings Based on use of Table Builder*. Canberra, Australia: Australian Bureau of Statistics.

Adams, V., Kaufman, S. R., Van Hattum, T. and Moody, S. (2011). Aging disaster: mortality, vulnerability, and long-term recovery among Katrina survivors. *Medical Anthropology*, 30, pp. 247–270.

Bambrick, H. J., Capon, A. G., Barnett, G. B., Beaty, R. M. And Burton, A. J. (2011). Climate change and health in the urban environment: adaptation opportunities in Australian cities. *Asia Pacific Journal of Public Health*, 23, pp. 67S–79S.

Barrett, A. E. (2003). Socioeconomic status and age identity: the role of dimensions of health in the subjective construction of age. *The Journals of Gerontology Series B: Psychological Sciences and Social Sciences*, 58, pp. S101–S109.

Benzie, M. (2014). Social justice and adaptation in the UK. *Ecology and Society*, 19, p. 39.

Boon, H. (2014). Investigation rural community communication for flood and bushfire preparedness. *The Australian Journal of Emergency Management*, 29, pp. 17–25.

Broder, J., Gao, C., Ikin, J., Berger, E., Campbell, T. C. H., Maybery, D., Tsoutsoulis, J., Mcfarlane, A. C., Abramson, M., Sim, M. R., Walker, J. and Carroll, M. (2020). The factors associated with distress following exposure to smoke from an extended coal mine fire. *Environmental Pollution*, 266, 115131.

Brown, L. M., Haun, J. N. and Peterson, L. (2014). A proposed disaster literacy model. *Disaster Medicine and Public Health Preparedness*, 8, pp. 267–275.

Burzynski, M., Deuster, C., Docquier, F. and De Melo, J. (2018). *Climate change, inequality and humamigration*. Technical report.

Canvin, K., Macleod, C. A., Windle, G. And Sacker, A. (2018). Seeking assistance in later life: how do older people evaluate their need for assistance? *Age & Ageing*, 47, pp. 466–473.

Chan, E. Y. Y., (ed.). (2019). *Disaster Public Health and Older People*. London: Routledge.

Deeny, P., Vitale, C. T., Spelman, R. and Duggan, S. (2010). Addressing the imbalance: empowering older people in disaster response and preparedness. *International journal of older people nursing*, 5, pp. 77–80.

Dialesandro, J. M., Wheeler, S. M. and Abunnasr, Y. (2019). Urban heat island behaviors in dryland regions. *Environmental Research Communications*, 1, p. 081005.

Duggan, S., Deeny, P., Spelman, R. and Vitale, C. T. (2010). Perceptions of older people on disaster response and preparedness. *International Journal of Older People Nursing*, 5, pp. 71–76.

Hansen, A., Bi, P., Nitschke, M., Pisaniello, D., Newbury, J. and Kitson, A. (2011). Older persons and heat-susceptibility: the role of health promotion in a changing climate. *Health Promotion Journal of Australia*, 22, pp. 17–20.

Howard, A., Blakemore, T. and Bevis, M. (2017). Older people as assets in disaster preparedness, response and recovery: lessons from regional Australia. *Ageing & Society*, 37, pp. 517–536.

Hughes, L., Rickards, L., Steffen, W., Stock, P. and Rice, M. (2016). *On the Frontline: Climate Change and Rural Communities*. Potts Point: Climate Council of Australia Limited.

Hutton, D. 2008. *Older People in Emergencies: Considerations for Action and Policy Development*. Geneva: World Health Organization.

IPCC. (2018). Global Warming of 1.5° C. An IPCC Special Report on the impacts of global warming of 1.5 C above pre-industrial levels and related global greenhouse gas emission pathways, in the context of strengthening the global response to the threat of climate change, sustainable development, and efforts to eradicate poverty. In: V. Masson-Delmotte, P. Z., H. O. Pörtner, D. Roberts, J. Skea, P.R. Shukla, A. Pirani, W., Moufouma-Okia, C. P., R. Pidcock, S. Connors, J. B. R. Matthews, Y. Chen, X. Zhou, M. I. Gomis, E. Lonnoy, and T. Maycock, M. T., T. Waterfield, eds., *Global Warming of 1.5° C. An IPCC Special Report*. Geneva: IPCC.

Jenkins, P., Laska, S. and Williamson, G. (2007). Connecting Future evacuation to current recovery: saving the lives of older people in the next catastrophe. *Generations*, 31, pp. 49–52.

Kinney, P. L., O'neill, M. S., Bell, M. L. and Schwartz, J. (2008). Approaches for estimating effects of climate change on heat-related deaths: challenges and opportunities. *Environmental Science and Policy*, 11, pp. 87–96.

Krawchenko, T., Keefe, J., Manuel, P. and Rapaport, E. (2016). Coastal climate change, vulnerability and age friendly communities: linking planning for climate change to the age friendly communities agenda. *Journal of Rural Studies*, 44, pp. 55–62.

Leaning, J. and Guha-Sapir, D. (2013). Natural disasters, armed conflict, and public health. *New England Journal of Medicine*, 369, pp. 1836–1842.

Loughnan, M. and Carroll, M. (2015). People who are elderly or have chronic conditions. In: R. Walker and W. Mason, eds., *Climate Change Adaptation for Health and Social Services*. Clayton, VIC: CSIRO Publishing, pp. 93–110.

Mayhorn, C. B. (2005). Cognitive aging and the processing of hazard information and disaster warnings. *Natural Hazards Review*, 6, pp. 165–170.

Mccann, D. G. C. (2011). A review of hurricane disaster planning for the elderly. *World Medical and Health Policy*, 3, pp. 1–26.

Moody, R. (2017). Elders and climate change: no excuses. *Public Policy and Aging Report*, 27, pp. 22–26.

Moser, S. C. (2017). Never too old to care: reaching an untapped cohort of climate action champions. *Public Policy and Aging Report*, 27, pp. 33–36.

Nitschke, N., Hansen, A., Bi, P., Tucker, G., Dal Grande, E., Avery, J., Slota-Ka, S. and Kelsall, L. (2013). *Adaptive Capabilities in Older People During Extreme Heat Events in Victoria: A Population Survey*. Melbourne, VIC: Victorian Department of Health.

Oven, K., Wistow, J. and Curtis, S. (2019). Older people and climate change. In: L. Mason and J. Rigg, eds., *People and Climate Change: Vulnerability, Adaptation, and Social Justice*. New York: Oxford University Press.

Pillemer, K. and Filiberto, D. (2017). Mobilizing older people to address climate change. *Public Policy & Aging Report*, 27, pp. 18–21.

Pillemer, K., Wells, N. M., Meador, R. H., Schultz, L., Henderson, C. R., Jr. and Cope, M. T. (2016). Engaging older adults in environmental volunteerism: the retirees in service to the environment program. *The Gerontologist*, 57, pp. 367–375.

Prohaska, T. R. and Peters, K. E. (2019. Impact of natural disasters on health outcomes and cancer among older adults. *The Gerontologist*, 59, pp. S50–S56.

Quast, T. and Feng, L. (2019). Long-term effects of disasters on health care utilization: hurricane Katrina and older individuals with diabetes. *Disaster Medicine and Public Health Preparedness*, 13, pp. 724–731.

Rafiey, H., Momtaz, Y. A., Alipour, F., Khankeh, H., Ahmadi, S., Khoshnami, M. S. and Haron, S. A. (2016). Are older people more vulnerable to long-term impacts of disasters? *Clinical Interventions in Aging*, 11, p. 1791.

Reher, D. and Requena, M. (2018). Living alone in later life: a global perspective. *Population and Development Review*, 44, pp. 427–454.

Rofi, A., Doocy, S. and Robinson, C. (2006). Tsunami mortality and displacement in Aceh province, Indonesia. *Disasters*, 30, pp. 340–350.

Rosenzweig, C., Solecki, W. D., Romero-Lankao, P., Mehrotra, S., Dhakal, S. and Ibrahim, S. A., (eds). (2018). *Climate Change and Cities: Second Assessment Report of the Urban Climate Change Research Network*. Cambridge: Cambridge University Press.

Sakauye, K. M., Streim, J. E., Kennedy, G. J., Kirwin, P. D., Llorente, M. D., Schultz, S. K. and Srinivasan, S. (2009). AAGP position statement: disaster preparedness for older Americans: critical issues for the preservation of mental health. *The American Journal of Geriatric Psychiatry*, 17, pp. 916–924.

Sawai, M. 2011. *Who is Vulnerable During Tsunamis? Experiences from the Great East Japan Earthquake 2011 and the Indian Ocean Tsunami 2004*. Geneva: United Nations Economic and Social Commission for Asia and the Pacific.

Smith, S. M., Tremethick, M. J., Johnson, P. and Gorski, J. (2009). Disaster planning and response: considering the needs of the frail elderly. *International Journal of Emergency Management*, 6, pp. 1–13.

Teague, B., Catford, J. and Petering, S. (2014). *Hazelwood Mine Fire Inquiry Report*. Available at: http://report.hazelwoodinquiry.vic.gov.au/ [Accessed 25 July 2016].

Teague, B., Mcleod, R. and Pascoe, S. (2010). 2009 *Victorian Bushfires Royal Commission: Final Report Summary*. Available at: www.royalcommission.vic.gov.au/Commission-Reports/Final-Report/Summary.html [Accessed 8 July 2016].

Tuohy, R., Stephens, C. and Johnston, D. (2014). Qualitative research can improve understandings about disaster preparedness for independent older adults in the community. *Disaster Prevention and Management*, 23, pp. 296–308.

Victorian Chief Health Officer. (2009). *January 2009 Heatwave in Victoria: An Assessment of Health Impacts*. Melbourne, VIC: Victorian Government Department of Human Services.

Walker, J., Carroll, M. and Chisholm, M. (2017). *Policy Review of the Impact of the Hazelwood Mine Fire on Older People: Final Report*. Melbourne, VIC: Hazelwood Health Study.

Warburton, J. and Lui, C. (2007). *Social Isolation and Loneliness among Older People: A Literature Review*. Brisbane, QLD: Australasian Centre on Ageing, The University of Queensland.

Warburton, J., Scharf, T. and Walsh, K. (2017). Flying under the radar? Risks of social exclusion for older people in rural communities in Australia, Ireland and Northern Ireland. *Sociologia Ruralis*, 57, pp. 459–480.

WHO (2017). *WHO Guidelines on Integrated Care for Older People (ICOPE)*. Geneva: World Health Organization.

WHO (2018). *The Global Network for Age-Friendly Cities and Communities: Looking Back Over the Last Decade, Looking Forward to the Next*. Geneva: World Health Organization.

Wolf, J., Adger, W. N., Lorenzoni, I., Abrahamson, V. And Raine, R. (2010). Social capital, individual responses to heat waves and climate change adaptation: An empirical study of two UK cities. *Global Environmental Change*, 20, pp. 44–52.

Part V

Conclusion

29 Towards a critical rural gerontology

Rachel Winterton, Kieran Walsh and Mark Skinner

Introduction

The primary objective of *Rural Gerontology: Towards Critical Perspectives on Rural Ageing* is to provide a comprehensive and critical foundation of knowledge about the interdisciplinary development, contemporary scope and critical prospects for rural gerontology. As outlined by Skinner et al. (Chapter 1), while research on rural ageing is by no means a new phenomenon, this book provides the first dedicated inquiry into a distinctive rural gerontology called for by Rowles (1988) more than 30 years ago, by providing a foundation for understanding the concepts, approaches and themes that distinguish the field of study. Through its multidisciplinary and global approach, this book provides an in-depth perspective on the complex intersection between rurality and the ageing experience, and what this means for rural ageing research, policy and practice. Rural gerontology scholars have long argued that rurality is salient to the experience of ageing, whether this be as a spatial or natural setting, an identity, an imagined community, or as an arbitrary categorisation that impacts on resource provision (Rowles, 1988; Keating and Phillips, 2008; Scharf et al., 2016). Our sincere hope is that this book assists its readers with better understanding this relationship, by providing a comprehensive foundation of conceptual and thematic tools that can be applied to interrogate the experiences of older adults in diverse rural settings. In doing so, we can truly understand why rurality matters for older people, in what circumstances it should matter and in what ways rural ageing knowledge can inform key gerontological debates.

Within this book, the key elements of rural gerontology are explored in four distinct parts. In Part I, an introduction to the field as well as an abridged historiography of key developments of the "rural turn" underway in the broader gerontological literature is provided by Skinner et al. (Chapter 1), with Part II drawing on the expertise of key leaders in the field to explain the foundational perspectives that underpin rural gerontology, including rural demography, rural studies, health studies, human ecology and social gerontology. The five chapters in Part II situated the book within the global contexts and, increasingly, critical fields of inquiry that characterise rural gerontology today. Part III illuminated the application of these foundational perspectives by exploring the scope of contemporary rural gerontological issues in 12 chapters that focused on key issues experienced by rural older people and ageing rural communities across diverse global, national/regional and community/individual levels, including retirement, migration, health care, housing, transportation, community development, ageing in place, social connectivity, inclusion and care. In Part IV, the various ways in which rural gerontology has engaged, more recently, with critical perspectives was examined, with an explicit focus on building a better theoretical and methodological understanding of the diversity of rural older people's experiences, and their interactions with and influence on the complexity of ageing societies. The ten chapters in Part IV revealed the potential for postcolonial, posthumanist and pragmatist traditions, as well as for emerging critical approaches to rural ageing dimensions of social exclusion, citizenship, voluntarism, poverty, mental health, technology and climate change.

Taken together, the preceding 28 chapters in this book go a long way towards meeting our foundational, scoping and critical aspirations for the field of rural gerontology. We cannot by any means claim that the book addresses the full complexity of experiences associated with rural ageing across global contexts. The lack of voices of under-considered older adult groups from culturally and linguistically diverse, Indigenous and older lesbian, gay, bisexual, transgender, intersex or queer (LGBTIQ) rural populations is of particular concern. However, we are confident that the contents of these chapters mark the first comprehensive attempt to address the scope of contemporary rural gerontological enquiry. Indeed, a major strength of this book is the international range of countries – both in the global north and global south – from which its contributors and contributions are drawn, including in non-English-speaking regions of the world where rural ageing scholarship is beginning to flourish. In doing so, the book provides a much-needed, if still incomplete, reflection on how rural ageing is experienced (and researched) across some developing, as well as developed nations.

Drawing together key debates and critiques developed across the four different sections of this book, this concluding chapter presents a new theoretically and conceptually informed framework that is representative of the twenty-first-century "rural turn" within gerontological studies. In doing so, it lays the foundations for a new critical rural gerontology that considers the diversity across and within rural settlements globally. This builds on early work that argued for and sought to define critical rural gerontology as a distinct field of enquiry (Scharf et al., 2016; Skinner and Winterton 2018a; see also Sidney, 2008). It also highlights key deficiencies and knowledge gaps within this emergent field, with the aim of setting a new research agenda for rural gerontological studies in the twenty-first century.

Characteristics of a distinctive rural gerontology

As is demonstrated throughout the pages of this edited collection, rural gerontology is informed by diverse theoretical and conceptual approaches, traditions and concepts. In some instances, as Scharf et al. (2016) note, rural acts merely as a setting for research inquiry or a context in which people age, with little reflection on how or why rurality is important. However, rural gerontologists generally employ rurality as an analytic lens by which to understand trends and experiences relating to ageing, across and within diverse locations and scales. In applying this lens, there are a diversity of orientations, which largely reflect differences in disciplinary backgrounds and the evolution of the field towards interdisciplinary studies of rural ageing.

A rural gerontological lens can be spatially applied, which encompasses the diverse scales, spaces and places within which rural people age (Skinner and Winterton, 2018b). Within this book, the chapter authors have explored trends and experiences related to ageing across diverse scales of enquiry. Rural gerontologists operate at a global scale (e.g., through examining global trends that impact on rural older people, global differences in relation to rural ageing issues and trends) and at a more regional or localised scale of enquiry. This localised scale encompasses the experiences of rural communities, stakeholders and older people in relation to these global trends. Rural gerontologists also explore different spaces. These might be material in nature, reflecting the rural natural, built or physical environments, or rural social, community or imagined spaces. In doing so, both spatial practices and spatial relationships are examined – so the ways that older people use, occupy and inhabit rural spaces and the affective relationships that rural older people have with these rural spaces and places. Spatial demographies also feature prominently within rural gerontology, with researchers interrogating factors related to population size, density and dispersion in understanding issues associated with rural ageing. Socio-cultural constructions of rurality are also employed, which encompass rural social structures, identities and norms that shape, or are shaped by, their ageing populations. Many contemporary rural gerontologists combine these conceptualisations of rurality to address core problems associated with rural ageing, which encompass rural community change and sustainability, rural service access and equity, in addition to drivers of social inclusion and exclusion. As Skinner and Winterton (2018a) have argued, a distinctly rural gerontology considers the mutually constitutive relationship between population ageing and rural sustainability and accounts for the sustained and rapid transformation of rural spaces. The multidisciplinary nature of rural gerontology is one of its greatest strengths, with contributions within this book drawn not only from within gerontology (social, cultural and geographical) but also from anthropology, geography, psychology, sociology, gender studies, planning and community development, social work and social policy, nursing and public health. It is this disciplinary diversity that prompts continued innovation within the field, in terms of applying new perspectives and approaches to complex rural ageing issues.

Across these varied traditions and approaches, perhaps the most important defining aspect of the twenty-first-century rural gerontological turn is the evolution of a critical approach. As Burholt and Scharf (Chapter 6) note, a critical lens provides insight into how the interplay between processes, structures and societal norms across different geographical scales influence experiences of ageing. Specifically, it enables interrogation of how older people become empowered or disenfranchised within certain contexts and how this occurs. In this context and taking into consideration sustained and ongoing calls from prominent rural scholars to interrogate what is specifically "rural" about rural ageing (Rowles, 1988; Scharf et al., 2016; Skinner and Winterton, 2018b), we highlight here some of the characteristics of a distinct critical rural gerontology moving forward. While there are many examples of excellent research within rural gerontology that do not engage with a critical approach, our aim is to highlight how a critical approach can build the existing body of research by shining a new light on contemporary issues and challenges.

Foundations for a critical rural gerontology

The various chapters within this book have identified a series of foundations that, when considered cumulatively, establish a critical framework for rural gerontology. It is posited here that by utilising a critical rural gerontological approach, we can better understand how and what circumstances, diverse elements of rurality are pertinent to experiences of ageing. Specifically, a critical framework enables us to situate the lived experience of rural ageing within a complex, multi-level set of processes and outcomes. This informs our understanding of the diverse everyday experiences of ageing in different rural places, in terms of how they may be positive or negative, or exclusionary or inclusionary. The application of a critical lens throughout this book offers much value as a means of reconciling these subjective and objective elements of the rural ageing experience. In doing so, it provides a nuanced set of insights that neither reinforce stereotypes of the rural idyll nor problematise rural ageing. This is important, as while the chapters in this book certainly testify to the challenges that rural ageing presents, they also highlight that in many cases, older people want to continue to reside in these locations. As elaborated below, there are three ways in which we see the foundation for a distinctly critical rural gerontology emerging. In line with the growing emphasis on ecological approaches within the critical rural gerontological literature (see Keating et al., Chapter 5), when considered in concert, these three elements reveal the interlinkages and reciprocal forces that occur across different levels of environmental context.

1 Critical rural gerontology interrogates the impact of broader macro-level trends and processes on ageing rural communities, and older people in rural communities

Numerous chapters in this book have noted the importance of the intersection of the global and the local in understanding rural ageing (see Hanlon and Poulin, Chapter 4, Keating et al., Chapter 5). Consequently, critical rural gerontology interrogates how macro-level processes and structures impact on rural community capacity to support older people. In doing so, it investigates how and why macro-level trends and processes have differential impacts on older people in rural locations, and challenges contemporary discourses and ideologies that relate to the needs and preferences of older people. It considers the intersection of these macro-level trends and processes with unique and diverse representations of rurality, including aspects of demography, scale and culture. As the authors in this book have described, this can encompass global discourses and national policies relating to ageing and/or rural development and broader global challenges or processes (e.g., globalisation, climate change, disaster events, financial crises). This enables an exploration of how situations of rural "precarity" are created over time. As Burholt and Scharf note in chapter 6, critical rural gerontology also interrogates how rural older people are represented within macro-level, mainstream policies that impact on them directly. By doing so, we can identify the dynamics that facilitate inequities across ageing rural communities (e.g., uneven geographies of care) and within rural communities (e.g., the differential impacts on diverse rural older adults).

2 Critical rural gerontology questions the construction or positioning of rurality within processes and structures that empower, or disenfranchise, diverse older people, while simultaneously considering how processes and constructions of demographic ageing shape rurality.

In applying a critical lens to rural gerontology, it is necessary to challenge notions of rurality as a broad contextual descriptor in understanding the ageing experience (Scharf et al., 2016) and instead move to identify which specific features of rurality are salient in structuring older people's use, occupancy and relationships with rural space. In doing so, it interrogates which specific elements of rurality create dynamics that empower or exclude older people. Are these elements spatial, physical or imagined, or are they culturally embedded norms related to rurality? How do these differ across communities, and how are they experienced differently by rural older adults with differential levels of resources and agency? In doing so, we can consider how specific elements of rurality impact on the representation of particular groups of older people within rural communities, and who is being othered, or how elements of rurality intersect with different social categorisations across the older adult life course. There is also a need to understand the scales at which influences of "rurality" occur, and how these intersect. For example, are salient elements of rurality locally distinctive (i.e., at a community level), do they manifest at a regional level, or are they reflective of macro-level discourses or structures related to rurality? Understanding the spatial and political relationships between different rural places, such as the relationships between small rural communities and regional towns, is also critical in understanding the dynamics that influence rural ageing, such as rural identities, resource distribution and modes of service provision.

Alternatively, as Keating et al. (Chapter 5) have provocatively suggested, critical rural gerontology also questions whether rurality is a useful descriptor within the ageing experience. There is increasing intersection between the global and the rural, due to increased population flows between metropolitan and rural areas, continued urban sprawl into rural and regional locations and the advent of new technologies that create new spaces and opportunities for engagement beyond the immediate local, rural environment (see Kosurko et al., Chapter 27). An additional challenge is the highly fluid boundaries in terms of defining rurality across and between locations – as Keating et al., (Chapter 5) have noted, traditional indicators that define the rural (e.g., population density, spatial distribution and economic status), are so different across countries, that use of a rural descriptor is rendered somewhat meaningless. Consequently, there is a need to closely examine how situated diverse older people are within their rural settings and the extent to which their experience of rural ageing "in place" is influenced by spaces and places beyond their immediate rural community. In an increasingly globalised world, this insight is critical in understanding how rural inequity is produced and reproduced in older age.

However, critical rural gerontology also examines the ways in which constructions and positionality of ageing, and processes associated with demographic ageing, are reshaping rurality. A key hallmark of rural gerontology is its acknowledgement of the mutually constitutive relationship between older people and sustainable rural communities (Skinner and Winterton, 2018a), and multiple chapters in this book highlight the capacity of rural ageing trends to impact on the economic, political and social fabric of rural life (e.g., Ryser et al., Chapter 13, Winterton and Warburton, Chapter 23). Applying a critical lens therefore provides insight into how both global and local processes related to population ageing create rural environments that can support, or marginalise, older adults.

3 Critical rural gerontology challenges and highlights the differential responsibilities and capacities of older people in producing, reproducing, or contesting rurality, and questions the scope and capacity of diverse rural communities to support population ageing.

Researchers have long argued that rural ageing is distinctive due to the mutually constitutive relationship between rural older people and their communities of residence (Keating and Phillips, 2008; Skinner and Winterton, 2018b). Consequently, in addition to identifying how rurality impacts on older people, critical rural gerontology also examines how older people produce or reproduce rurality, through processes such as active citizenship or place integration processes, or performances of rurality. In doing so, it also examines how they contest rurality, through active and passive engagement with macro and community-level structures that seek to disempower rural people through restricting access to resources or spaces that are salient to the rural ageing experience.

While it is acknowledged that older people produce rurality simply through being "in place" and through their everyday practices within rural spaces and places, critical rural gerontology simultaneously seeks to challenge both macro and community-level stereotypes and expectations relating to older people's desire and capacity to produce rurality through engagement as active rural citizens and mobilise in support of the rural (see Winterton and Warburton, Chapter 23, Colibaba et al., Chapter 24). In doing so, it brings into question the agency of certain groups of older adults to respond to dynamics that impact on spaces of rural ageing. It moves beyond conceptualisations of rural ageing populations as homogenous and considers how intersectionality impacts levels of agency, considering factors such as socio-economic status, gender, sexual orientation, cultural diversity and life-course experience with the rural (for example, see Hansen et al., Chapter 12). It also highlights differences within communities, and across diverse rural communities, in relation to these factors. However, in doing so, it also considers how the production or performance of rurality, or the contestation of rurality by certain groups of older adults, may create dynamics that marginalise other groups of older adults within rural communities. This is particularly important in the consideration of global-local relations, where older adults with higher levels of resources can rally against rural resource constraints by using non-local services, or by engaging in non-local volunteer activity. In cases such as this, the implications for older adults who are reliant on local, rural spaces, must be considered.

Future possibilities and challenges for critical rural gerontology

In considering how critical rural gerontology might move forward as a field of study, there is scope to address several opportunities and challenges. At a conceptual level, the multidisciplinary scope of critical rural gerontology will mean that scholars working in this field will need to be cognisant of the composition of disciplinary approaches that underpin a truly critical, rural approach to gerontology. As the chapters in this book illustrate, approaches to critical rural gerontology encompasses conceptual or theoretical approaches from geography, gerontology and/or rural studies, and consequently scholars will need to actively engage with multiple disciplinary fields. This poses an ongoing challenge in relation to keeping in touch with disciplinary advances and turns across each field. However, it also presents a significant opportunity, from the perspective of developing truly innovative, multidisciplinary models and approaches and a shared language of sorts to underpin models of enquiry. To that end, inclusivity must be the agenda of critical rural gerontologists, in terms of adopting insights from fields that perhaps are not their own.

From a content perspective, given the focus on marginalisation within critical rural gerontology, a key gap within this book is its lack of attention to LGBTIQ populations, or culturally diverse and Indigenous older people's experiences of ageing in rural settings. In capturing the voices of older people who have been silenced within contemporary rural ageing debates, a key opportunity (and simultaneous challenge) for rural gerontologists will be to explore the experiences of older people who are marginalised and/or "out of place", which may also encompass economic migrants (for instance, see Cheng et al., Chapter 9), socially excluded older adults (for instance, see Walsh et al., Chapter 22) or older adults in rural environments experiencing significant environmental changes (for instance, see Carroll and Walker, Chapter 28). While a key strength of this book is the inclusion of research conducted in LMICs, a key challenge for the field will be to move beyond the predomination of accounts of rural ageing from the global North and ensure that the voices of older adults (and scholars) in developing nations continue to be heard within contemporary rural ageing debates.

Critical rural gerontologists also have a significant role to play in interrogating the relevance of contemporary global or macro-level movements aimed at improving the lives of older people, such as initiatives relating to age-friendly or dementia-friendly communities. Current research commentary on these types of initiatives is relatively uncritical, and as multiple chapters in this book have noted, application of these programmes in the rural context is not without implications for older people and their rural communities. Subsequently, there is a need to critically scrutinise discourses that impact on rural ageing (for an example, see Glasgow and Doebler, Chapter 10). At a macro-level, many policy and practice discourses pertinent to the rural ageing experience are firmly rooted in either rural, or ageing spheres. This ensures that the representation of rural older people's experiences, needs and voices is limited. Critical rural gerontology can add much value in terms of interrogating how representative policies are in relation to the intersectionality of both rural and ageing discourses. The impact of global challenges and events related to environmental sustainability and climate change (as highlighted by Carroll and Walker, Chapter 28) and contemporary global health challenges such as the COVID-19 pandemic (as highlighted by Colibaba et al., Chapter 24), should also be interrogated closely in relation to their impacts on rural older people and the capacity of rural communities to support their older residents.

The employment of temporal approaches to understanding rural ageing is also a significant opportunity for the field and a relative gap within the rural ageing literature. While life-course approaches to rural ageing have been prevalent within the field for some time, employment of longitudinal perspectives to explore processes and practices that influence the ageing experience would shed additional light on how older people become marginalised within rural contexts. Specifically, this would enable exploration of how individual and community changes intersect to facilitate positive or negative outcomes for older people, and this is particularly important against a backdrop of changing rural dynamics. At a spatial level, the increasing use of new technologies in providing care and resource support to older adults is an area of enquiry that critical rural gerontologists must engage with, particularly in determining how these impact upon rural identities and relationships with place in older age. While chapters within this book have touched on the intersection between rural migration patterns and ageing (e.g., Cheng et al., Chapter 9), at both the temporal and spatial level there is considerable scope to explore the complex relationship between rural migration trends (e.g., counter-urbanisation, immigration, return migration and circular migration) and the rural ageing experience. This may include, but is not limited to, place identity and attachment, community sustainability, social connectedness and access to care and services.

As always, a challenge for rural gerontologists will be to ensure that the lived experiences of diverse rural ageing populations, across different types of rural communities, are captured in a way that support higher levels of representation, greater levels of knowledge translation, better models of service delivery and more sensitive policies and practices. In ensuring the voices of diverse rural older people and places are captured, critical rural gerontologists must explore new and innovative methods to engage rural older people in developing and producing research that is about them, through participatory methods and co-production and to involve traditionally marginalised groups. Addressing these shortfalls within the field more generally is important in ensuring that the voices of rural older adults are representative. Further, while many of the chapters in this book have acknowledged the considerable diversity among rural places, a more active analysis of different kinds of rural places is required, beyond factors such as population size and scale. Future research within critical rural gerontology should investigate the constraints and opportunities associated with ageing in less traditionally represented types of rural settings, such as urban fringe, remote, resource, mountainous, coastal, island and desert communities around the world.

Concluding comments

This book has provided a foundation for the emergence of a distinct rural gerontology, characterised by a multidisciplinary critical approach to understanding the experience of ageing in rural areas. In doing so, the book recognises the contributions of rural ageing scholars across the preceding decades, while profiling contemporary approaches employed within the field today and proposing new directions. As such, *Rural Gerontology: Towards Critical Perspectives on Rural Ageing* provides a conceptual and theoretical platform for current and future generations of rural gerontologists. It is our aspiration that this will prompt empirical and theoretical insights over time that contribute to addressing inequity and marginalisation among rural older adults across diverse international contexts. We hope that rural gerontologists are inspired to meet this challenge and look forward to witnessing and contributing to the future evolution of rural gerontology.

References

Keating, N., and Phillips, C. (2008). A critical human ecology perspective on rural ageing. In: N. Keating, ed., *Rural Ageing: A Good Place to Grow Old?* Bristol: Policy Press, pp. 1–10.

Rowles, G. D. (1988). What's rural about rural aging? An Appalachian perspective. *Journal of Rural Studies*, 4(2), pp. 115–124.

Scharf, T., Walsh, K., and O'Shea, E. (2016). Ageing in rural places. In: M. Shucksmith and D. Brown, eds., *Routledge International Handbook of Rural Studies*. Oxon: Routledge, pp. 80–91.

Sidney, I. (2008). Towards a critical rural gerontology: a multi-disciplinary journey to a place where others age. *Generations Review*, July. Available at: www.britishgerontology.org/printer-friendly/DB/gr-issues-2007-to-present/generations-review/towards-a-critical-rural-gerontology-a-multi-disci

Skinner, M. W., and Winterton, R. (2018a). Interrogating the contested spaces of rural aging: implications for research, policy, and practice. *The Gerontologist*, 58(1), pp. 15–25.

Skinner, M. W., and Winterton, R. (2018b). Rural ageing: contested spaces, dynamic places. In: M. Skinner, G. Andrews and M. Cutchin, eds., *Geographical Gerontology*. London Routledge, pp. 136–148.

Index

Page numbers in *italics* indicate figures and in **bold** indicate tables on the corresponding pages.